BIOLOGICAL AND MEDICAL PHYSICS, BIOMEDICAL ENGINEERING

BIOLOGICAL AND MEDICAL PHYSICS, BIOMEDICAL ENGINEERING

The fields of biological and medical physics and biomedical engineering are broad, multidisciplinary and dynamic. They lie at the crossroads of forntier research in physics, biology, chemistry, and medicine. The Biological & Medical Physics/Biomedical Engineering Series is intended to be comprehensive, covering a broad range of topics important to the study of the physical, chemical and biological sciences. Its goal is to provide scientists and engineers with textbooks, monographs, and reference works to address the growing need for information.

For other volumes in this series, go to:
www.springer.com/series/3740

Hans Frauenfelder

The Physics of Proteins

An Introduction to Biological Physics and Molecular Biophysics

Editors: Shirley S. Chan and Winnie S. Chan

With contributions from:
Robert H. Austin, Charles E. Schulz,
G. Ulrich Nienhaus, and Robert D. Young

Hans Frauenfelder
Los Alamos National Laboratory
Theory Division
Los Alamos NM 87545
USA

ISSN 1618-7210
ISBN 978-1-4419-1043-1 e-ISBN 978-1-4419-1044-8
DOI 10.1007/978-1-4419-1044-8
Springer New York Dordrecht Heidelberg London

Library of Congress Control Number: 2010927467

Printed on acid-free paper

Springer is part of Springer Science+Business Media (www.springer.com)

PHYSICS OF PROTEINS

From the Editors: This book is a compilation of the notes that Prof. Hans Frauenfelder used in his pioneering course in Biological Physics at the University of Illinois at Urbana-Champaign with an emphasis on proteins and protein dynamics. We have tried to maintain as best we can the spirit of these lectures in these notes. This is not a conventional textbook, but rather a guide to the lectures which were written on something now sadly obsolete, the large chalkboard. Whenever possible, we have used Prof. Frauenfelder's hand-drawn figures as they would appear on the board. We believe that, while some may lack in detail, they more than make up for in their charm and elegance. Too bad we could not capture his unique Swiss-German accent! The editors (and Robert Austin) take no credit for the joy of these notes but take any blame for what is left unclear and possibly even found to be wrong.

Since these notes were compiled some years after Prof. Frauenfelder "retired" from Urbana, the editors recruited a few colleagues of Prof. Frauenfelder to fill in some gaps in the discussion. We also moved his wonderful notes on experimental techniques and ancillary topics to the end of the book so that the flow of ideas as Prof. Frauenfelder develops the concept of the protein's free energy landscape can move as smoothly as possible. We urge the reader to consult those chapters because, as Prof. Rob Phillips of Caltech likes to say, Biological Physics without numbers is Biology.

And now, Professor Frauenfelder speaks as we settle into our chairs in Loomis Laboratory of Physics:

Life is based on biomolecules. Biomolecules determine how living systems develop and what they do. They store and propagate information, build the systems and execute all processes, from transport of energy, charge, and matter to catalysis. A knowledge of the structure and function of biomolecules is essential for biology, biochemistry, biophysics, medicine, and pharmacology, and it even has technological implications.

The present course should provide a first introduction to the structure and function of biomolecules, particularly proteins. Emphasis is on aspects of interest to physicists. The number of unsolved problems is very large. Any new development in physical tools usually leads to exciting advances in our understanding of biomolecules. Tools and concepts of experiments will therefore be stressed.

In the present lectures we will discuss only a few of the physical tools, and we neglect chemical ones. We try to avoid techniques that are already well established and concentrate on those that may make major new steps possible. New tools appear regularly; laser, synchrotron radiation, muons, proton radiology, holography, gamma-ray lasers (?), and inner-electron spectroscopy are some of the ones that either have recently been introduced into biomolecular physics or are likely to become important.

Our approach is that of physicists. In recent years it has become customary to distinguish **biophysics** and **biological physics**. In biophysics, physics is the servant and the goal is unambiguously to understand the biology of living systems. In biological physics, the situation is not so clear, but one goal is to describe the physics of biological systems, to discover physical models, and possibly even to find new laws that characterize biological entities.

Progress in physics has often followed a path in which three areas are essential: structure, energy levels, and dynamics. Of course, progress is not linear and usually occurs in all three areas at once. Moreover, experimental results and theoretical understanding are both needed for progress. Nevertheless, the three steps can often be seen clearly and frequently they can be related to specific names. The present deep understanding of atoms and atomic structure is linked to a chain of discoveries and theories. Every student is familiar with the Balmer series, the Rutherford atom, the Bohr model, and the theories of Schrödinger, Heisenberg, Pauli, and Dirac. Similar discoveries, models, and theories have elucidated solids (Einstein, Laue, Debye, ...), nuclei, and particles. We try to follow a similar path here. In Part I, we give a brief and superficial introduction to biomolecules. In Part II, we describe the structure of two main classes of biomolecules, proteins, and nucleic acids, and we treat the relevant methods. In Part IIII, we discuss the energy levels or, more properly, the energy landscape, of proteins. Since proteins are complex systems, their energy can no longer be described by a simple level diagram, but requires more general concepts. In Part IV, the heart of the book, we treat the dynamics of proteins. We show how in particular two types of fluctuations, alpha and beta, control protein motions and functions. We also apply the concepts to some selected biological problems. In Part V, we collect some of the background information which can be useful for understanding what has been described in the earlier parts.

The sequence structure–energy landscape–dynamics suffices for "simple" systems such as atoms and nuclei and for "passive" complex systems such as glasses and spin glasses. "Active" complex systems, such as biomolecules, computers, or the brain, perform functions. We do not discuss biological function in detail, but we consider some specific examples together with dynamics.

As we will describe briefly in Chapter 2, the number of biological molecules is extremely large, and indeed, the literature, while covering only a small fraction of existing systems, is vast. Since we are interested in concepts and general features, we cover only a very restricted set of biomolecules but hope to nevertheless cover many of the essential ideas.

Contents

Part III THE ENERGY LANDSCAPE AND DYNAMICS OF PROTEINS

Part IV FUNCTION AND DYNAMICS

Part V APPENDICES: TOOLS AND CONCEPTS FOR THE PHYSICS OF PROTEINS

About the Author

Hans Frauenfelder

Hans Frauenfelder received his Dr. Sc. Nat. (Ph.D.) in 1950 from the Swiss Federal Institute of Technology for the study of surface physics with radioactivity. In 1951 he discovered perturbed angular correction of nuclear radiation (PAC), an effect that is still used today. In 1952 he moved to the University of Illinois where he stayed for 40 years. His research there included parity violation in the weak interaction and the study of nuclear energy levels. After the discovery of recoilless emission of nuclear gamma rays by Rudolf Mössbauer, he and his group applied this tool to a variety of problems. Of particular interest was the application of the Mössbauer effect to biological systems. The early investigations were focused on spectroscopy, the study of spectral lines and bands in proteins as they relate to the protein functions. Crucial for this research was (and still is) the close interaction with the biological community. With another physicist, Peter Debrunner, a close collaboration with a biochemist, I. C. Gunsalus, assured that the proteins under investigation were active and were worth studying. At a workshop attended by physicists and biochemists, Frauenfelder realized that the topics studied were all essentially static and asked whether anyone investigated protein motions. He was told that dynamics was too difficult. He and his group were challenged by this statement and used a typical physics approach to explore dynamics, namely measuring the response of proteins over broad temperature and time range after excitation by a light pulse. This approach led to much of the materials covered in the present book. It demonstrated that proteins are exceedingly complex and can assume a very large number of different conformations. Frauenfelder left the University of Illinois in 1990 because of a mandatory retirement age. He continues his research at the Los Alamos National Laboratory as a Laboratory Fellow. Frauenfelder has been elected to the National Academy of Sciences, the American Academy of Arts and Sciences, the Academy Leopoldina and the American Philosophical Society. He also is the recipient of numerous prestigious scientific fellowships and honors.

Part I

BIOMOLECULES

In Part I, we sketch some of the main features of biomolecules. The discussion is short and is meant to introduce the language and the concepts. More complete treatments can be found in a number of excellent books, which can be read by physicists without previous biological knowledge.

Further readings

1. R. E. Dickerson and I. Geis. *The Structure and Action of Proteins.* Benjamin/Cummings, Menlo Park, CA, 1969.
2. J. M. Berg, J. L. Tymoczko, and L. Stryer. *Biochemistry,* 6th edition. W. H. Freeman, New York, 2006.
3. C. R. Cantor and P. R. Schimmel. *Biophysical Chemistry.* W. H. Freeman, San Francisco, 1980. 3 vols.
4. C. Brändén and J. Tooze. *Introduction to Protein Structure.* Garland Science, New York., 1991.
5. G. E. Schulz and R. H. Schirmer. *Principles of Protein Structure.* Springer, New York, 1996.
6. G. A. Petsko and D. Ringer. *phProtein Structure and Function. New Science Press, London, 2004.*

1

The Hierarchy of Living Things

The chain from atoms to organisms in Fig. 1.1 consists of a number of clearly distinguishable systems. The complexity increases with increasing number of atoms. At the present time it is impossible to predict the behavior of an organism starting from the individual properties of the atoms. An understanding can only be reached by breaking the chain in pieces and studying, for instance, how the properties of biomolecules depend on the properties of its building blocks.

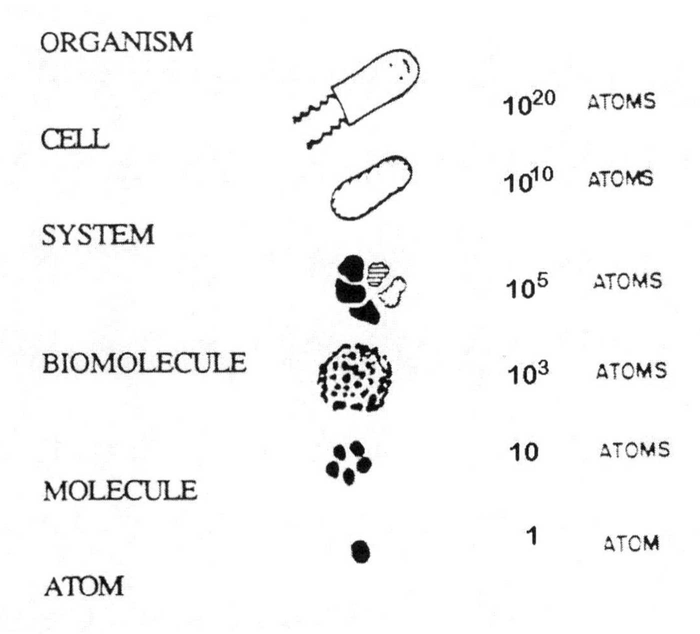

Fig. 1.1. A survey: from atoms to organisms. The number of atoms is of course only approximate but gives an idea of the size of the system.

H. Frauenfelder, *The Physics of Proteins*, Biological and Medical Physics, Biomedical Engineering, DOI 10.1007/978-1-4419-1044-8_1,

More detail concerning the hierarchy of living systems is given in Table 1.1.

Table 1.1. The Hierarchy of Living Systems.

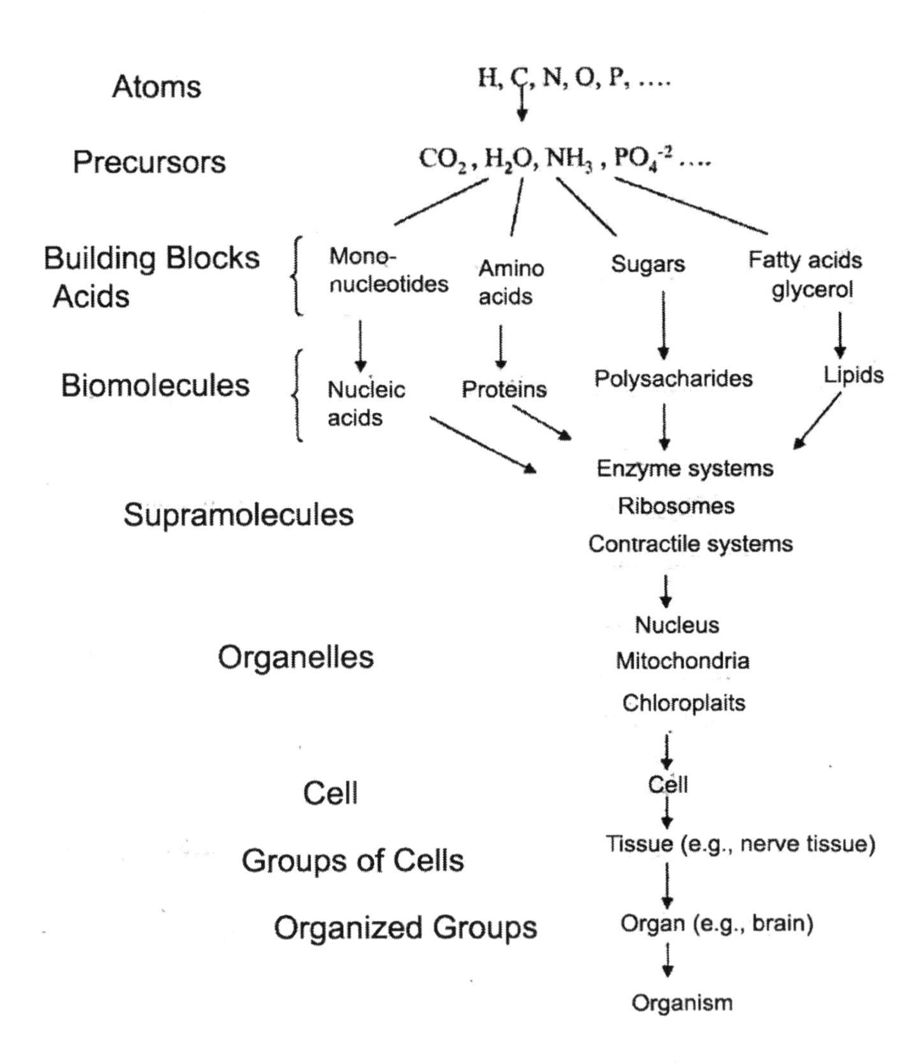

The phenomena that occur at the various levels change remarkably as one goes from the atom to the more complex systems; each new level permits exploration of processes that cannot be studied at lower ones. Some typical processes are listed in Table 1.2.

Table 1.2. Molecular Organization and Characteristic Phenomena.

Atom	Quantum mechanics
Diatomic molecule	Vibrations and rotations
Polyatomic molecule	Molecular chirality, radiationless transitions
Macromolecules	Conformations, phase transitions, tunneling, phonons, solitons, catalysis
Macromolecular complexes	Collective modes, cooperative phenomena, multiple excitations

It is likely that further investigation of biomolecules and biomolecular complexes will lead to the discovery and understanding of other phenomena that cannot be seen in simpler systems.

2

Information and Function

2.1 Information and Construction

If we want to build a low-cost computer, we buy components and build it. Three essential features are involved: (i) instructions for how to assemble, (ii) parts, and (iii) assembler. In a biological system, these components are also present, but the situation is more complex: The assembler must be part of the system, and the system must be self-reproducing. The structure of such a system is schematically shown in Fig. 2.1.

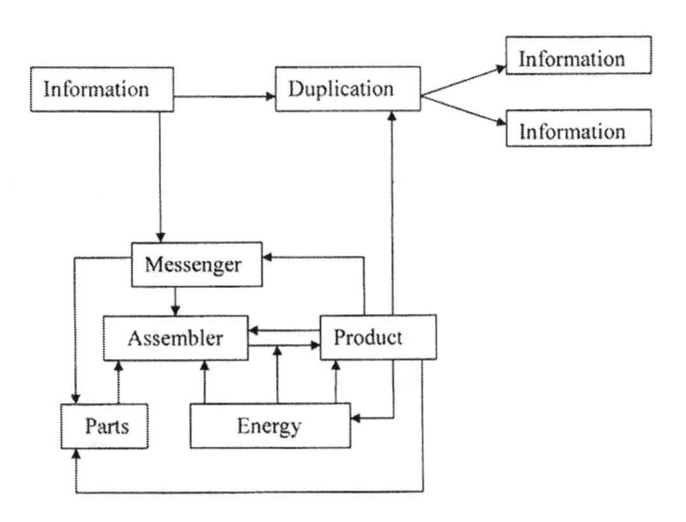

Fig. 2.1. Self-reproducing system.

H. Frauenfelder, *The Physics of Proteins*, Biological and Medical Physics, Biomedical Engineering, DOI 10.1007/978-1-4419-1044-8_2,

The stored information directs the assembly of new parts. These are needed in the reading and transfer of the information and in the assembly. The entire system thus involves a very large number of components and multiple feedback loops.

2.2 Information Content

Since information storage and transfer are crucial in all life processes, a brief description of some relevant concepts is in order. Consider a system where g, the *basis*, gives the number of different "letters" and n, the *digits*, the number of units of the information carrier (number of letters in a word). The *information capacity* $N_{\lambda\nu}$ is given by

$$N_{\lambda\nu} = \lambda^{\nu}. \tag{2.1}$$

The *information content* is defined by

$$I = \ln N_{\lambda\nu} / \ln 2 = \nu \ln \lambda / \ln 2. \tag{2.2}$$

The unit of I is 1 *bit* (binary digit). As an example, consider four-letter words in the English language, proteins with 100 building blocks, and nucleic acids with 10^7 units. The English language has 26 letters, proteins are built from 20 amino acids, and nucleic acids have 4 building blocks. We therefore have the values of $N_{\lambda\nu}$ and I given in Table 2.1. The essential components of the system as it occurs in living things are shown in Fig. 2.2. The actual arrangement is far more complex and involves many feedback loops. The basic principles, however, are contained in Fig. 2.2.

Table 2.1. Information Capacity and Content.

System	Basis λ	Digits ν	Information Capacity $N_{\lambda\nu}$	Information Content I (bits)
English words	26	4	26^4	18.8
Protein	20	100	20^{100}	432
Nucleic acid	4	10^7	4^{107}	2×10^7

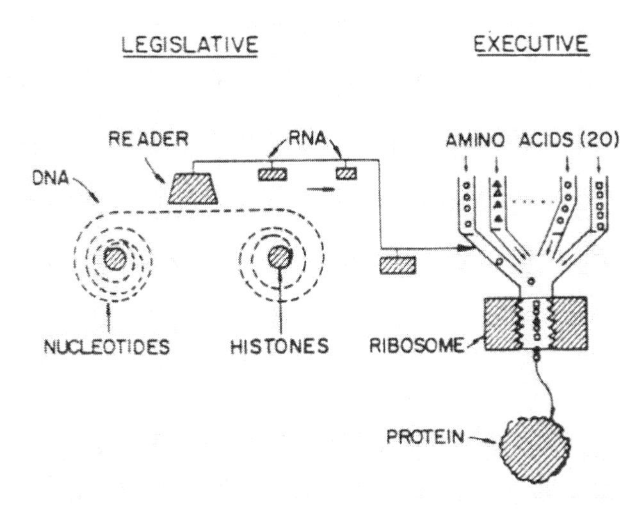

Fig. 2.2. Biomolecules: Legislative and executive. Nucleic acids store and transport information and direct the assembly of proteins. Proteins, assembled from amino acids (AA), are the machines of life. The information is stored on DNA and transported by RNA.

3

Biomolecules, Spin Glasses, Glasses, and Solids (R. H. Austin[1])

Before discussing the building blocks of life in more detail, we compare a few many-body systems [1, 2]. Figure 2.2 shows that biomolecules are linear systems that fold into their three-dimensional structure. The folding is so compact that the density of a protein, for instance, is nearly the same as the density of a crystal made out of one amino acid. Schrödinger [3] actually called proteins "aperiodic crystals." Indeed, proteins and solids have a number of similar properties. However, proteins have features that distinguish them from solids. In a sense, they can be considered a separate state of matter.

Since these features are crucial for the functioning of proteins, we compare biomolecules, glasses, spin glasses, and solids:
(i) Crystalline solids possess a periodic spatial structure, whereas glasses and protein possess a nonperiodic type. While the disorder in glasses is random, it has been carefully selected by evolution in proteins. The "disorder" in proteins is closer to the "disorder" in Beethoven's Gross Fuge (Op.133) or a Picasso painting, both of which are not periodic. Nonperiodicity consequently describes the situation in a protein better than disorder.
(ii) In solids, glasses, and spin glasses, the "strong" forces that hold the atoms together are essentially equally strong in all three directions. In proteins, however, the bonds are "strong" (covalent) along the backbone, but the cross connections are "weak" (hydrogen bonds, disulfide bridges, van der Waals forces). A solid is "dead" and an individual atom can, as a rule, only vibrate around its equilibrium position. In contrast, the weak bonds in a biomolecule can be broken by thermal fluctuations. A biomolecule can therefore execute large motions; it can breathe and act as a miniature machine.
(iii) Solids are spatially homogeneous, apart from surface effects and defects. In glasses, inhomogeneities are random and minor. Biomolecules, in contrast, are spatially inhomogeneous; properties such as density charge and dipole moment change from region to region within a protein.

[1] Department of Physics, Princeton University, Princeton, NJ 08544, USA.

H. Frauenfelder, *The Physics of Proteins*, Biological and Medical Physics, Biomedical Engineering, DOI 10.1007/978-1-4419-1044-8_3,

(iv) Solids or glasses cannot be modified on an atomic or molecular scale at a particular point: modifications are either periodic or random. In contrast, a protein can be changed at any desired place at the molecular level: Through genetic engineering, the primary sequence is modified at the desired location, and this modification leads to the corresponding change in the protein.

The difference between a crystalline solid on the one hand, and glasses, spin glasses, and biomolecules on the other hand, can be expressed in another way. A crystal possesses essentially a unique structure and, as stated in (ii), each atom remains near its unique equilibrium position. The energy surface of a crystal, that is, its energy as a function of its conformation, is nondegenerate and has a single minimum as shown in Fig. 3.1(a). (We can, in principle, describe the conformation of a system by giving the coordinates of all atoms. The energy hypersurface then is a function of all of these coordinates. The curves in Fig. 3.1 are one-dimensional cross sections through the hypersurface.) The energy surface of a complex system such as a glass, spin glass, or biomolecule is sketched in Fig. 3.1(b). The surface shows a very large number of minima, corresponding to many slightly different conformations that a complex system can assume. We will discuss the energy landscape in detail in Part III.

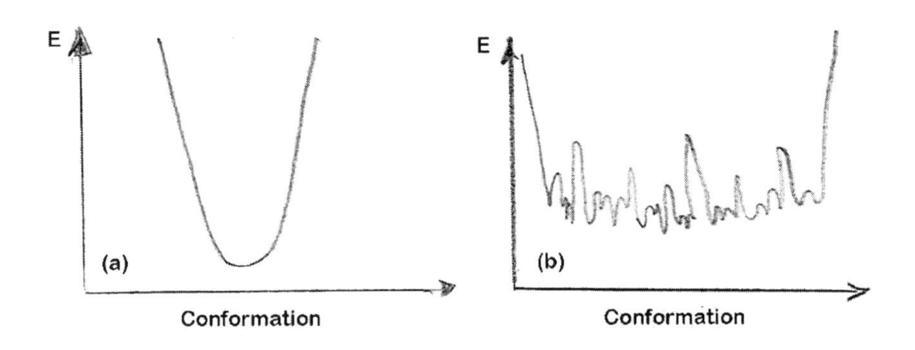

Fig. 3.1. Energy landscape of (a) a simple system (crystal), and (b) a complex system.

A central concept in spin glass physics is the concept of *frustration* [4]. The concept of frustration is extremely important in understanding why a spin glass forms a disordered state at low temperatures and must play a crucial role in the protein problem as well. A clear discussion of frustration, as well as a very readable discussion of many of the problems in spin glass physics can be found in the article by Fisher et al. [5].

Frustration arises when there are competing interactions of opposite signs at a site. Frustration is easy to demonstrate as we show in Fig. 3.2; if we con-

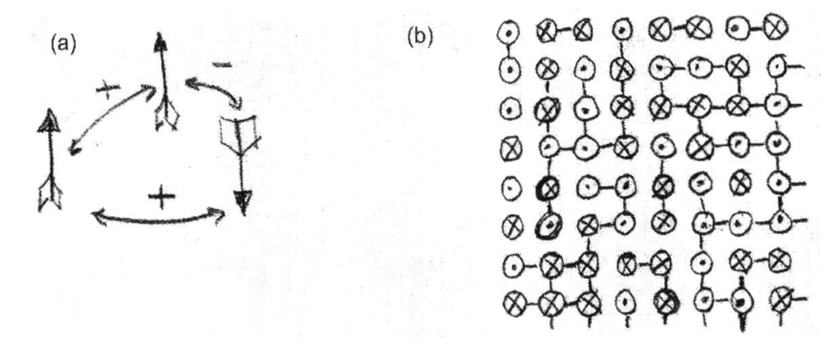

Fig. 3.2. (a) A simple frustrated lattice. The two spins at the end of the chain have equal energies with spins parallel or antiparallel. (b) The ground state of a more complicated lattice, where the sign of the interaction changes randomly. Black lines indicate a negative interaction. From http://www.informatik.uni-koeln.de

sider a chain of spins that close in a circle. With an odd number of antiparallel couplings there is not a unique ground state because of frustration. Frustration generally implies that there is no global ground energy state but rather a large number of nearly isoenergetic states separated by large energy barriers. In other words: imagine that you take a large interacting system and split it into two parts. Minimize the energy of the two separate parts. Now bring the two parts back together. A frustrated system will not be at the global energy minimum because of the interactions across the surface of the two systems.

There are a number of physical consequences that arise from frustration. The primary one that we wish to stress here is *the presence of a dense multitude of nearly isoenergetic states separated by a distribution of energy barriers between the states.* This is due to the presence of frustrated loops that cannot be easily broken by any simple symmetry transformation. This complexity inherently leads to distributions of relaxation times.

The different energy landscapes for crystals and complex systems lead to a profound difference in the dynamic behavior. The motions in a crystal are predominantly elastic; atoms vibrate about their equilibrium positions but very rarely move from place to place. The conformation of the crystal remains unchanged. Protein and glasses, however, show both elastic and conformational (plastic) motions. In the conformational motions, the entire structure can change. The conformational motions, important for the function, will be discussed in Part IV.

Comparison of Fig. 3.1(a) and Fig. 3.1(b) makes it clear that the exploration of the energy landscape of a protein is far more complex than the study of the energy levels of an atom or a crystal. Two more estimates make it even clearer that the field is rich.

(v) The number of "possible" biomolecules is incredibly large. Consider a medium-sized protein, constructed from 150 amino acids. Since there exist 20 amino acids, the number of possible combinations is $(20)^{150} \approx 10^{200}$. If

we produce one copy of each combination and fill the entire universe, we need 10^{100} universes to store all combinations. Thus, new proteins cannot be constructed successfully by randomly linking amino acids. The problem is even more glaring for DNA.
(vi) A protein with a given primary sequence can fold into a very large number of slightly different conformational substates. Each individual building block can, on average, assume 2 to 3 different configurations with states of approximately equal energy. The entire protein thus possesses about $(2 \text{ to } 3)^{150}$ states of approximately equal energy.

The properties alluded to in (i) to (vi) imply that biomolecules are complex many-body systems. Their size indicates that they lie at the border between classical and quantum systems. Since motion is essential for their biological function, collective phenomena play an important role. Moreover, we can expect that many of the features involve nonlinear processes. Function, from storing information, energy, charge, and matter, to transport and catalysis, is an integral characteristic of biomolecules. The physics of biomolecules stands now where nuclear, particle, and condensed matter physics were around 1930. We can expect exciting progress in the next few decades.

References

1. P. G. Wolynes. Aperioidic crystals: Biology, chemistry and physics in a fugue with stretto. In *Proceedings of the International Symposium on Frontiers in Science*, AIP conference proceedings, v. 108. New York, 1988, pp. 39–65.
2. D. Stein, editor. *Spin Glasses and Biology.* World Scientific, Singapore, 1992.
3. E Schrödinger. *What Is Life?* Cambridge Univ. Press, London, 1944. Well work reading.
4. G. Toulouse. Theory of the frustration effect in spin glasses. I. *Commun. Phys.*, 2:115–19, 1977.
5. D. S. Fisher, G. M. Grinstein, and A. Khurana. Theory of random magnets. *Physics Today*, 14:56–67, 1988.

4

Proteins

A protein is a linear chain built from 20 amino acids [1]–[6]. (A few rare amino acids occur occasionally, but we will not discuss them.) The chain contains at the order of 100 to 200 amino acids. Of particular interest are the globular proteins, which act, for instance, as enzymes (catalysis). In the proper solvent, these systems fold into the native protein, as sketched in Fig. 4.1.

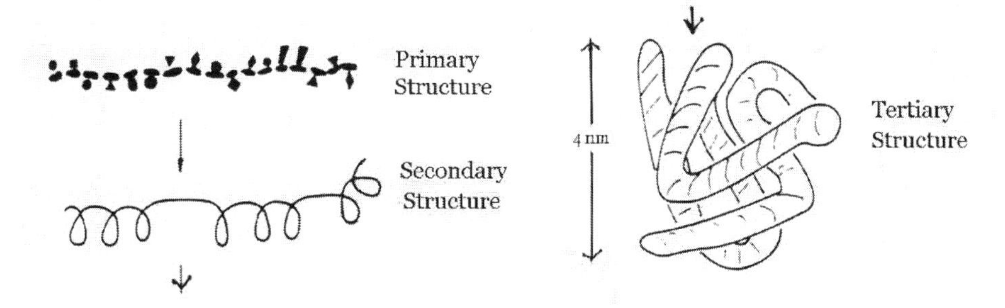

Fig. 4.1. The linear polypeptide chain (primary sequence) folds into the final tertiary structure with a diameter of about 4 nm.

In this chapter we describe the building blocks of proteins and make some remarks about protein structure.

4.1 Amino Acids, the Building Blocks

The building blocks from which proteins are constructed are *amino acids.* To a physicist, such organic molecules are usually a nightmare. We suggest forgetting at the beginning that they are organic molecules and simply consider them building blocks with specified properties. As they appear again and again, their properties become familiar.

H. Frauenfelder, *The Physics of Proteins*, Biological and Medical Physics, Biomedical Engineering, DOI 10.1007/978-1-4419-1044-8_4,

All amino acids are built around a carbon atom (Fig. 4.2).

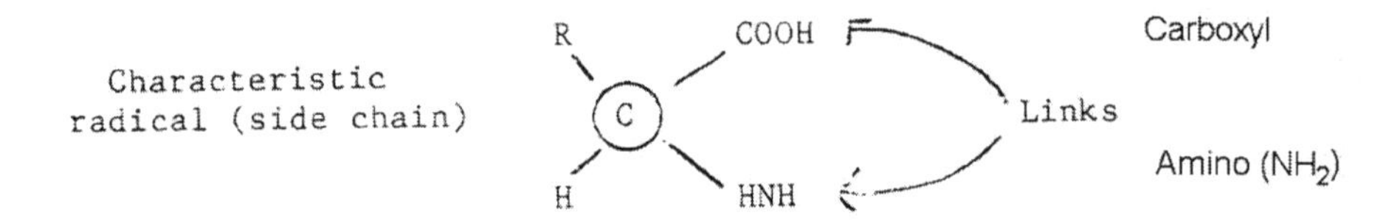

Fig. 4.2. General structure of an amino acid.

Two of the four ligand positions (valences) connect the amino acid to the other building blocks in the chain, and the third carries a hydrogen atom. The fourth position binds the side chain or residue R that determines the specific properties of the amino acid. The individual building blocks in a protein are linked together by *peptide bonds.* The bond has covalent character and is so strong that it is practically never broken by thermal effects. In the formation of peptide bonds, water is eliminated and a bond is formed: COOH$\cdots$HNH $\longrightarrow$ –NH–CO– + H_2O. The resulting *polypeptide chain* thus looks as in Fig. 4.3. This amino acid chain consists of a polypeptide backbone with one sidechain (R) per amino acid residue. The backbone is nonspecific; the entire specificity is in the side chain. In addition to the components shown, proteins contain two terminal groups:

the amino terminal $-NH_3^+$ denoted by –N
the carboxy terminal $-CO_2^-$ denoted by –C

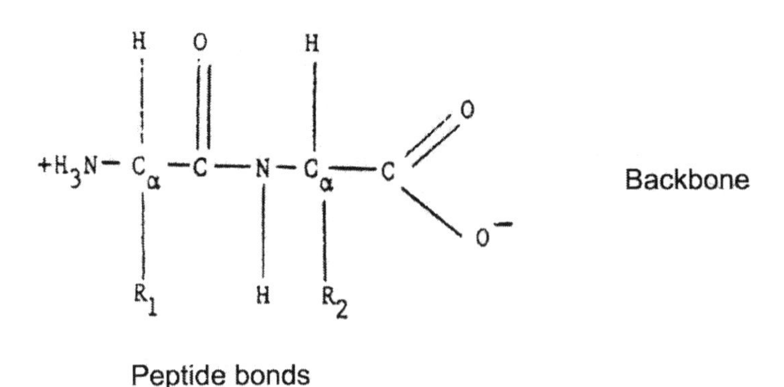

Fig. 4.3. A polypeptide chain formed from individual amino acids through peptide bonds.

The twenty common amino acids are given in Table 4.1.

Table 4.1. Properties of protein building blocks. The length (L) is for the side chain only. The molecular weight is for the entire amino acid. Subtract 17.9 (molecular weight of water) to obtain the net molecular weight of the residue. The polarity indicates whether the amino acid is nonpolar (NP) or polar with a net positive, negative, or neutral charge at pH = 6.

Amino Acid	Symbol	Molecular Weight (amu)	L (nm)	Polarity	Side chain (X = benzene)	One-Letter Symbol
Alanine	ALA	89	0.28	NP	—C	A
Arginine	ARG	174	0.88	+	—C—C—C—N—C=N, N	R
Asparagine	ASN	132	0.51	0	—C—C=O, N	N
Aspartic acid	ASP	133	0.50	—	—C—C=O, O	D
Cysteine	CYS	121	0.43	0	—C—S	C
Glutamine	GLN	146	0.64	0	—C—C—C=O, N	Q
Glutamic acid	GLU	147	0.63	—	—C—C—C=O, O	E
Glycine	GLY	75	0.15	0	—H	G
Histidine	HIS	155	0.65	+	—C—C=C, N, N, C	H
Isoleucine	ILE	131	0.53	NP	—C—C—C, C	I
Leucine	LEU	131	0.53	NP	—C—C—C, C	L
Lysine	LYS	146	0.77	+	—C—C—C—C—N	K
Methionine	MET	149	0.69	NP	—C—C—S—C	M
Phenylalanine	PHE	165	0.69	NP	—C—X	F
Proline	PRO	115		NP	C—C, C, N—C	P
Serine	SER	105	0.38	0	—C—O	S
Threonine	THR	119	0.40	0	—C—C, O	T
Tryptophan	TRP	204	0.81	NP	—C—C—C, N	W
Tyrosine	TYR	181	0.77	0	—C—X—O	Y
Valine	VAL	117	0.40	NP	—C—C, C	V

The side chains are shown schematically in Fig. 4.4.

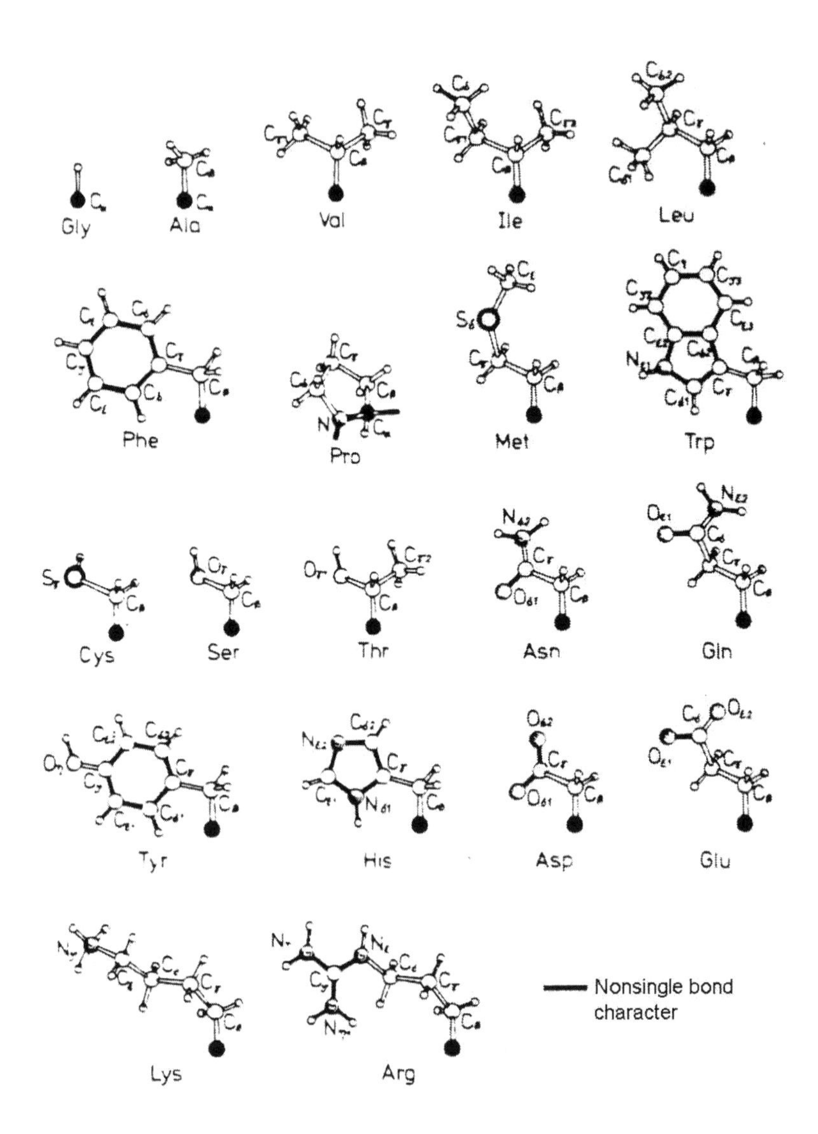

Fig. 4.4. The 20 standard amino acid side chains. For proline part of the main chain is inserted because of bonding to C_α and N_α. All other side chains are shown as they emerge from the C_α-atom in the main chain. The residue names are given as three-letter symbols. Atom names are those given in the IUPAC-IUB recommendations of 1969. The main chain in Pro is indicated by solidly drawn bonds. All C_α-atoms are black. (After Schulz and Schirmer [5]).

The amino acids are divided into four groups:

I. Amino acids with *nonpolar (hydrophobic)* residues. These amino acids are less soluble in water than the polar amino acids.

II. Amino acids with *uncharged polar* residues. The polar R groups of these amino acids can hydrogen-bond with water, and they are thus more soluble in water than those of group I.

III. Amino acids with *negatively charged (acidic)* R groups. The members of this class possess a negative charge at pH 6–7.

IV. Amino acids with *positively charged (basic)* R groups. The basic amino acids have a positive charge at pH 7. Histidine is a borderline case; at pH 7 only about 10% have a positive charge; at pH 6, the percentage is about 50.

An excellent discussion of the properties of each amino acid can be found in Schulz and Schirmer [5].

All amino acids can exist in two different forms, related to each other by a mirror reflection (Fig. 4.5). *D* and *L* stand for *dextro* (right) and *levulo* (left). The two forms rotate the polarization of light in different directions. In nature, only the *L*-amino acids are biologically active. In the laboratory, both forms can be produced. Some experiments, probably wrong, provide food for speculation. Did the *L* dominance come about by accident or through a deeper cause? Neutrinos and beta particles emitted in the weak interaction have a handedness. Could this be the agent to produce predominantly left-handed amino acids? Could one of the two forms be energetically favored [7]?

In describing a sequence, the various amino acids are designated by either a three-letter or a one-letter abbreviation. The abbreviations are listed in Table 4.1.

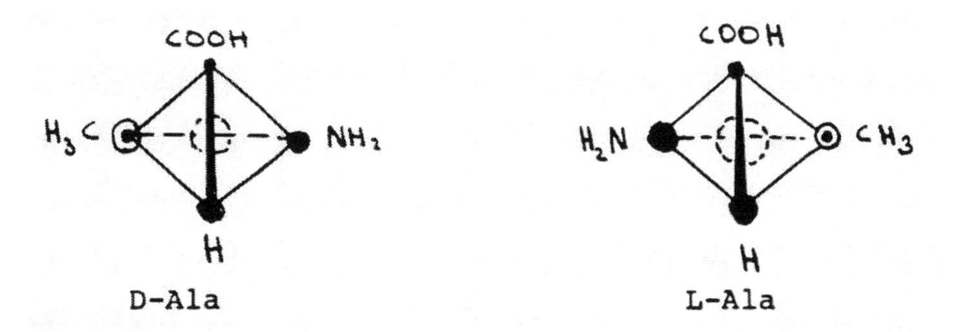

Fig. 4.5. *D* and *L* forms of alanine.

4.2 Amino Acid Composition and Sequence

Two questions arise: In a given protein, what is the number of each type of amino acid, and how are the amino acids arranged?

To determine the amino acid composition, a protein is first decomposed (hydrolyzed or digested) into a mixture of amino acids. (*Hydrolysis* means that in water a proton is transferred from a molecule HA to a water molecule, $HA + H_2O \longrightarrow A^- + H_3O^+$.) Commercial amino acid analyzers are available. They are based on ion-exchange chromatography.

At this point we can make an analogy between proteins and words. Assume that the 20 common amino acids correspond to 20 letters of a complete alphabet set, *a*, *b*, *c*. The primary sequence of a protein then corresponds to a word, for instance,

b i o m o l e c u l a r

The composition analysis then tells us how often a particular letter appears in a word, but gives no information about the order. To establish the sequence, the word must be read. The strategy is based on Sanger's pioneering work on insulin: The protein is first degraded into a mixture of overlapping peptides. The preceding word would, for instance, be broken into

biomo molecu cular

Each of the peptide groups is then sequenced, and the peptides are then reassembled. An individual peptide group is read by degrading it sequentially, starting from both ends.

A technique of sequencing that appeals to physicists is mass spectroscopy [8, 9].

A more convenient way to establish the primary sequence of a protein is to determine the arrangement of nucleic acids in the corresponding DNA or mRNA.

4.3 The Primary Sequence

Proteins consist of about 100 to 1000 amino acids bound together to form a polypeptide chain. Investigation of the primary sequence already reveals fascinating features. Consider, for instance, the primary sequence of cytochrome *c*, an electron transport enzyme. (R. Dickerson, *Scientific American*, April 1973; reprinted in *The Chemical Basis of Life* [10]). The primary sequence of this protein has been determined for a large number of systems. Table 4.2 gives part of the sequence for some. We notice that this enzyme has remarkably little changed in the course of about 10^9 years. Cyt *c* is essentially the same in bacteria, fungi, plants, and humans. In particular, some positions are *invariant*; they are occupied by the same amino acid in every sequence. A mutation in such a position is lethal.

Table 4.2 can be used to draw a family tree: The differences between any two chains are counted. Man and rhesus monkey, for instance, differ by one residue. The difference table then provides us with a family tree. Table and family tree are shown in Fig. 4.6.

Table 4.2 presents only a very small part of the sequence information that is available today. Sequencing has become a major and rapidly growing industry. Information can be found on the Web [11].

Table 4.2. Partial Sequence of Cytochrome *c*.

	1	2	3	4	5	6	7	8	9	10	11	12	13	14	15	16	17	18	19	20	21	22	23	24	25	26	27	28	29	30	31	32	33	34	35	36	37	38	39	40
Human	-	-	-	-	-	-	-	-	G	D	V	E	K	G	K	K	I	F	I	M	K	C	S	Q	C	H	T	V	E	K	G	G	K	H	K	T	G	P	N	L
Rhesus monkey	-	-	-	-	-	-	-	-	G	D	V	E	K	G	K	K	I	F	I	M	K	C	S	Z	C	H	T	V	Z	K	G	G	K	H	K	T	G	P	N	L
Horse	-	-	-	-	-	-	-	-	G	D	V	E	K	G	K	K	I	F	V	Q	K	C	A	Q	C	H	T	V	E	K	G	G	K	H	K	T	G	P	N	L
Pig, bovine, sheep	-	-	-	-	-	-	-	-	G	D	V	E	K	G	K	K	I	F	V	Q	K	C	A	Q	C	H	T	V	E	K	G	G	K	H	K	T	G	P	N	L
Dog	-	-	-	-	-	-	-	-	G	D	V	E	K	G	K	K	I	F	V	Q	K	C	A	Q	C	H	T	V	E	K	G	G	K	H	K	T	G	P	N	L
Gray whale	-	-	-	-	-	-	-	-	G	D	V	E	K	G	K	K	I	F	V	Q	K	C	A	Q	C	H	T	V	E	K	G	G	K	H	K	T	G	P	N	L
Rabbit	-	-	-	-	-	-	-	-	G	D	V	E	K	G	K	K	I	F	V	Q	K	C	A	Q	C	H	T	V	E	K	G	G	K	H	K	T	G	P	N	L
Kangaroo	-	-	-	-	-	-	-	-	G	D	V	E	K	G	K	K	I	F	V	Q	K	C	A	Q	C	H	T	V	E	K	G	G	K	H	K	T	G	P	N	L
Chicken, turkey	-	-	-	-	-	-	-	-	G	D	I	E	K	G	K	K	I	F	V	Q	K	C	S	Q	C	H	T	V	E	K	G	G	K	H	K	T	G	P	N	L
Penguin	-	-	-	-	-	-	-	-	G	D	I	E	K	G	K	K	I	F	V	Q	K	C	S	Q	C	H	T	V	E	K	G	G	K	H	K	T	G	P	N	L
Pekin duck	-	-	-	-	-	-	-	-	G	D	V	E	K	G	K	K	I	F	V	Q	K	C	S	Q	C	H	T	V	E	K	G	G	K	H	K	T	G	P	N	L
Pigeon	-	-	-	-	-	-	-	-	G	D	I	E	K	G	K	K	I	F	V	Q	K	C	S	Q	C	H	T	V	E	K	G	G	K	H	K	T	G	P	N	L
Snapping turtle	-	-	-	-	-	-	-	-	G	D	V	E	K	G	K	K	I	F	V	Q	K	C	A	Q	C	H	T	V	E	K	G	G	K	H	K	T	G	P	N	L
Rattlesnake	-	-	-	-	-	-	-	-	G	D	V	E	K	G	K	K	I	F	T	M	K	C	S	Q	C	H	T	V	E	K	G	G	K	H	K	T	G	P	N	L
Bullfrog	-	-	-	-	-	-	-	-	G	D	V	E	K	G	K	K	I	F	V	Q	K	C	A	Q	C	H	T	C	E	K	G	G	K	H	K	V	G	P	N	L
Tunafish	-	-	-	-	-	-	-	-	G	D	V	A	K	G	K	K	T	F	V	Q	K	C	A	Q	C	H	T	V	E	N	G	G	K	H	K	V	G	P	N	L
Dogfish	-	-	-	-	-	-	-	-	G	D	V	E	K	G	K	K	V	F	V	Q	K	C	A	Q	C	H	T	V	E	N	G	G	K	H	K	T	G	P	N	L
Lamprey	-	-	-	-	-	-	-	-	G	D	V	E	K	G	K	K	V	F	V	Q	K	C	S	Q	C	H	T	V	E	K	A	G	K	H	K	T	G	P	N	L
Fruit fly	-	-	-	-	G	V	P	A	G	D	V	E	K	G	K	K	L	F	V	Q	R	C	A	Q	C	H	T	V	E	A	G	G	K	H	K	V	G	P	N	L
Screw-worm fly	-	-	-	-	G	V	P	A	G	D	V	E	K	G	K	K	I	F	V	Q	R	C	A	Q	C	H	T	V	E	A	G	G	K	H	K	V	G	P	N	L
Silkworm moth	-	-	-	-	G	V	P	A	G	N	A	E	N	G	K	K	I	F	V	Q	R	C	A	Q	C	H	T	V	E	A	G	G	K	H	K	V	G	P	N	L
Tobacco horn worm moth	-	-	-	-	G	V	P	A	G	N	A	D	N	G	K	K	I	F	V	Q	R	C	A	Q	C	H	T	V	E	A	G	G	K	H	K	V	G	P	N	L
Wheat	A	S	F	S	E	A	P	P	G	N	P	D	A	G	A	K	I	F	K	T	K	C	A	Q	C	H	T	V	D	A	G	A	G	H	K	Q	G	P	N	L
Neurospora crassa	-	-	-	-	G	F	S	A	G	D	S	K	K	G	A	N	L	F	K	T	R	C	A	E	C	H	G	E	G	G	N	L	T	Q	K	I	G	P	A	L
Baker's yeast	-	-	-	T	E	F	K	A	G	S	A	K	K	G	A	T	L	F	K	T	R	C	E	L	C	H	T	V	E	K	G	G	P	H	K	V	G	P	N	L
Candida krusei	-	-	P	A	P	F	E	Q	G	S	A	K	K	G	A	T	L	F	K	T	R	C	A	E	C	H	T	I	E	A	G	G	P	H	K	V	G	P	N	L
Common									G					G				F				C			C	H									K		G	P		L
Alternatives	-	-	-	-	-	-	-	-		D	V	E	K		K	K	I		V	Q	K		A	Q			T	V	E	K	G	G	K	M		T			N	
	A	S	F	S	G	V	P	A		N	A	K	N		A	T	L		K	T	R		S	E			G	C	D	A	A	A	P	Q		V			A	
			P	T	E	F	S	P		S	I	D	A			N	V		I	M			E	L				E	G	N	N	L	T			Q				
				A	P	A	K	Q			P	A					T		T									I		G			G			E				
							E				S																													

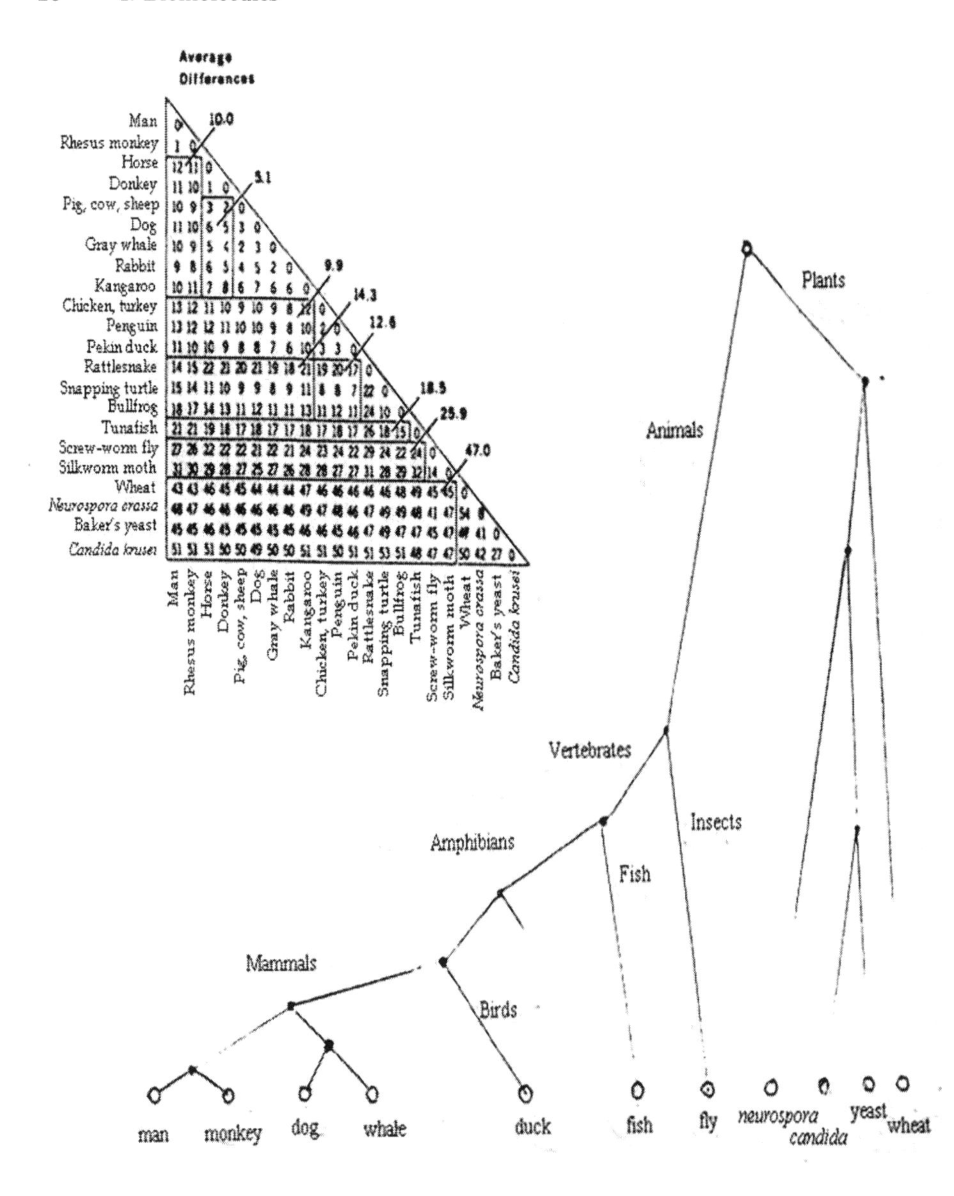

Fig. 4.6. The family tree of the cytochromes *c*. The species differences are given in the insert. The tree is shown inverted in order to agree with the concept of hierarchy treated later. Only the "bottom line" corresponds to actually existing species; earlier vertices simply indicate where branching occurred. The tree is hierarchical, and the species are lumped in clusters. (After Dickerson and Geis [1]).

4.4 Molecular Weight

The molecular weight of a protein or a nucleic acid gives a clue to overall size and complexity. It is therefore usually the first property of a newly discovered species that is given.

One tool to determine molecular weights of biomolecules is the *ultracentrifuge*. The molecules in a solution are spun at rotational speeds of up to 70,000 rpm. The biomolecules, originally distributed uniformly through the rotating cell, move toward the outside. From the time dependence of the motion of the boundary between the regions where the biomolecules concentrate and where they deplete, M can be computed. The boundary is observed with a schlieren optical system. After an equilibrium has been reached (which may take days), the equilibrium distribution also permits finding M. The actual calculations of the expected effect can be done as a problem for qualifying exams.

A second method for determining the molecular weight is *gel electrophoresis*. Thin polyacrylamide gel of a given cross linkage is casted between two glass plates. (These gels are commercially produced insoluble polymers, cross linking to form a porous matrix.) The biomolecules are dissolved in the proper buffer and applied to the top of the gel well, together with a tracking dye (e.g., bormophenol blue). A small DC current is applied for a few hours. The gel is removed from the plates and the positions of the various proteins on the gel are made visible by staining. It turns out that the molecular weight is related to d, the distance traversed by the protein

$$M \propto \exp(-\text{const } d). \tag{4.1}$$

The presence of an anionic detergent masks the charge difference between different proteins so that the mobility depends only on the size of the protein. This simple method permits determination of the molecular weight of an unknown protein with respect to a known one. It also allows separation of a mixture into components.

Some typical molecular weights are:

Cytochrome *c*	13.4 kD
Lysozyme	14.3 kD
Myoglobin	17.8 kD
Hemoglobin	65 kD
Catalase	250 kD
Myosin	500 kD
Tobacco mosaic virus	40 MD

4.5 Some Typical Proteins

Proteins perform an amazing number of functions; we only mention a major subdivision here between structural and functional proteins. Structural proteins use repeated units, as in a mass-produced building. Functional proteins are tailor-made for a specific job, such as storing energy or matter (oxygen) or catalyzing a reaction. The basic building blocks are the same, but they are used differently. Moreover, nonprotein subunits are sometimes incorporated into functional proteins. The nonpolypeptide part is called the *prosthetic group*, the protein without it *apoprotein*. As shown in Fig. 4.1, a protein first folds into a secondary structure and then into a tertiary one. The geometry of the amino acids permits only a small number of secondary structures. The two that are most often used are the alpha helix and the antiparallel beta pleated sheet. Their basic structures are shown in Fig. 4.7.

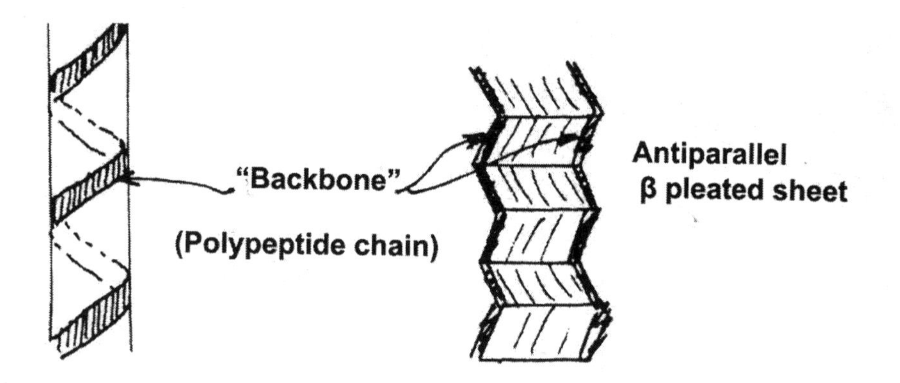

Fig. 4.7. Alpha helix and antiparallel beta pleated sheet.

Globular proteins are largely built from pieces of alpha helices, interrupted by groups that permit bending and twisting, and from beta pleated sheets. Structural proteins also use both units, but in a periodic arrangement.

4.5.1 Heme Proteins

Nature uses good tricks over and over again. In heme proteins, an organic molecule (heme) is embedded into different proteins ("globins"). The protein modifies the properties of the heme group so that the entire system performs a given task. A heme protein thus looks schematically as sketched in Fig. 4.8. Before discussing the heme group and some heme proteins in more detail, we list in Table 4.3 a number of heme proteins and their functions.

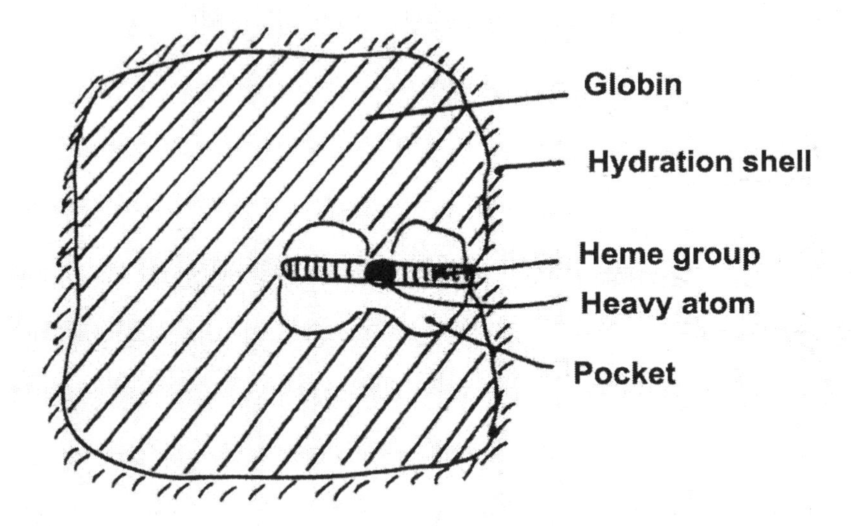

Fig. 4.8. A cross section through a heme protein.

Table 4.3. Selected Metalloporphyrin Proteins and Function.

Protein	Function	Central Atom
Myoglobin	Storage of O_2, NO catalysis	Fe
Hemoglobin	Transport of O_2	Fe
Cytochrome *c*	Transport of electron	Fe
Cytochrome P450	Hydroxilation	Fe
Peroxidase	Oxidation with H_2O_2	Fe
Chlorophyll	Transformation of light into chemical energy	Mg

The *metalloporphyrin group* [12]–[14], shown in Fig. 4.9, consists of an organic part and a central heavy atom. If the metal is iron, it is called the *heme group*. For the chemist, the organic part, protoporphyrin, is made of four pyrrole groups linked by methene bridges to form a tetrapyrrole ring. Four methyl, two vinyl, and two propionate side groups are attached to the tetrapyrrole ring. The side groups can be arranged in 15 different ways, but only one arrangement, protoporphyrin IX, is commonly found in biological systems. The iron atom binds covalently to the four nitrogens in the center of the protoporphyrin ring. The iron can form two additional bonds, one on either side of the heme plane.

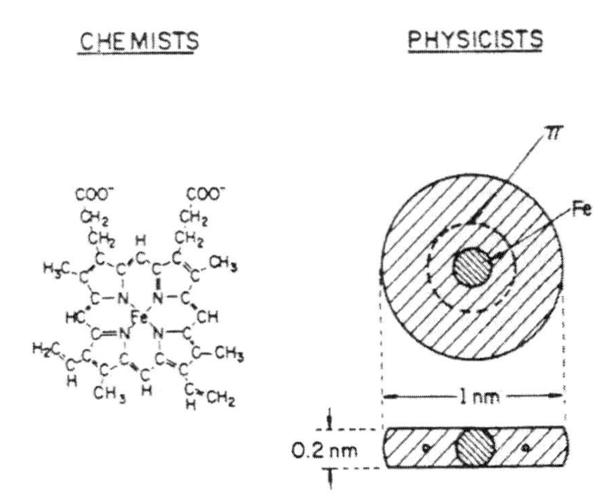

Fig. 4.9. The heme group as seen by chemists and physicists. π indicates the pi-electron ring.

For physicists, who are usually afraid of any molecular structure with more than two atoms, the heme group can be shown as a disk, about 1 nm in diameter and 0.2 nm thick. The disk has an iron atom with two free valences in the center and a one-dimensional electron ring (pi electrons) surrounds the iron.

4.5.2 Myoglobin (Mb)

Myoglobin [15] is a protein that consists of 153 amino acids and has a molecular weight of about 18 kD. It is responsible for the red color in muscle tissue. The main function is oxygen storage, but it also transports oxygen in the cell and may carry energy. Moreover, at low pH, it catalyzes the reaction [16]

$$\mathrm{MbO_2} + \mathrm{NO} \rightarrow \mathrm{NO_3^-} + \mathrm{metMb^+}.$$

Schematically, the Mb molecule is as sketched in Fig. 4.8: The heme group is embedded in the globin (folded polypeptide chain) with pockets on either side of the heme group. The entire molecule is surrounded by the hydration shell, a layer of water about 0.4 nm thick with properties different from liquid water.

The primary sequence of sperm whale myoglobin is given in Fig. 4.10. The three-dimensional structure, elucidated by X-ray diffraction (Chapter 25) is shown in Fig. 4.11. The impression given by Fig. 4.11 is misleading; the entire

Val-Leu-Ser-Glu-Gly-Glu-Trp-Gln-Leu-Val- 10
NA1 NA2 A1 A2 A3 A4 A5 A6 A7 A8

Leu-His-Val-Trp-Ala-Lys-Val-Glu-Ala-Asp- 20
A9 A10 A11 A12 A13 A14 A15 A16 AB1 B1

Val-Ala-Gly-His-Gly-Gln-Asp-Ile-Leu-Ile- 30
B2 B3 B4 B5 B6 B7 B8 B9 B10 B11

Arg-Leu-Phe-Lys-Ser-His-Pro-Glu-Thr-Leu- 40
B12 B13 B14 B15 B16 C1 C2 C3 C4 C5

Glu-Lys-Phe-Asp-Arg-Phe-Lys-His-Leu-Lys- 50
C6 C7 CD1 CD2 CD3 CD4 CD5 CD6 CD7 CD8

Thr-Glu-Ala-Glu-Met-Lys-Ala-Ser-Glu-Asp- 60
D1 D2 D3 D4 D5 D6 D7 E1 E2 E3

Leu-Lys-Lys-His-Gly-Val-Thr-Val-Leu-Thr- 70
E4 E5 E6 E7 E8 E9 E10 E11 E12 E13

Ala-Leu-Gly-Ala-Ile-Leu-Lys-Lys-Lys-Gly- 80
E14 E15 E16 E17 E18 E19 E20 EF1 EF2 EF3

His-His-Glu-Ala-Glu-Leu-Lys-Pro-Leu-Ala- 90
EF4 EF5 EF6 EF7 EF8 F1 F2 F3 F4 F5

Gln-Ser-His-Ala-Thr-Lys-His-Lys-Ile-Pro- 100
F6 F7 F8 F9 FG1 FG2 FG3 FG4 FG5 G1

Ile-Lys-Tyr-Leu-Glu-Phe-Ile-Ser-Glu-Ala- 110
G2 G3 G4 G5 G6 G7 G8 G9 G10 G11

Ile-Ile-His-Val-Leu-His-Ser-Arg-His-Pro- 120
G12 G13 G14 G15 G16 G17 G18 G19 GH1 GH2

Gly-Asn-Phe-Gly-Ala-Asp-Ala-Gln-Gly-Ala- 130
GH3 GH4 GH5 GH6 H1 H2 H3 H4 H5 H6

Met-Asn-Lys-Ala-Leu-Glu-Leu-Phe-Arg-Lys- 140
H7 H8 H9 H10 H11 H12 H13 H14 H15 H16

Asp-Ile-Ala-Ala-Lys-Tyr-Lys-Glu-Leu-Gly- 150
H17 H18 H19 H20 H21 H22 H23 H24 HC1 HC2

Tyr-Gln-Gly 153
HC3 HC4 HC5

Fig. 4.10. Amino acid sequence of sperm whale myoglobin. The label beneath each residue in the sequence refers to its position in an α-helical region or a nonhelical region. For example, B4 is the fourth residue in the B helix; EF7 is the seventh residue in the nonhelical region between the E and F helices [11, 17].

structure is nearly close-packed and contains few cavities and no channels for the exit and entry of small molecules [18].

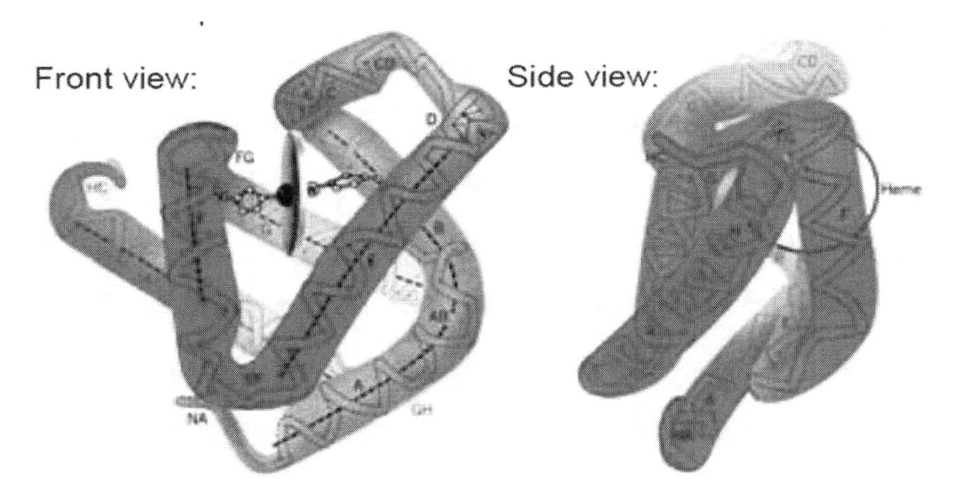

Fig. 4.11. The myoglobin molecule is built up from eight stretches of α helices that form a box for the heme group. Histidines interact with the heme to the left and right, and the oxygen molecule sits at point W. Helices E and F build the walls of the box for the heme; B, G, and H are the floor; and the CD corner closes the open end. After Dickerson and Geis [1].

The iron in the active Mb is in the ferrous state and is formally written as Fe^{++}. Oxygen moves into the pocket and binds to the ferrous iron, *without oxidation*. In most organic heme compounds, oxygen snatches an electron from the iron; changing O_2 into O_2^- and oxidizing the iron from +2 to +3 ("rust"). The oxygen then is not easily available, and binding is essentially irreversible. The protein thus produces the right environment to achieve reversible oxygen storage and protects the binding site; it also provides for specificity and keeps unwanted molecules away. Moreover, the surface of the protein provides the recognition sites for antibodies [19, 20].

4.5.3 Hemoglobin (Hb)

Oxygen transport in vertebrates is performed by hemoglobin [1, 2, 15, 21]–[24], a molecule with molecular weight of about 64 kD and containing four heme groups. Hb is an aggregate of four Mb-like molecules. The Hb tetramer is built from two α chains and two β chains. What distinguishes Hb from Mb? Hb shows *collective behavior*; its oxygen-binding properties depend on whether some O_2 is already bound. Each Hb molecule can bind four O_2; the first one binds slowest, the later ones bind faster. This collective behavior ensures that oxygen transport works like a mandatory car-pool: An Hb molecule will load all four seats before leaving. Loading also depends on the pH; high oxygen affinity at high pH, low affinity at low pH (Bohr effect). These properties, and the difference between Hb and Mb, can be seen in a plot of percent saturation versus oxygen pressure in Fig. 4.12. Since pH is lower in rapidly metabolizing tissues than in the lungs, the Bohr effect promotes the release of O_2 where it is needed.

Christian Bohr, 1855–1911

Niels Bohr, 1885–1962

Aage Bohr, 1922–2009

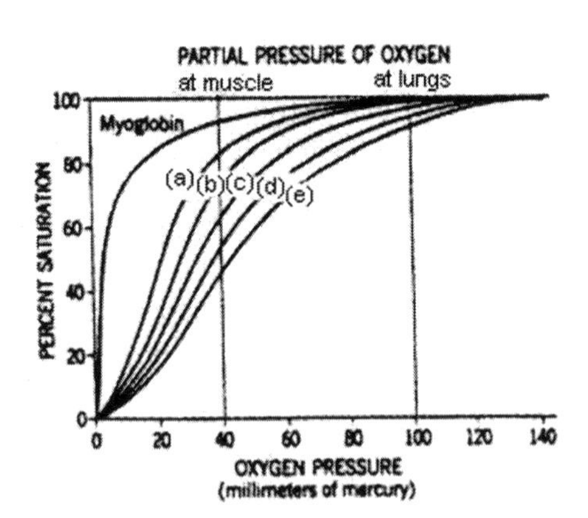

Fig. 4.12. Oxygen binding curves for myoglobin and for hemoglobin at **(a)** pH 7.6, **(b)** pH 7.4, **(c)** pH 7.2, **(d)** pH 7.0, and **(e)** pH 6.8.

So far we have talked about Mb and Hb as if there was a structure common to all mammals. Comparison of myoglobins and hemoglobins from many different sources shows, however, that selection does occur and that species have proteins that are adapted to their lifestyle [25]–[27].

4.5.4 Structural Proteins

In contrast to the globular proteins, which use sophisticated arrangements of primary sequences to arrive at tailor-made tertiary and quarternary structures, structural proteins involve simple repetition of a fixed pattern. Keratins are the protective covering of all land vertebrates. Actin and myosin build the muscles. Silks and insect fibers form another class. Collagen forms connective ligaments in tendons and hides. In all these structures, the main building elements are the alpha helix, the antiparallel beta pleated sheet, and the triple helix. As an example, the beta sheet is shown again in Fig. 4.13.

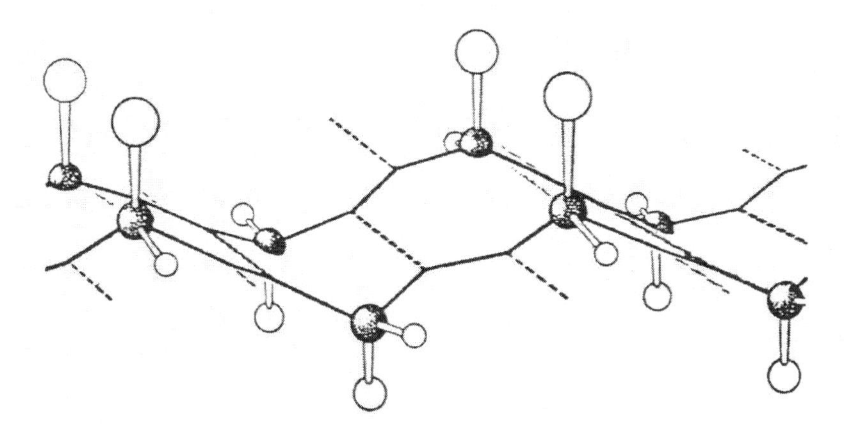

Fig. 4.13. The antiparallel beta pleated sheet. Dashed lines denote hydrogen bonds.

References

1. R. E. Dickerson and I. Geis. *The Structure Action of Proteins.* Benjamin/Cummings, Menlo Park, CA, 1969.
2. J. M. Berg, J. L. Tymoczko, and L. Stryer. *Biochemistry*, 6th edition. W. H. Freeman, New York, 2006.
3. C. R. Cantor and P. R. Schimmel. *Biophysical Chemistry.* W. H. Freeman, San Francisco, 1980. 3 vols.
4. C. Brändén and J. Tooze. *Introduction to Protein Structure.* Garland Science, New York, 1991.
5. G. E. Schulz and R. H. Schirmer. *Principles of Protein Structure.* Springer, New York, 1996.
6. G. A. Petsko and D. Ringe. *Protein Structure and Function.* New Science Press, London, 2004.
7. For a review see L. Keszthelyi, Origin of the asymmetry of biomolecules and weal interaction, *Origins of Life*, 8:299–340, 1977.
8. C. Fenselau. Beyond gene sequencing: Analysis of protein structure with mass spectrometry. *Ann. Rev. Biophys. Biophys. Chem.*, 20:205–20, 1991.
9. P. L. Ferguson and R. D. Smith. Proteome analysis by mass spectrometry. *Ann. Rev. Biophys. Biomol. Struct.*, 32:399–424, 2003.
10. P. C. Hanawalt and R. H. Hanes, editors. *The Chemical Basis of Life.* W. H. Freeman, San Francisco, 1973. Readings from *Scientific American.*
11. http://www.rcsb.org/pdp/.
12. K. M. Smith, editor. *Porphyrins and Metalloporphyrins.* Elsevier, New York, 1975.
13. D. Dolphin, editor. *The Porphyrins.* Academic Press, New York, 1979. 7 vols.
14. A. B. P. Lever and H. B. Gray, editors. *Iron Porphyrins.* Addison-Wesley, New York, 1983. 3 vols.
15. R. E. Dickerson and I. Geis. *Hemoglobin: Structure, Function, Evolution and Pathology.* Benjamin-Cummings, Menlo Park, CA, 1983.
16. H. Frauenfelder, B. H. McMahon, R. H. Austin, K. Chu, and J. T. Groves. The role of structure, energy landscape, dynamics, and allostery in the enzymatic function of myoglobin. *Proc. Natl. Acad. Sci. USA*, 98:2370–74, 2001.
17. http://www.ncbi.nlm.nih.gov/RefSeq/.
18. F. M. Richards. Areas, volumes, packing, and protein structure. *Ann. Rev. Biophys. Bioeng.*, 6:151–76, 1977.

19. J. A. Berzofsky. Intrinsic and extrinsic factors in protein antigenic structure. *Science*, 229:932–40, 1985.
20. H. Lodish, D. Baltimore, A. Berk, S. L. Zipursky, P. Matsudaira, and J. Darnell. *Molecular Cell Biology*, 3rd edition. W. H. Freeman, New York, 1995.
21. M. F. Perutz. The hemoglobin molecule. *Sci. Amer.*, 211(11):2–14, 1964.
22. M. F. Perutz. Hemoglobin structure and respiratory transport. *Sci. Amer.*, 239(6):92–125, 1978.
23. M. Weissbluth. *Hemoglobin.* Springer, New York, 1974.
24. M. F. Perutz. Mechanisms of cooperativity and allosteric regulation in proteins. *Q. Rev. Biophysics*, 22:139–236, 1989. Reprinted as *Mechanisms of Cooperativity and Allosteric Regulation in Proteins*, Cambridge Univ. Press, Cambridge, 1990.
25. R. A. Bogardt, B. N. Jones, F. E. Dwulet, W. H. Garner, L. D. Lehman, and F. R. N. Gurd. Evolution of the amino acid substitution in the mammalian myoglobin gene. *J. Mol. Evol.*, 15:197–218, 1980.
26. M. F. Perutz. Species adaptation in a protein molecule. *Mol. Bio. Evol.*, 1:1–28, 1983.
27. A. C. T. North and J. E. Lydon. The evolution of biological macromolecules. *Contemp. Phys.*, 25:381–93, 1984.

5

Nucleic Acids

We have already pointed out that nucleic acids [1]–[3] are responsible for information storage and transfer and that they direct the synthesis of proteins. Three properties guarantee life over many periods of replication:

i) the information transfer is highly accurate [4], and
ii) self-replication and
iii) self-repair occur.

Without these three properties, life could not exist.

Two different types of nucleic acids dominate:

DNA (deoxyribonucleic acid) carries the genetic information, and
RNA (ribonucleic acid) transfers the information and directs protein synthesis.

DNA has a molecular weight of between 10^7 and 10^9 D. The amount of DNA in a system increases with system complexity. Basically, DNA is a double-stranded linear helical molecule. RNA comes in three forms:

tRNA transfer RNA	about 25 kD,
mRNA messenger RNA	100–4000 kD,
rRNA ribosomal RNA	40–1600 kD.

tRNA and mRNA are single-stranded; rRNA exists as a nucleo-protein in the ribosome. tRNA transports the correct amino acids to the growing protein chain, mRNA is involved in the protein synthesis, and rRNA is the catalytic center for the protein synthesis.

The most important function of nucleic acids is storage and transfer of information. To store information, an *alphabet* and a *dictionary* are needed. We first describe the building blocks and then the arrangement of complete nucleic acids. In the next chapter, we discuss the alphabet and the dictionary.

5.1 General Arrangement of Nucleic Acids

Nucleic acids contain four building blocks:

H. Frauenfelder, *The Physics of Proteins*, Biological and Medical Physics, Biomedical Engineering, DOI 10.1007/978-1-4419-1044-8_5,

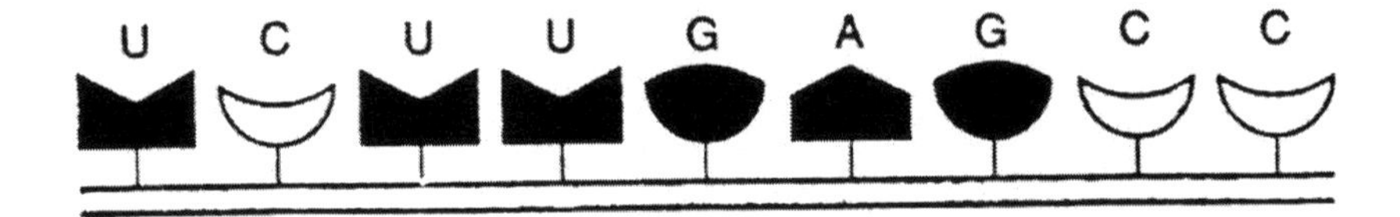

Fig. 5.1. Schematic arrangement of building blocks in nucleic acids.

DNA: A T G C
RNA: A U G C

Here, A is adenine, T thymine, U uracil, G guanine, and C cytosine. The building blocks are hooked onto a backbone, as shown schematically in Fig. 5.1. The four building blocks are matched in pairs, and the interaction between the partners of a pair is stronger than between two nonpartners. Matched partners are

DNA	RNA
A = T	A = U
G ≡ C	G ≡ C.

Here we have symbolically indicated that A and T are connected by two bonds, G and C by three. Matching thus is forced by the number of hydrogen bonds: A binds well to T (or U) via two bridges; G binds strongly to C via three bridges. The matching is shown schematically in Fig. 5.2. Through the matching, AT (or AU) and GC form base pairs that are nearly identical in size and form.

A single strand of nucleic acids is most stable if it can fold such that a maximum number of matched pairs are produced. For short single strands, the result is a "hairpin"; for longer ones, a "clover leaf." Bonding between two strands is particularly strong if each base of the sequence on one strand is bound to a complementary base on the other strand. Such an arrangement is

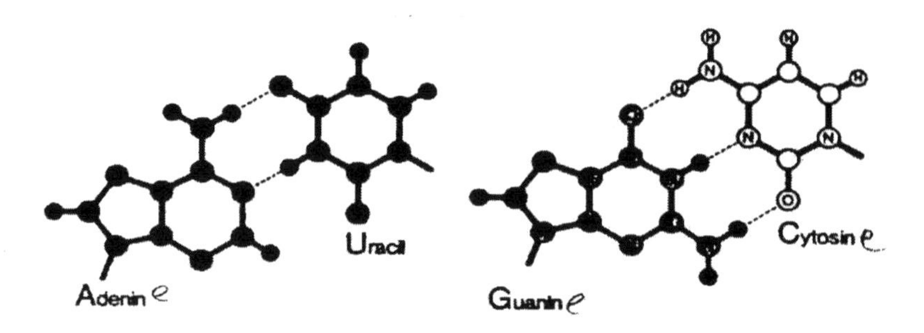

Fig. 5.2. Matching of base pairs.

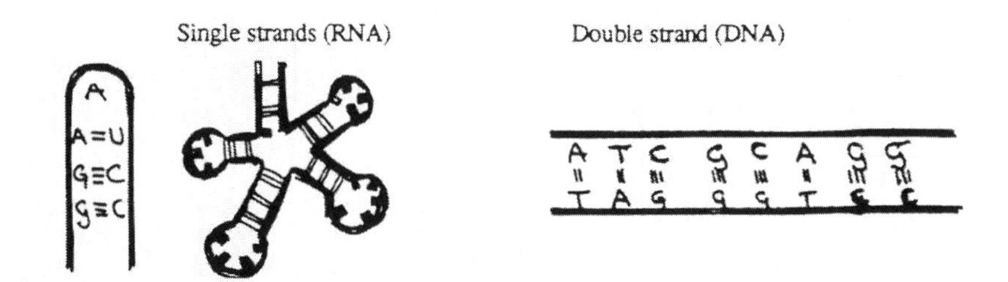

Fig. 5.3. Particularly stable forms for single- and double-stranded nucleic acids.

obtained if two separate complementary sequences bind. This arrangement is most stable in the famous Watson-Crick double helix. The three basic forms, hairpin, cloverleaf, and complementary strands are shown schematically in Fig. 5.3.

5.2 Mononucleotides—The Building Blocks

The four building blocks of nucleic acids are sketched schematically in Fig. 5.2. These nucleotides consist of a complex containing base + sugar + phosphate, as indicated in Fig. 5.4. The base is one of the logic building blocks and distinguishes the four units; phosphate and sugar are nonspecific and provide the links for the backbone.

The *bases* are formed from purines and pyrimidines; the five most important ones are given in Fig. 5.5.

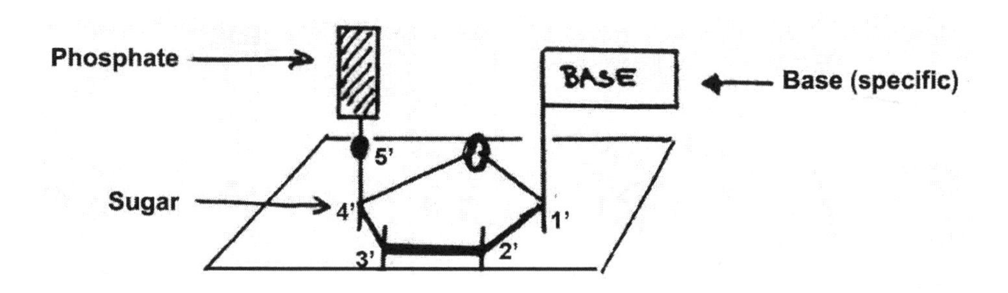

Fig. 5.4. Mononucleotides consist of the structural units sugar and phosphate and the specific base.

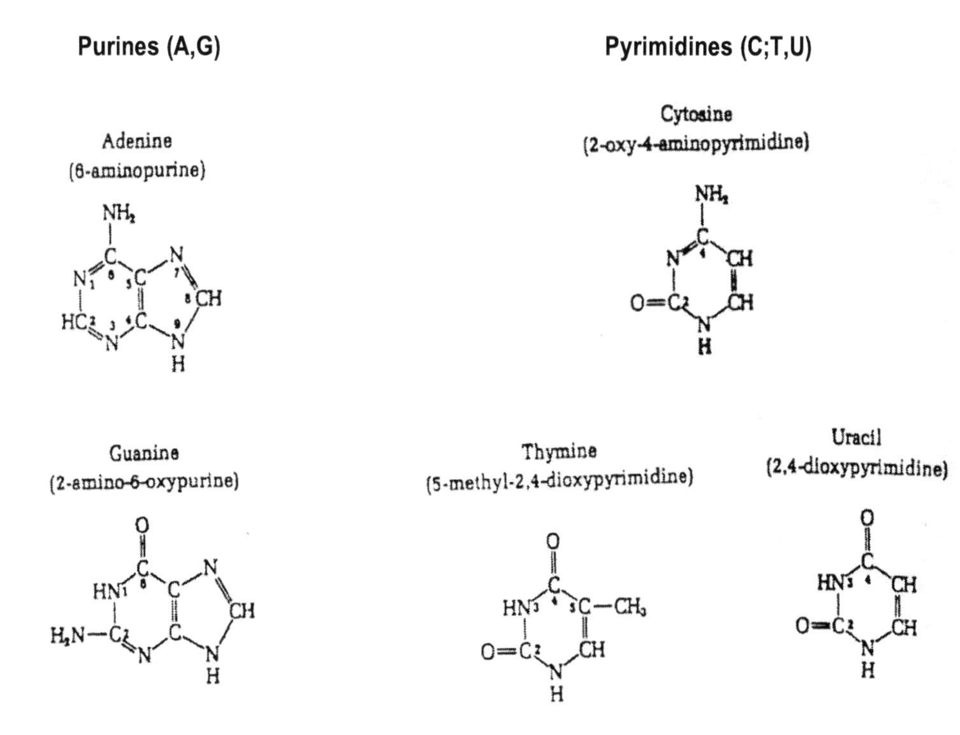

Fig. 5.5. The five important bases.

Names: A *nucleoside* is a purine or pyrimidine base linked to a pentose.
A *nucleotide* is a phosphate ester of a nucleoside.

In passing we mention that nucleotides are not only extremely important as the building blocks of RNA and DNA, but they occur also in other roles; energy-carrying coenzymes, coenzymes in redox reactions, and in transfer reactions.

5.3 Polynucleotides

DNA and RNA are linear polymers of successive mononucleotide units in which one mononucleotide is linked to the next by a phosphodiester bridge between the 3′-hydroxyl group of one nucleotide and the 5′-hydroxyl group of the next. The basic structure of such a polynucleotide is thus as in Fig. 5.6. Phosphates and sugars form the nonspecific structure; the bases are the specific letters of the genetic message. The message has a direction. One end of the chain has a 5'-OH group, the other a 3′-OH group, neither linked to other nucleotides. By definition, a base sequence (ACGAG..) is written in the $5' \to 3'$ direction. ($1'$ to $5'$ labels the carbon atoms of the sugar ring.)

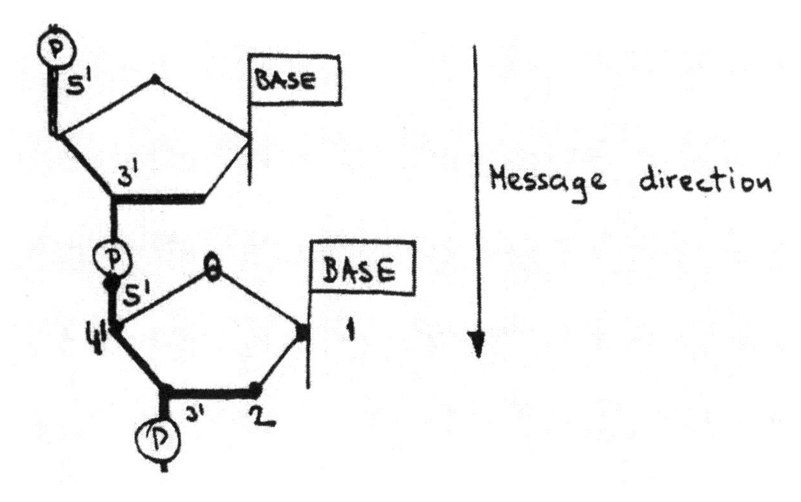

Fig. 5.6. DNA chain—basic arrangement.

5.4 The Double Helix

Solutions of purified natural DNA [5] can be very viscous, suggesting long and rigid rather than compact and folded DNA molecules. The elucidation of the actual structure was based on X-ray data and an understanding of the base pairing. Paired bases are already shown in Fig. 5.2; a more detailed figure is in Fig. 5.7.

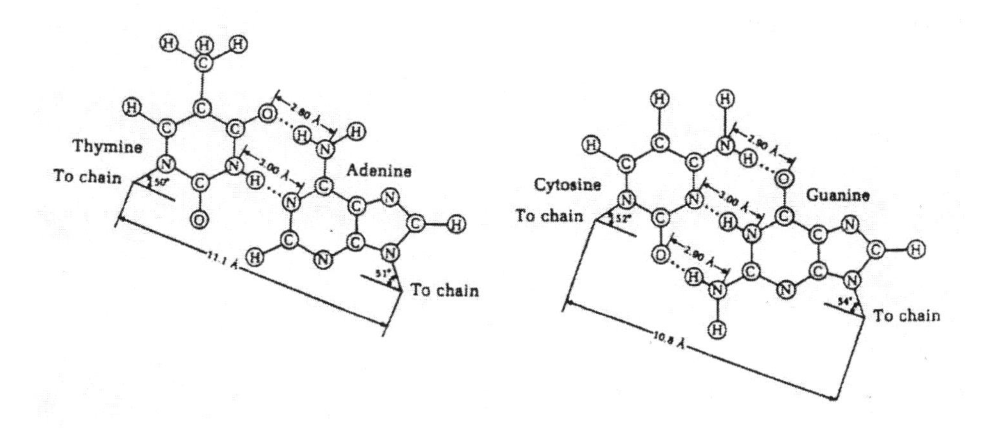

Fig. 5.7. The two base pairs.

Figure 5.7 shows that the two base pairs have nearly identical dimensions. They can thus be arranged in arbitrary sequence, without straining the backbone too much. X-ray data indicate a helical structure, with two periodicities, a major one of 3.4 Å and a minor one of 34 Å. This information is incorporated in the double helix of Watson and Crick.

The Watson-Crick model [6] postulates two right-handed helical nucleotide chains coiled around the same axis to form a double helix. The two strands are antiparallel; their $3'$-$5'$-phosphodiester bridges run in opposite directions. The arrangement is shown schematically in Fig. 5.8. The coiling is such that the two chains cannot be separated unless the coils unwind. The two strands are bonded together by the hydrogen bonds of complementary bases. The 3.4-Å periodicity is explained by assuming that the bases are stacked perpendicular to the long axis at a center-to-center distance of 3.4 Å. The 34-Å periodicity is explained by assuming that there are ten nucleotide residues in each complete turn of the helix.

Stability of the double helix is obtained through the hydrogen bonds between bases *and* through the proper arrangement of the components. The hydrophobic bases inside the double helix are shielded from the solvent water. The hydrophilic sugar residues and electrically charged phosphate groups are at the periphery, exposed to water. Initially only one form of DNA was recognized, the Watson-Crick double helix. It is now known that there are three different forms of double helices [7, 8]:

A-DNA is double-helical DNA that contains about 11 residues per turn. The planes of the base pairs in this right-handed helix are tilted 20 degrees away from the perpendicular to the helical axis. A-DNA is formed by dehydration of B-DNA.

B-DNA is the classical Watson-Crick double helix with about 10 residues per turn. The helix is right-handed and the base planes are perpendicular to the helical axis.

Z-DNA is a left-handed double helix containing about 12 base pairs per turn.

5.5 Some Properties of DNA

DNA is a remarkable biomolecule. Although the diameter is always fixed to about 2 nm by the base pairing, the length depends on the number of base pairs and can exceed centimeters. DNA thus is extremely asymmetric. In Table 5.1, a few data are collected.

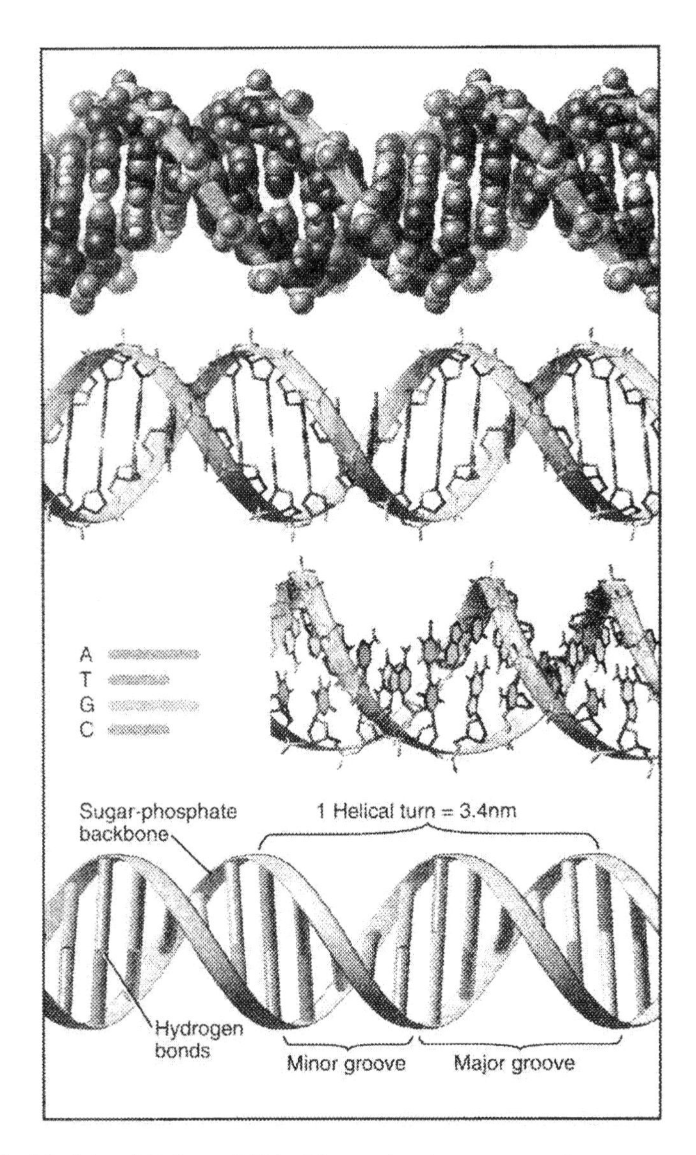

Fig. 5.8. Models of B-form DNA. The molecule consists of two complementary antiparallel strands arranged in a right-handed double helix with the backbone on the outside and stacked pairs of hydrogen-bonded bases on the inside. Top: Space-filling model. Middle: Stick figures, with the lower figure rotated slightly to reveal the faces of the bases. Bottom: Ribbon representation [9].

DNA molecules have been seen directly by electron microscopy. Many form closed loops. Some DNA molecules interconvert between linear and circular forms.

Table 5.1. Properties of Some DNA.

Source	Number of Base Pairs	Length	Molecular Weight	Information Content (bit)
Polyoma virus	4,600	1.6 μm	3×10^6	9×10^3
T2 phage	185,000	63 μm	122×10^6	3.7×10^5
E.coli bacteria	3.4×10^6	1.2 mm	2.3×10^9	6.8×10^6
Drosophila	6×10^7	2.1 cm	43×10^9	1.2×10^8
Human cell	$\sim 3 \times 10^9$	1 m	2×10^{12}	6×10^9

5.6 Replication

The Watson-Crick model leads directly to an understanding of the way in which information is replicated [6]. The bases on one strand unambiguously determine the bases on the complementary strand. A DNA molecule thus can duplicate as sketched in Fig. 5.9.

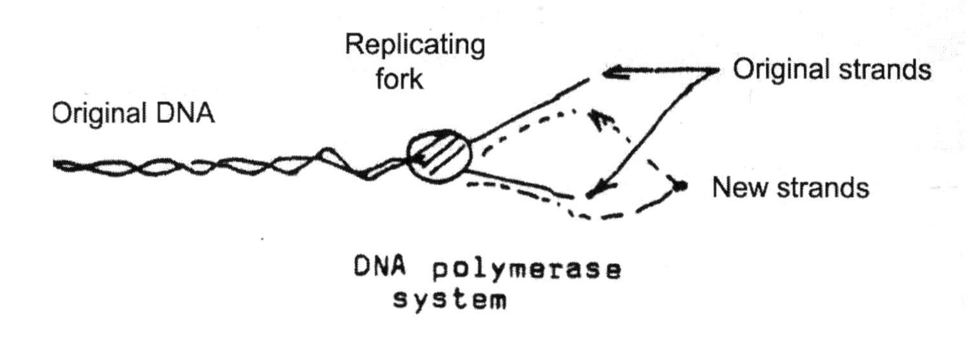

Fig. 5.9. DNA replication.

In a protein system (DNA polymerase), the original double strand is unwound. Each single strand then serves as a template for a new complementary strand. The complete system after duplication thus consists of two double strands, each identical to the original DNA. Actually, replication occurs in fragments which are then joined by DNA ligase. Replication involves a proofreading mechanism because replicating mistakes can happen [10, 11].

5.7 Storage and Retrieval of Information

The information encoded on the DNA is useless unless it can be stored and retrieved efficiently [2, 9, 12]–[14]. These tasks are extremely difficult to achieve, as one can realize quickly. The human DNA consists of about 6.6×10^9 base pairs. Each base pair has a linear dimension of about 3.4 Å. The DNA, stretched, thus is about 2 m long, but only about 1 nm in diameter. In the cell, it is compacted into a region with linear dimensions of about 10 μm. How is this compaction achieved, and how can information be extracted from the compact core? It is clear that the packaging must occur hierarchically in a number of steps. Indeed, the first step is reasonably well understood, but the processes beyond the first step, reduction in size by about a factor of ten, are largely unknown. The organization of the first step is sketched in Fig. 5.10.

In the first level of folding, the DNA coils around proteins called *histones* to form the nucleosome, shortening the DNA by about a factor of 7. The string of nucleosomes then folds into a filament called 30-nm fiber, further shortening the DNA by a factor of about 40. Further folding, not shown, finally results in the compact 10 μm structure. Access to a particular stretch of the DNA in such a tight complex is obviously very difficult, and this state is therefore termed "silent." If a stretch has to be read, the 30-nm fiber partially unfolds, leading to the active form shown on the left.

This crude picture already makes it clear that a great deal of research will be needed to obtain a quantitative picture of the structure and dynamics of the information storage and retrieval in cells.

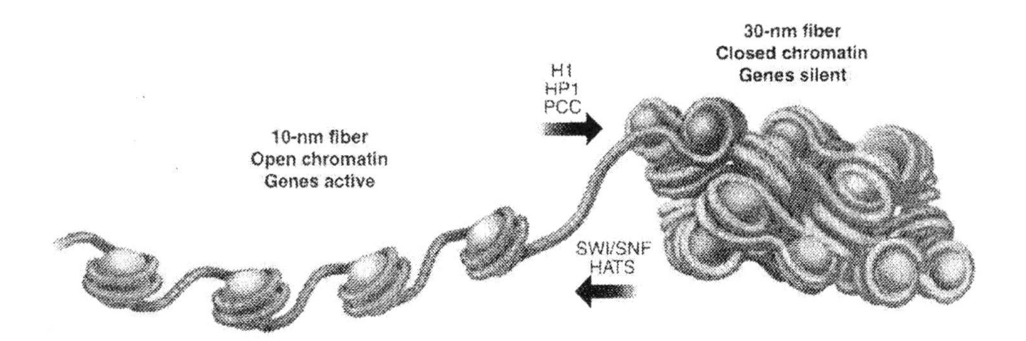

Fig. 5.10. The first step in the organization of the DNA into a compact form. The DNA is wound around proteins (histones) to form nucleosomes. The entire structure is called *chromatin*. The structure can assume an open form, left, or a compact form, right. The closed chromatin then can further fold into larger structures, not shown (after Mohd-Sarip and Verrijzer [14]).

References

1. J. D. Watson. *Molecular Biology of the Gene*, 4th edition. Benjamin/Cummings, Menlo Park, CA, 1987.
2. J. D. Watson, T. A. Baker, S. P. Bell, A. Gann, M. Levine, and R. Losick. *Molecular Biology of the Gene*, 5th edition. Benjamin Cummings, Menlo Park, CA, 2003.
3. J. D. Watson and A. Berry. *DNA*. Knopf, New York, 2003.
4. M. Eigen and L. de Maeyer. Chemical means of information storage and readout in biological systems. *Naturwiss.*, 53:50–7, 1966.
5. The double helix has inspired literature: J. H. Watson, *The Double Helix*. R. Olby, *The Path to the Double Helix*. H. Judson, *The Eighth Day of Creation*, Simon and Schuster, 1979. F. Rick. How to Live with a Golden Helix, *The Sciences*, 9:6–9, 1979.
6. J. D. Watson and F. H. C. Crick. Genetical implications of the structure of deoxyribonucleic acid. *Nature*, 171(4361):964–7, 1953.
7. A. H.-J. Wang, G. J. Quigley, F. J. Kolpak, J. L. Crawford, J. H. van Boom, G. van der Marel, and A. Rich. Molecular-structure of a left-handed double helical DNA fragment at atomic resolution. *Nature*, 282(5740):680–6, 1979.
8. R. E. Dickerson, H. R. Drew, B. N. Conner, R. M. Wing, A. V. Fratini, and M. L. Kopka. The anatomy of A-, B-, and Z-DNA. *Sci. Amer.*, 216:475–85, 1982.
9. T. D. Pollard and W.C. Earnshaw. *Cell Biology*. Saunders, Philadelphia, 2002.
10. J. J. Hopfield. Kinetic proofreading—new mechanism for reducing errors in biosynthetic processes requiring high specificity. *Proc. Natl. Acad. Sci. USA*, 71:4135–9, 1974.
11. M. Gueron. Enhanced selectivity of enzymes by kinetic proofreading. *Amer. Scientist*, 66:202–8, 1978.
12. H. Lodish et al. *Molecular Cell Biology*, 5th edition. W. H. Freeman, New York, 2004.
13. S. C. R. Elgin and J. L. Workman, editors. *Chromatin Structure and Gene Expression*. Oxford Univ. Press, Oxford, 2000.
14. A. Mohd-Sarip and C. P. Verrijzer. Molecular biology: A higher order of silence. *Science*, 306(5701):1484–5, 2004.

6

The Genetic Code

The next problem is the connection between information and function: How does DNA direct protein synthesis? The basic idea is straightforward and was indicated in Fig. 2.2: The information is written in the DNA linearly; it is read and transferred to the assembler. Proteins are involved in reading, selecting, and assembling. The question thus arises: What came first, nucleic acid or protein information or function? The solution to this chicken-and-egg problem is not yet fully known.

6.1 Transcription and Translation

The information contained in DNA is not used directly to control biosynthesis; the flow of information occurs as follows:

$$\text{DNA} \underset{\text{transcription}}{\longrightarrow} \text{RNA} \underset{\text{translation}}{\longrightarrow} \text{protein.}$$

In transcription, mRNA (messenger RNA) acts as a template and takes the information required for the synthesis of a particular protein from the master, the DNA. In translation, the information on the mRNA is used to synthesize the protein: The action occurs at the ribosome, where the rRNA (ribosomal RNA) and the tRNA (transfer RNA) play the major roles.

Transcription occurs at the double-stranded DNA, but only one strand is used. The DNA thus unwinds partially so that the information contained on a certain length portion (typically 600–900 bases) is transcribed onto the particular mRNA that corresponds to the protein to be produced.

6.2 The Code

The information on the DNA is coded in terms of the four bases A, T, C, and G. A given amino acid must be described by more than one base (letter);

H. Frauenfelder, *The Physics of Proteins*, Biological and Medical Physics, Biomedical Engineering, DOI 10.1007/978-1-4419-1044-8_6,

otherwise only four amino acids could be targeted. Two-letter words, such as AT or AG, would suffice for at most 16 amino acids. Since at least 20 must be specified, and start and stop signals are also needed, the code must contain at least three-letter words. Indeed, a series of brilliant experiments in the early 1960s established that the genetic code uses three-letter words in mRNA sequence, with the dictionary given in Table 6.1.

Table 6.1 shows that the code is degenerate; most amino acids correspond to more than one triplet. Degenerate triplets are called *synonyms*. Most synonyms differ only in the last base of the triplet. In particular XYC and XYU always, and XYG and XYA usually, code for the same amino acid. Degeneracy minimizes the effect of deleterious mutations: If the code were not degenerate, 20 codons would specify amino acids, and 44 would signify chain terminations (stop). Most mutations would lead to chain terminations and thus to inactive proteins.

UAA, UAG, and UGA designate chain termination. AUG or GUG is part of the initiation signal and the conformation around the corresponding mRNA determines whether the signal is to be read as a codon for an amino acid or as a start signal.

Table 6.1. The Genetic Code.

First Position (5′ end)	Second Position				Third position (3′ end)
	U	C	A	G	
	Phe	Ser	Tyr	Cys	U
U	Phe	Ser	Tyr	Cys	C
	Leu	Ser	Stop	Stop	A
	Leu	Ser	Stop	Trp	G
	Leu	Pro	His	Arg	U
C	Leu	Pro	His	Arg	C
	Leu	Pro	Gln	Arg	A
	Leu	Pro	Gln	Arg	G
	lle	Thr	Asn	Ser	U
A	lle	Thr	Asn	Ser	C
	lle	Thr	Lys	Arg	A
	Met	Thr	Lys	Arg	G
	Val	Ala	Asp	Gly	U
G	Val	Ala	Asp	Gly	C
	Val	Ala	Glu	Gly	A
	Val	Ala	Glu	Gly	G

6.3 Codon and Anticodon

A triplet on mRNA specifying a particular amino acid is called a codon. The codon is recognized by a corresponding anticodon on tRNA. The tRNA is the adaptor molecule; at least one tRNA exists for each amino acid. Each contains an amino acid attachment site and a template recognition site, the anticodon. A tRNA carries a specific amino acid to the site of protein synthesis.

Protein synthesis was schematically described in Fig. 2.2. More detail is shown in Fig. 6.1. DNA can either replicate or direct protein synthesis via mRNA. The synthesis occurs at the ribosome, a nucleoprotein complex consisting of nucleic acids and proteins. The tRNAs pick up the required amino acids and transfer them to the ribosome. There, the anticodons on the tRNA match the codons on the mRNA to determine the order on the polypeptide chain. In all operations, proteins are involved. Nucleic acids and proteins thus are fully interwoven.

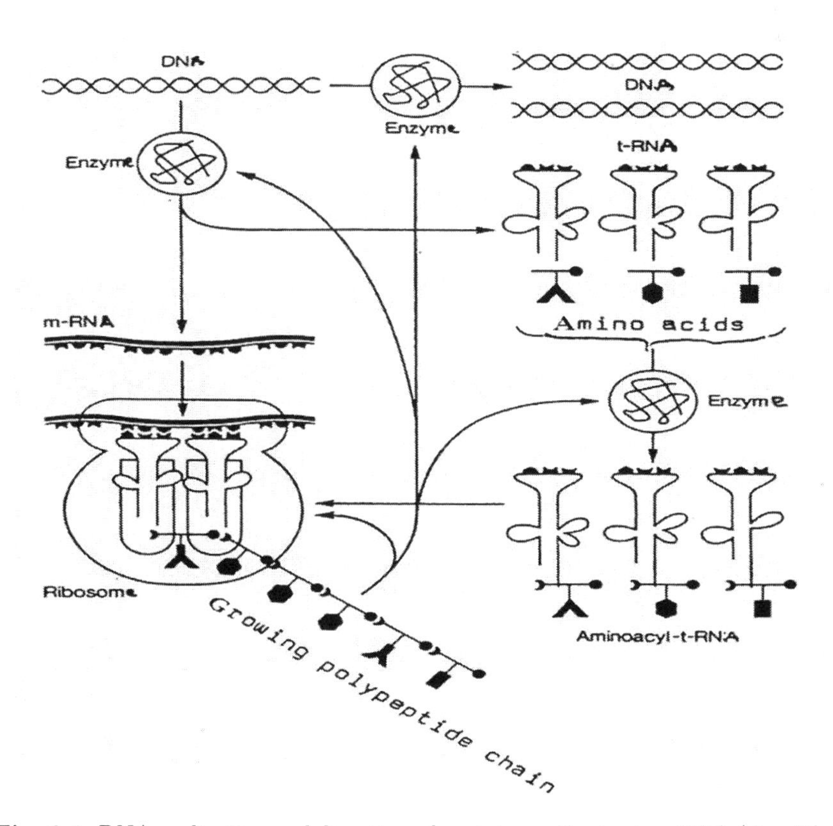

Fig. 6.1. DNA replication and direction of protein synthesis via mRNA (simplified).

One final remark concerns the physical location of DNA and protein synthesis. DNA is located in the cell nucleus; the synthesis occurs in the ribosomes, as indicated in Fig. 6.2.

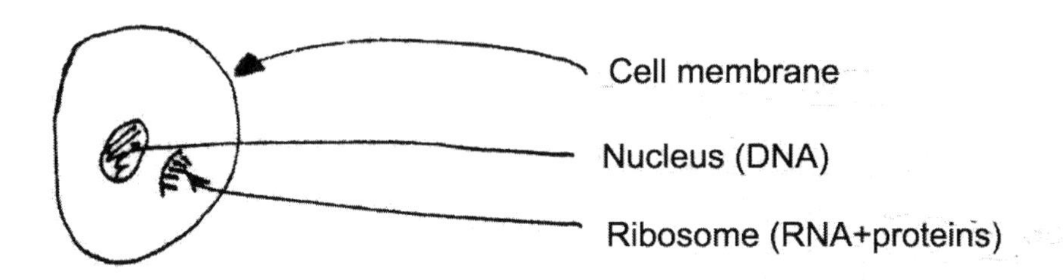

Fig. 6.2. The most important components of the cell.

Actually, the chromosomes in the cell nucleus contain the DNA; a single chromosome contains a single DNA molecule. Genes, which were originally defined genetically, thus can be identified with DNA.

7

Lipids and Membranes

Membranes are organized assemblies composed mainly of lipids and proteins. They are highly selective barriers, contain pumps and gates, and are involved in the two most important energy conversion processes, photosynthesis and oxidative phosphorylation. In photosynthesis, light is transformed into chemical energy. In oxidative phosphorylation, energy is gained by the oxidation of energy-rich materials. We will not discuss the chemical functions of membranes, but only sketch some features of their building blocks and their construction and describe some interesting physical properties. Membranes are crucial for the function of cells; an excellent description of the detailed membrane structure and function can be found in texts on cell biology, for instance [1, 2]. The physics of membranes is treated in, for instance, [3]–[6].

A membrane with an embedded protein is shown schematically in Fig. 7.1. The membrane is a sheetlike structure with a thickness between 5 and 10 nm, and consists mainly of lipids and proteins. The lipids and proteins are held together by cooperative noncovalent forces. The figure indicates three areas to be discussed: the characteristics of the membrane building blocks (lipids), the properties of the membranes alone without proteins, and the membrane-property phenomena.

7.1 Lipids

Lipids are water-insoluble biomolecules that are highly soluble in organic solvents. They are *amphipathic*: they contain both a hydrophilic and a hydrophobic moiety as indicated in Fig. 7.1. The polar head group can be either charged or neutral (zwitterion). The nonpolar tail usually consists of one or two hydrocarbon chains. In biological membrane, phospholipids are abundant. Most phospholipids are derived from glycerol, a three-carbon alcohol, and consist of a glycerol backbone, two fatty acid chains, and a phosphorylated alcohol. (Alcohol is obtained from a hydrocarbon by replacing one hydrogen atom by a hydroxyl group, –OH.) These phospholipids are called phosphoglycerides.

H. Frauenfelder, *The Physics of Proteins*, Biological and Medical Physics, Biomedical Engineering, DOI 10.1007/978-1-4419-1044-8_7,

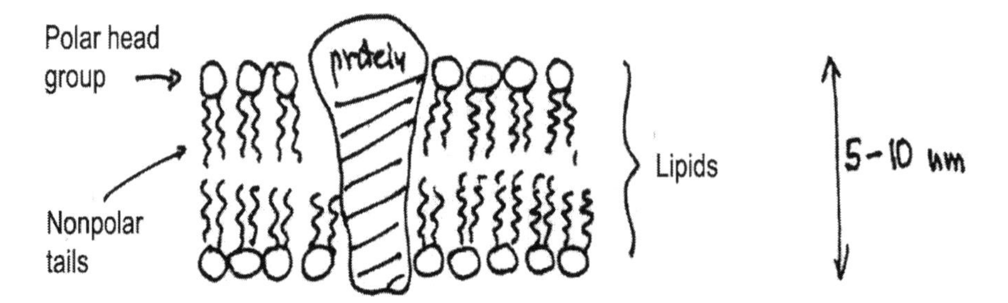

Fig. 7.1. Schematic representation of a bilayer membrane with an embedded protein.

The schematic structure is shown in Fig. 7.2, and a specific example is shown in Fig. 7.3. The fatty acid chains usually contain an even number of carbon atoms, most frequently 16 and 18. They may be saturated or unsaturated. The length and degree of unsaturation of the chains has a strong effect on membrane fluidity.

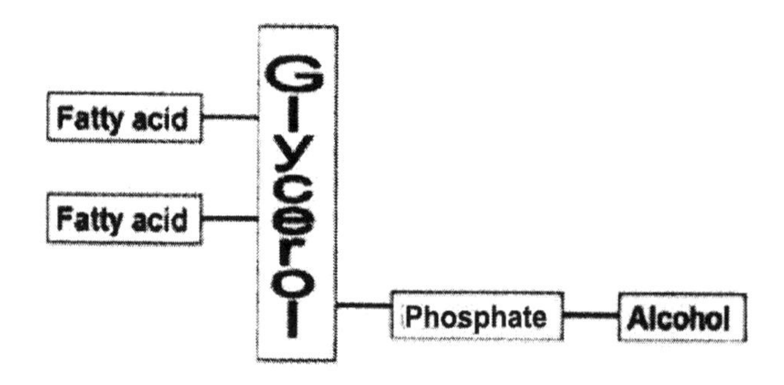

Fig. 7.2. Components of a phosphoglyceride.

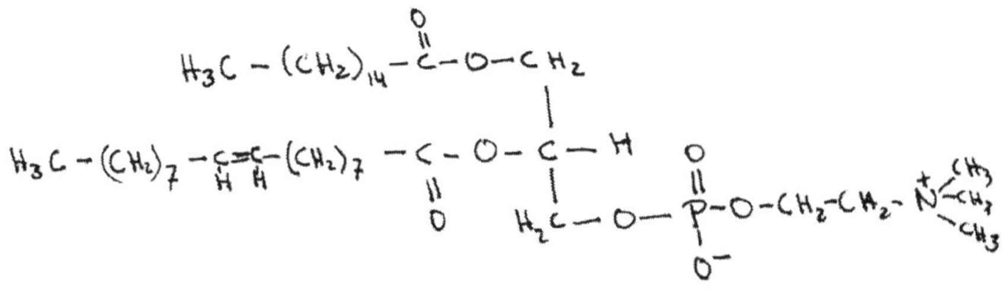

$H_3C-(CH_2)_{14}-C(=O)-O-CH_2$

$H_3C-(CH_2)_7-CH=CH-(CH_2)_7-C(=O)-O-C-H$

$H_2C-O-P(=O)(O^-)-O-CH_2-CH_2-N^+(CH_3)_3$

Fig. 7.3. Phosphatidyl choline, a frequently occurring phosphoglyceride.

7.2 Membranes

In aqueous solutions, the polar head groups of lipids like to stay in water, the hydrophilic hydrocarbon tails try to avoid it. There are various structures, shown in Fig. 7.4, that can be formed by lipids to lower the Gibbs energy, micelles, bilayer membranes, and closed bilayers or liposomes.

In water, lipids assemble and form bilayer membranes spontaneously and rapidly. The driving forces are the hydrophobic interaction and van der Waals interaction between tails. Membranes thus form noncovalent cooperative structures.

Figure 7.4 shows that membranes form essentially two-dimensional systems. At physiological temperatures, lipid membranes are fluid systems; at lower temperatures, they become solidlike. The physics of membranes, still in its infancy, displays many phenomena similar to those in fluids and solids.

Structure. The basic structure of a membrane is shown in Fig. 7.1. Real membranes are obviously far more complex. The detailed arrangement of the lipids and proteins depends very much on characteristics of all components, and the phase diagram of a membrane system can be complicated [3, 5].

Energy Landscape. We would expect that the energy of a membrane as a function of its conformation (structure) displays a rugged landscape as sketched in Fig. 3.1(b). Not enough is known, however, to give details.

Dynamics—Fluidity. Motion in membranes are highly anisotropic, with a diffusion anisotropy of the order of 10^7. If r denotes a linear distance, diffusion gives

$$\langle r^2 \rangle = 4Dt \tag{7.1}$$

for the mean square lateral displacement after a time t. A typical diffusion coefficient is $D \approx 10^{-8}\text{cm}^2/\text{s}$. In one second, a lipid can travel about 2 μm. The corresponding rotational diffusion expression,

$$\langle \Theta^2 \rangle \approx 2D_r t \tag{7.2}$$

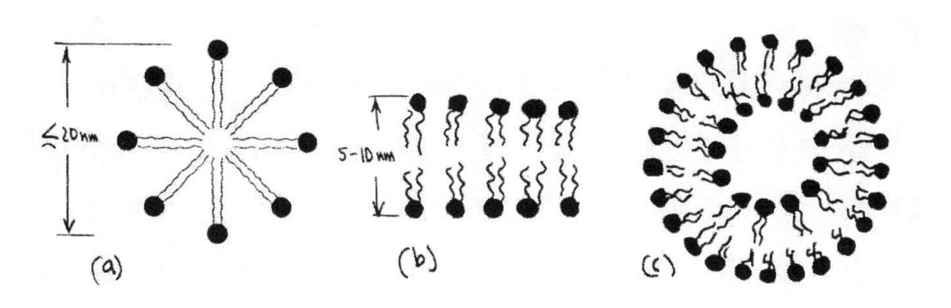

Fig. 7.4. Schematic arrangements – the cross sections of **(a)** a micelle, **(b)** a bilayer membrane, and **(c)** a liposome.

with $D_r \approx 10^8$, s^{-1} shows that a lipid rotates about once around its long axis while it travels over its own diameter.

Dynamics—Phase Transitions [5]–[7]. All lipid bilayers show a first-order phase transition associated with the melting of the hydrocarbon chains. Below the transition, in the crystalline (gel) phase, the hydrocarbon chains form a lattice. Above the transition, the membrane is fluid (liquid crystal) like. The transition is accompanied by a large expansion of the bilayer area, a thinning of the membrane, and a large increase of the entropy, $\delta S/k_B \approx 15$/molecule. A very large number of degrees of freedom, of the order of 10^6, is activated during the transition.

Function—Permeability. Since membranes form the boundaries between biological systems, their permeability for various molecules is crucial for function and control. In general, lipid bilayers have a low permeability for ions and most polar molecules, with the exception of water.

7.3 Biomembranes and Inclusions

Biomembranes are usually not simple lipid bilayers. They contain proteins, polypeptides, and cholesterol. Four arrangements of proteins in membranes are shown in Fig. 7.5. The type and amount of proteins depend on function; the content varies from less than 20% to about 75%.

Mosaic Model. So far we have treated membranes as if they were uniform systems. A real biomembrane can exist as a two-dimensional lipid mixture. According to the model of Singer and Nicolson [8], membrane proteins may be incorporated in regions of rigid lipids; the biomembrane then assumes a mosaic structure.

Membrane Asymmetry. The membrane components are asymmetrically distributed between the two surfaces. For proteins, the asymmetry is absolute;

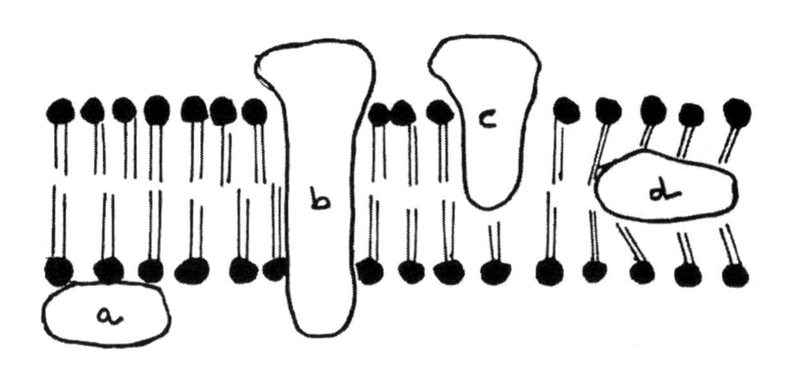

Fig. 7.5. Membrane proteins. **(a)** is a peripheral protein, **(b)**–**(d)** are integral proteins.

every copy has the same orientation in the membrane. Almost every type of lipid is present on both sides, but in different amounts.

Protein Diffusion. The diffusion coefficients for proteins in membranes are considerably smaller than those for the lipids itself; the coefficients in Eqs. (7.1) and (7.2) are of the order $D \approx 10^{-10}\text{cm}^2/\text{s}$ and $D_r \approx 10^4\text{s}^{-1}$. Transverse diffusion is very slow for lipids, and the time for "flip-flop" is of the order of hours. Transverse diffusion (rotation) of membrane proteins has not been observed.

Lipid–Protein Interactions. How do membranes affect the structure and function of the membrane proteins? How do proteins affect the membranes? We will not discuss these questions here but only state that dynamics will be an important aspect of the solution. For reviews, see [7, 9].

References

1. T. D. Pollard and W. C. Earnshaw. *Cell Biology.* Saunders, Philadelphia, 2002.
2. B. Alberts, A. Johnson, J. Lewis, M. Raff, K. Roberts, and P. Walter. *Molecular Biology of the Cell*, 4th edition. Garland Science, New York, 2002.
3. R. Lipowsky and E. Sackmann, editors. *Structures and Dynamics of Membrances.* Handbook of Biological Physics, Vol. 1A and Vol. 1B. Elsevier, Amsterdam, 1995.
4. P. Nelson. *Biological Physics.* W. H. Freeman, New York, 2004.
5. E. Sackmann. In W. Hoppe and R. D. Bauer, editors, *Biophysics.* Springer, Berlin, 1983, pp. 425–57.
6. O. G. Mouritsen. In D. Baeriswyl, M. Droz, A. Malaspinas, and P. Martinoli, editors, *Physics in Living Matter* (Lecture Notes in Physics, 284). Springer, Berlin, 1987. pp. 77–109.
7. O. G. Mouritsen and M. Bloom. Models of lipid-protein interactions in membranes. *Ann. Rev. Biophys. Biomol. Struct.*, 22:145–71, 1993.
8. S. J. Singer and G. L. Nicolson. The fluid mosaic model of the structure of cell membranes. *Science*, 173:720–31, 1972.
9. J. P. Cartailler and H. Luecke. X-ray crystallographic analysis of lipid-protein interactions in the bacteriorhodopsin purple membrane. *Ann. Rev. Biophys. Biomol. Struct.*, 32:285–310, 2003.

Part II

SPATIAL STRUCTURE OF PROTEINS: MEASUREMENT AND CONSEQUENCE

In Part II, we focus our attention on how we obtain spatial structure of biomolecules, in particular, proteins, and some earlier known examples of how spatial structure controls protein function. The composition and structure of biomolecules were not elucidated quickly. It took many years before it was established that biomolecules were separate entities and not complexes of simpler organic molecules. More years passed before the primary sequence of the first protein was unraveled, the three-dimensional structure of myoglobin became known, and the double helix became a household word. The discussion here is centered on the physical tools that have led to the present state of understanding.

8

The Secondary Structure

In principle we could expect that X-ray diffraction provides a complete picture of any protein that can be crystallized, with the exact position of each atom. Indeed, models displayed in showcases give the impression that the protein structure is fully known and has been found *ab initio*. In reality, however, stereochemical information is used in the elucidation of the structure of a protein. The knowledge of the structure of the building elements is consequently essential. Moreover, X-rays do not "see" hydrogen atoms well; the charge density of these is often determined by using the well-known structure of the amino acids. We give here a brief discussion of the secondary structure; a beautiful and clear treatment is given by Dickerson and Geis [1].

8.1 The Amino Acids

We return to the building blocks, amino acids. The backbone of the polypeptide chain consists of repeating units, $\cdots$ N—C_α—C—N—C_α—C $\cdots$. A single amino group is planar as shown in Fig. 8.1, where the dimensions are given. What structures can be formed from these building blocks? The problem was first attacked by Pauling and collaborators [2, 3], who proposed the α-helix and the parallel and antiparallel pleated sheets. These structures indeed occur in all proteins.

The amino group and its links to the neighboring residues are sketched in Fig. 8.2. Two degrees of freedom exist, namely rotation about the axes C_α—N(ϕ) and C_α—$C(\psi)$. The angles ϕ and ψ are shown in Fig. 8.3. A list of all pairs (ϕ, ψ) completely describes the geometry of the chain backbone.

If the angles at every C_α are the same, the polypeptide chain forms a helix, as shown for some cases in Fig. 8.3. Pitch, p, number of elements per turn, n, and rise, d, are defined in Fig. 8.3. For given ϕ and ψ, n and d are determined.

H. Frauenfelder, *The Physics of Proteins*, Biological and Medical Physics, Biomedical Engineering, DOI 10.1007/978-1-4419-1044-8_8,

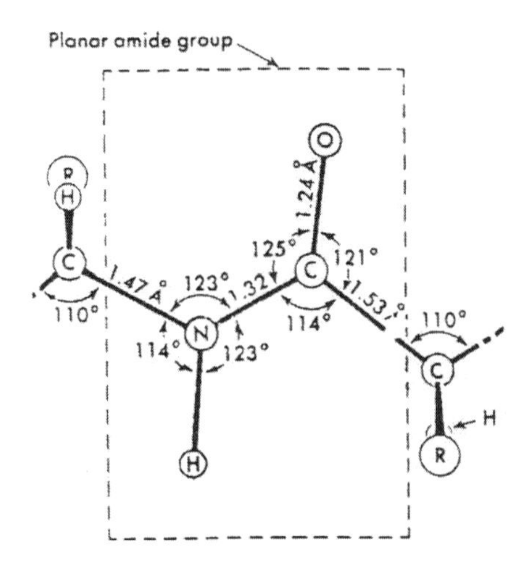

Fig. 8.1. Structural parameters describing the amino group [4].

8.2 The Ramachandran Plot

Some conformations will be natural, some will be impossible to achieve because atoms come too close together. The problem was studied in detail by Ramakrishnan and Ramachandran [5] and Ramachandran and Sasisekharan [6]. The result of the studies is usually expressed in the Ramachandran map, given in crude outline in Fig. 8.4.

This plot shows region of ϕ and ψ that can be occupied. A number of possible building elements have been found, but only the three most important ones are indicated in Fig. 8.4. Properties of two of the helices are given in Table 8.1. Note that the numbers are not unique. On the line $n = 3.6$, the α-helix can be realized by more than one pair (ϕ, ψ).

8.3 The α-Helix

The α-helix is particularly stable because it is maximally hydrogen bonded, as can be seen from Fig. 8.5. Each peptide bond participates in the hydrogen bonding and the hydrogen bonds are so oriented that they give nearly maximal bond strength. Not all amino acids form α-helices equally easily, as shown in Table 8.2.

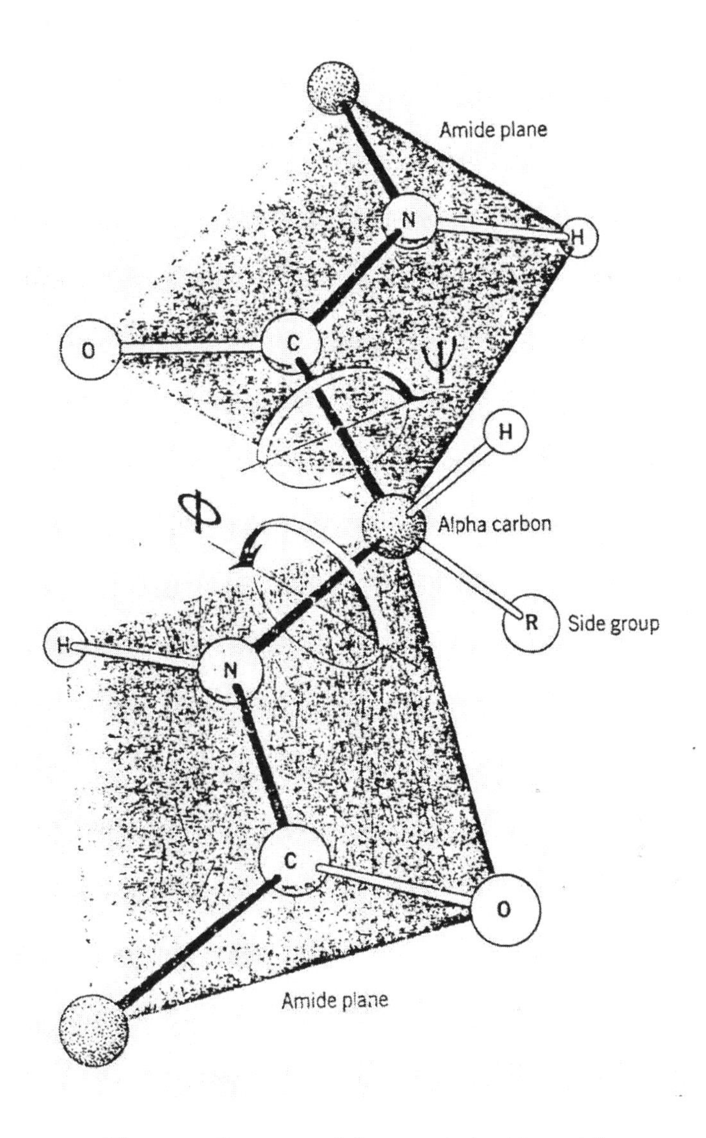

Fig. 8.2. Geometry of the protein backbone [1].

Reading textbooks leaves the impression that the α-helix is always a well-defined structure, with each turn having identical characteristics. Reality, however, is different. The number of turns, $n = 3.61$, is not determined by a symmetry law, but by an optimum in hydrogen bonding. Destabilizing residues in a helix and interaction of the helix with other protein components change the

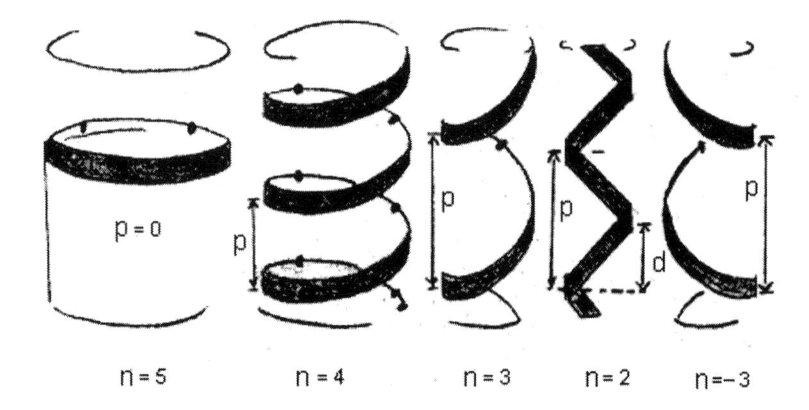

Fig. 8.3. The polypeptide chain can form different helices (after Dickerson and Geis [1]).

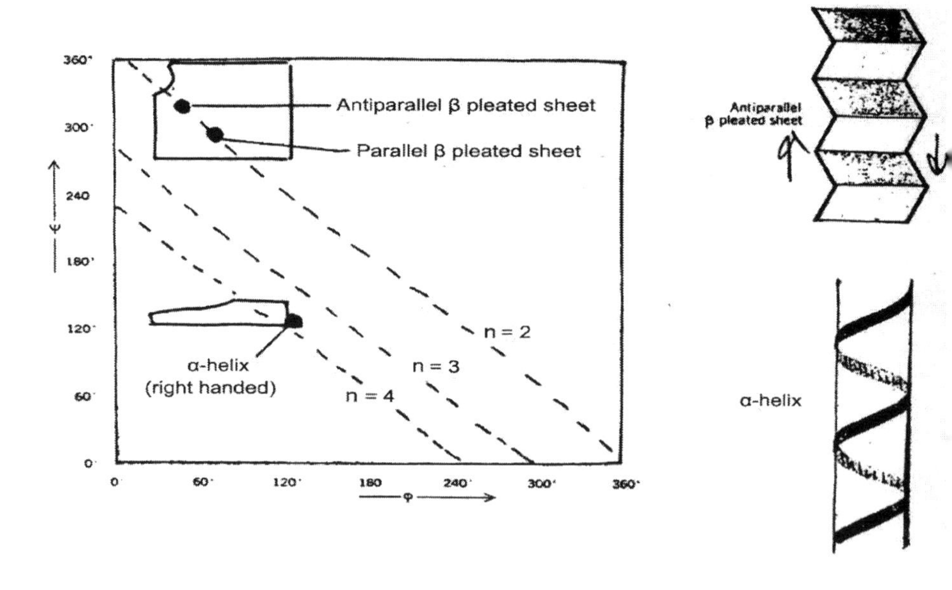

Fig. 8.4. Ramachandran map. The two small fields give the main allowed regions. Only the positions of the three most important helices are given.

optimum position. In a typical α-helix in Mb, about 40% of the residues are of the destabilizing type. The numbers of units per turn can therefore vary from about 3 to more than 4.

Table 8.1. Properties of Two Helices.

	α-Helix	Antiparallel β Pleated Sheet
ϕ	132°(113°)	40°
ψ	123°(136°)	315°
n	3.61	2.00
d (Å)	1.50	3.47
p (Å)	5.41	6.95

Table 8.2. Formation of α-Helices.

skip 0.1in

Allow	α-helix	Ala, Leu, Phe, Tyr, Try, Cyd, Met, His, Asn, Gln, Val
Destabilize	α-helix	Ser, lle, Thr, Glu, Asp, Lys Arg, Gly
Break	α-helix	Pro, Hydroxypro

8.4 β Pleated Sheets

In addition to the helices, β pleated sheets are often found in globular proteins. While the α-helix is formed by a continuous structure, the β pleated sheets are obtained by the interaction between stretched segments (Fig. 8.6).

The backbone is stretched nearly as much as is possible, with $n = 2$. Two such stretched helices can have hydrogen bonds between NH and CO groups in different strands, thus forming a pleated sheet. The bond arrangement in

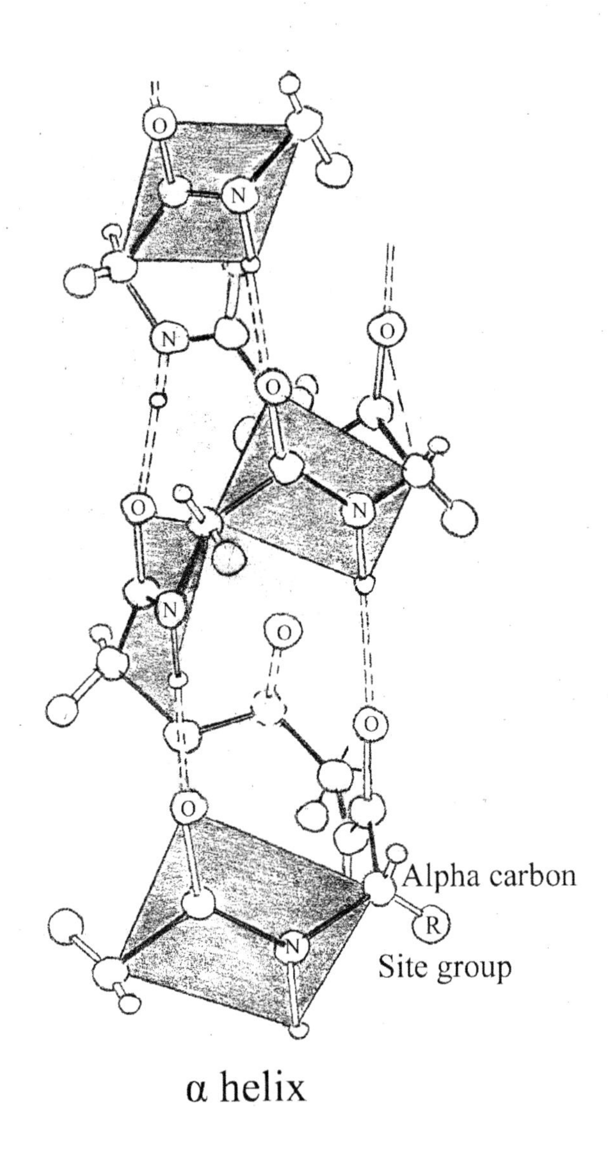

Fig. 8.5. Hydrogen bonding is nearly optimal in the α-helix.

one antiparallel sheet is shown in Fig. 8.7. The hydrogen bond patterns for parallel and antiparallel β sheets are sketched in Fig. 8.8.

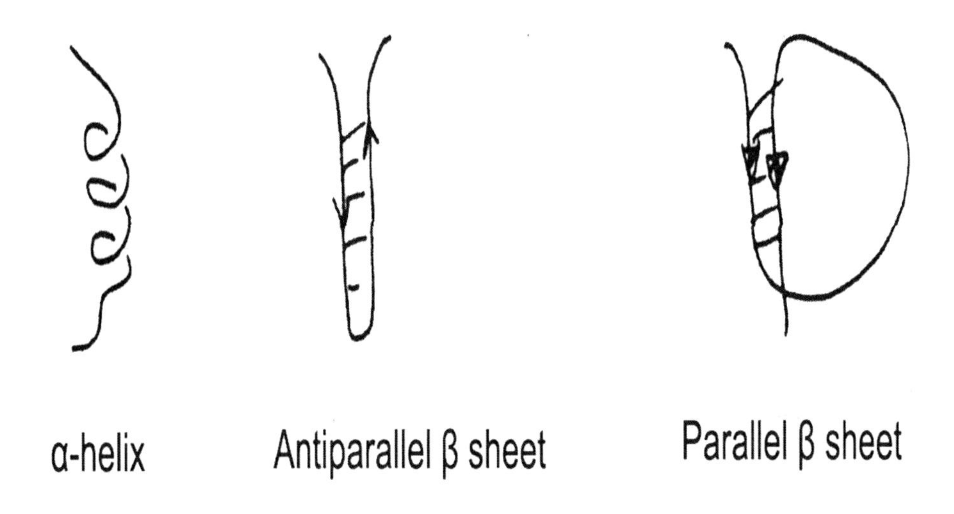

Fig. 8.6. Geometry of the backbone in an α-helix and the β pleated sheets.

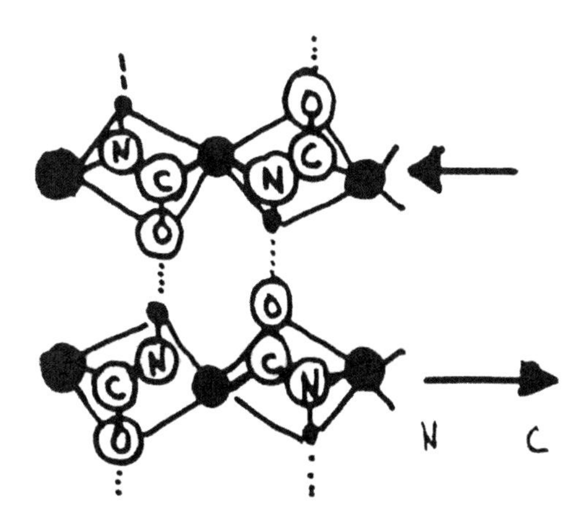

Fig. 8.7. Hydrogen bonds between CO and NH in the antiparallel β pleated sheet.

8.5 Secondary Structure Prediction

In the goal of determining the tertiary working structure of proteins from their primary sequence, the prediction of secondary structures is the first

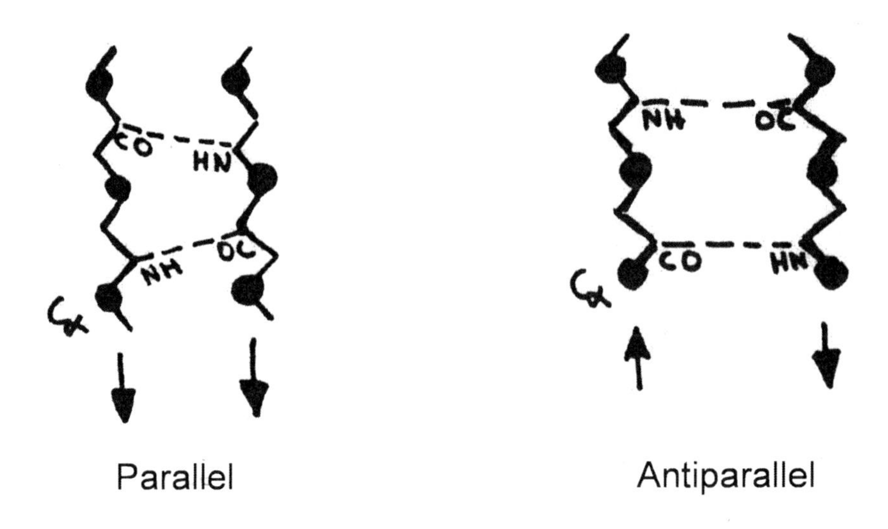

Fig. 8.8. Hydrogen bond geometry in parallel and antiparallel β pleated sheets.

waystation. A number of different approaches are being pursued, for instance, statistical analysis, energy minimization, and stereochemical methods [7, 8]. (One of the Web search engines has about tens of million entries on "secondary structure prediction.")

References

1. R. E. Dickerson and I. Geis. *The Structure and Action of Proteins.* Benjamin/Cummings, Menlo Park, CA, 1969.
2. L. Pauling, R. B. Corey, and H. R. Branson. The structure of proteins: Two hydrogen-bonded helical configurations of the polypeptide chain. *Proc. Natl. Acad. Sci. USA*, 37:205–11, 1951.
3. L. Pauling and R. B. Corey. Atomic coordinates and structure factors for two helical configurations of polypeptide chains. *Proc. Natl. Acad. Sci. USA*, 37:235–85, 1951.
4. L. Pauling. *The Nature of the Chemical Bond*, 3rd edition. Cornell Univ. Press, Ithaca, NY, 1960, p. 498.
5. C. Ramakrishnan and G. N. Ramachandran. Stereochemical criteria for polypeptide and protein chain conformations. II. Allowed conformations for a pair of peptide units. *Biophys. J.*, 5:909–33, 1965.
6. G. N. Ramachandran and V. Sasisekharan. Conformation of polypeptides and proteins. *Adv. Protein Chem.*, 23:283–438, 1968.
7. G. E. Schulz. A critical evaluation of methods for prediction of protein secondary structures. *Ann. Rev. Biophys. Biophys. Chem.*, 17:1–21, 1988.
8. G. A. Petsko and D. Ringe. *Protein Structure and Function.* New Science Press, London, 2004.

9

Tertiary Structure of Proteins

In Chapter 4, we gave a brief introduction to proteins. The structures of a very large number of proteins have been determined and it is possible to ask fundamental questions: Given the primary sequence, what is tertiary structure? How does the protein fold into the final structure? This "folding problem" has attracted a great deal of attention, and it has become an industry. (One of the Web search engines has more than 10^6 entries.) We will not treat the folding problem here, but refer to review articles for more information [1]–[7]. Here we discuss a simpler problem, how the main secondary structures, α-helices and β pleated sheets, combine to form globular proteins. The folding problem is also treated in Section 17.3.

9.1 Packing of α-Helices and β Pleated Sheets

Three principles dominate the way secondary structures associate [8]–[12].

1. Residues that become buried in the interior of a protein pack closely together; they occupy a volume similar to that in crystals of their amino acids. 2. Associated secondary structures retain a conformation close to the minimum free energy conformation of the isolated secondary structure. Bond lengths, bond angles, and torsion angles have the same standard values as in small molecules.

3. Almost all atoms that are acceptors and donors of H-bonding pair up to form hydrogen bonds.

The principles imply that the secondary structures found in proteins interact to give the maximum van de Waals attraction and the minimum steric strain.

With these rules and the well-known structures of the α-helix and the β pleated sheets, the various possibilities of packing have been explored with computer graphics and numerical calculation. Some results can simply be stated.

H. Frauenfelder, *The Physics of Proteins*, Biological and Medical Physics, Biomedical Engineering, DOI 10.1007/978-1-4419-1044-8_9,

Consider first *helix-helix packing.* The α-helix has ridges and grooves. The models indicate that packing occurs so that ridges fit into grooves. Three classes of interaction occur, as shown in Fig. 9.1. Each class can be characterized by Ω, the angle between the strands of the pleated sheet and/or helix axes when projected onto their plane of contact. The angles given in Fig. 9.1 are for two identical helices with 3.6 residues per turn and radius (5 Å). Examples of all three classes have been found in proteins, and the observed angles are in the neighborhood of the ones given in Fig. 9.1.

Helix-sheet packing is shown in Fig. 9.2 [8]. An α-helix packs onto a β pleated sheet with its axis parallel to the strands of the sheet. The angle Ω then is zero. In a more detailed model, Ω will depend on the twist in the sheet.

Face-to-face packing of two β sheets contains two classes: One class is formed by the packing of two large and independent sheets, the second by the folding over of single sheets. In the proteins studied, the sheets have a right-hand twist when viewed in the direction of the strands. We consider here only the first class as an example. Computer model building then shows that the best packing of two sheets is obtained if

$$\Omega = -2(|T_1 - T_2|)$$

where T_1 and T_2 are the twists of the two sheets.

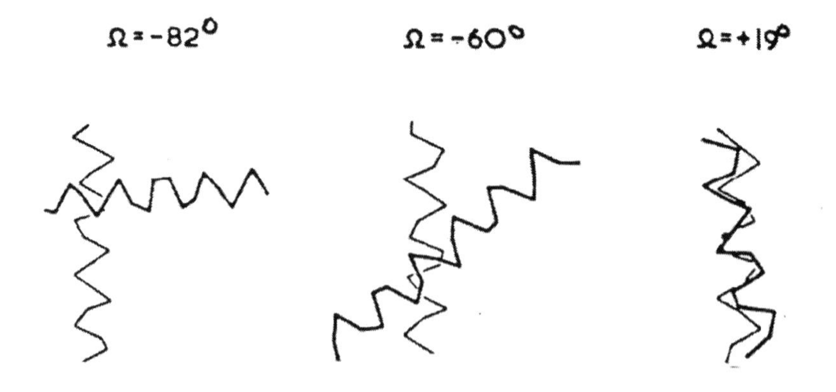

Fig. 9.1. Helix-helix packing.

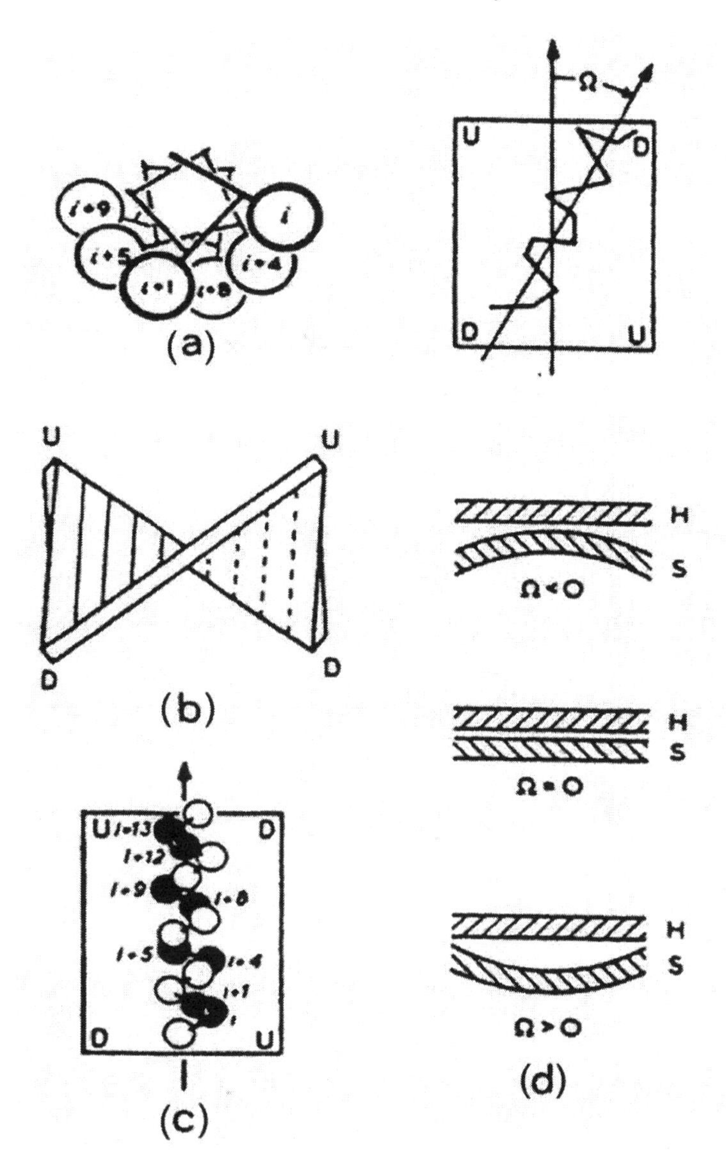

Fig. 9.2. The model for helix-sheet packing. (a) The helical residues form a surface with a right-hand twist that is complementary to the right-hand twist of a pleated sheet shown in (b). (c) The helix is shown on top of the twisted sheet. The corners marked U are above the plane of the page, and those marked D are below it. (d) Sections showing idealized helix-sheet interfaces for different values of Ω. After Chothia et al. [8].

9.2 Structural Patterns

The ultimate goal of structure research is ambitious: how to achieve a desired function by specifying only the primary sequence. We are far from that goal (and it may be unreachable), but a systematic study of many proteins reveals some patterns. Since proteins are rather complicated it is not even easy to compare two or more of them; some new classification techniques are required. References [11]–[13] give the beginning of such schemes. (Earlier work is cited in these papers.) Proteins are characterized by topology/packing diagrams. We will not describe this approach in detail but give in Fig. 9.3 three simple examples:

Analysis of many proteins using topology/packing diagrams indicates that very often residues that are close together in the primary sequence are also in close contact in the tertiary structure. Thus folding subunits are formed that may well make the initial folding easier. In protein folding, pieces of secondary structure would first diffuse together to form folding units. These would then associate to form the native structure.

More insight comes from a detailed study of the proteins that contain β pleated sheets [12, 13]. Many patterns that are observed may be similar not because of evolution, but because the arrangement is topologically simple and

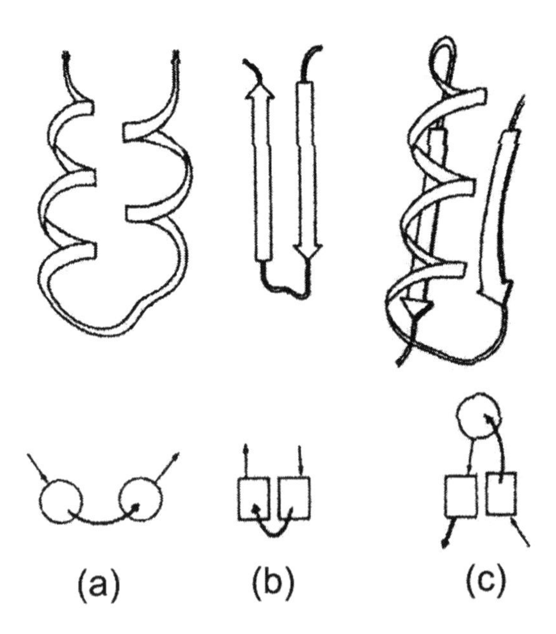

Fig. 9.3. The folding of the polypeptide chain in the three commonly occurring folding units: $(\alpha\alpha)$, $(\beta\beta)$, and $(\beta\alpha\beta)$. The ribbon illustrates the path of the backbone with the arrows directed from the N to C terminals. After Levitt and Chothia [11].

thus easy to arrive at. Examples of some such patterns are given in Fig. 9.4. Some other cases are presented in Fig. 9.5.

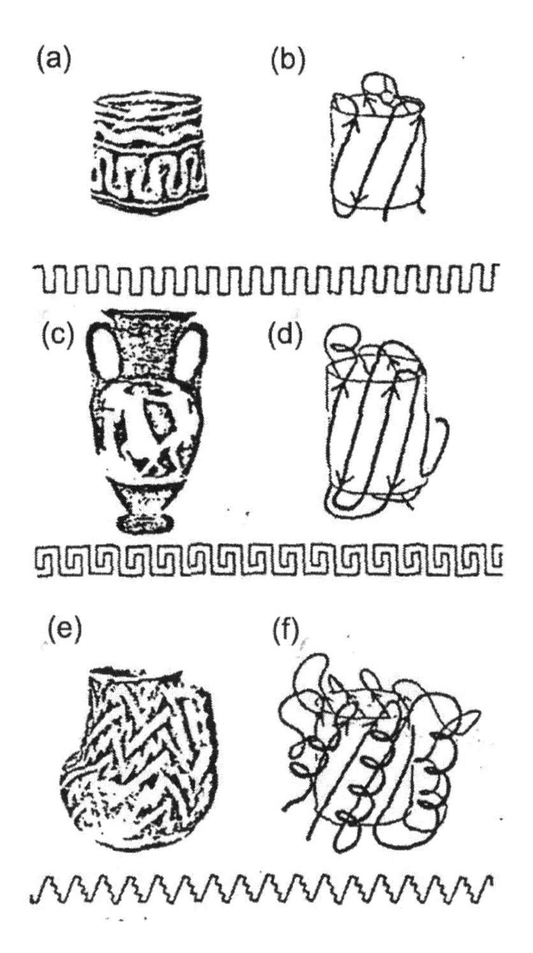

Fig. 9.4. Comparison of geometric motifs common on Greek and American Indian weaving and pottery with the backbone folding patterns found for cylindrical β sheet structures in globular proteins. (a) Indian polychrome cane basket, Louisiana. (b) Polypeptide backbone of rubredoxin. (c) Red-figured Greek amphora showing Cassandra and Ajax (about 450 B.C.). (d) Polypeptide backbone of prealbumin. (e) Early Anasazi Indian redware pitcher, New Mexico. (f) Polypeptide backbone of triose phosphate isomerase. Photographs (a) and (e) courtesy of the Museum of the American Indian, Heye Foundation; photograph (c) courtesy of the Metropolitan Museum of Art, Fletcher Fund, 1956. After Chothia [10].

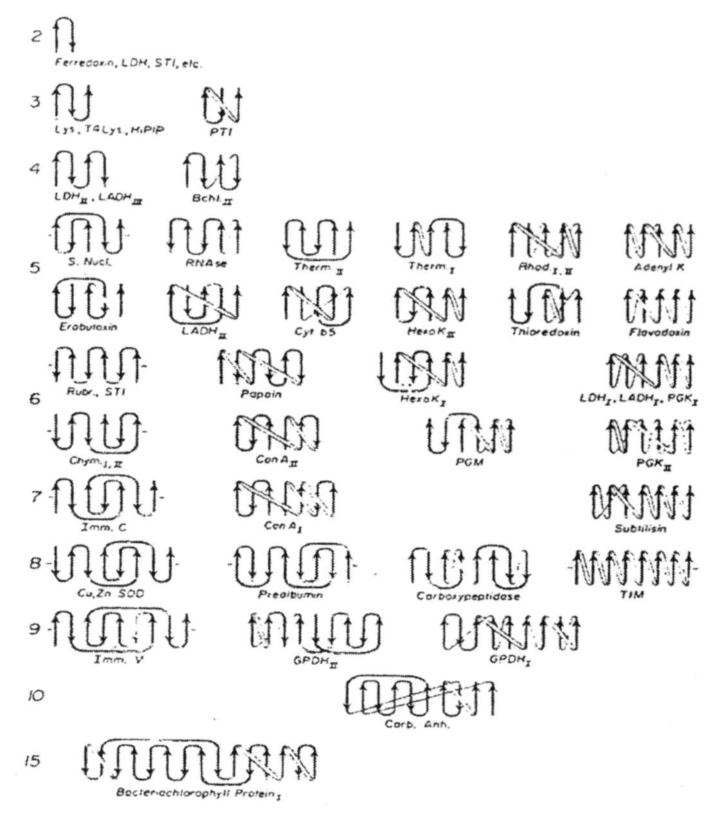

Fig. 9.5. Schematic diagrams for all the topologically distinct β pleated sheets found in proteins of known three-dimensional structure. Arrows indicate the strands of the β sheet, considered to be in the plane of the paper; crossover connections shown wide are above the sheet and thin ones are below it. Only topological connectivity is represented, with no attempt to indicate the length or conformation of the connections. The β sheets are sorted from top to bottom by number of strands, and from left to right according to their percentage of antiparallel as against parallel β structure. In cases of proteins with more than one β sheet, Roman numeral subscripts are added. Those β sheets that can be considered to form cylindrical barrels are flanked by hyphens and are shown as viewed from the outside. Abbreviations for the 33 proteins included (in decreasing order of β sheet size) are as follows: bacteriochlorophyll protein (*Bchl.*), carbonic anhydrase C (*Carb. anh.*), glyceraldehyde phosphate dehydrogenase (*GPDH*), immunoglobulin variable (Imm. V) and immunoglobulin constant (Imm. C) domains, Cu,Zn superoxide dismutase (*Cu, Zn SOD*), prealbumin, carboxypeptidase *A*, triose phosphate isomerase (*TIM*), concanavalin A (*Con A*), subtilisin, chymotrypsin (*Chym.*), phosphoglycerate mutase (*PGM*), phosphoglycerate kinase (*PGK*), rubredoxin (*Rubr.*), soybean trypsin inhibitor (*STI*), papain, hexokinase (*HexoK*), lactate dehydrogenase (*LDH*), liver alcohol dehydrogenase (*LADH*), erabutoxin, cytochrome b5 (*Cyt. b5*), thioredoxin, flavodoxin, staphylococcal nuclease (*S. nucl.*), pancreatic ribonuclease (*RNAse*), thermolysin (*Therm.*), rhodanese or thiosulphate sulphurtransferase (*Rhod.*), adenyl kinase (*AdenylK*), egg white lysozyme (*Lys.*), T4 phage lysozyme (*T4 Lys.*)), high-potential iron protein (*HiPIP*), pancreatic trypsin inhibitor (*PTI*), ferredoxin. After Richardson [12].

References

1. F. M. Richards, D. S. Eisenberg, and P. S. Kim. *Protein Folding Mechanisms.* Academic Press, San Diego, 2000.
2. R. Bonneau and D. Baker. Ab initio protein structure prediction: Progress and prospects. *Ann. Rev. Biophys. Biomol. Struct.*, 30:173–89, 2001.
3. S. S. Plotkin and J. N. Onuchic. Understanding protein folding with energy landscape theory: Part I: Basic concepts. *Q. Rev. Biophys.*, 35:111–67, 2002.
4. S. S. Plotkin and J. N. Onuchic. Understanding protein folding with energy landscape theory: Part II: Quantitative aspects *Q. Rev. Biophys.*, 35:205-86, 2003.
5. C. Dobson. Principles of protein folding, misfolding, and aggregation. *Seminars in Cell and Developmental Biology*, 15:3–16, 2004.
6. J. N. Onuchic and P. G. Wolynes. Theory of protein folding. *Curr. Opin. Struct. Bio.*, 14:70–5, 2004.
7. M. Oliverberg and P. G. Wolynes. The experimental survey of protein folding energy landscapes. *Q. Rev. Biophys.*, 38:245-86, 2005. Published online June 19, 2006.
8. C. Chothia, M. Levitt, and D. Richardson. Structure of proteins: packing of alpha-helices and pleated sheets. *Proc. Natl. Acad. Sci. USA*, 74:4130–4, 1977.
9. D. A. D. Parry and E. N. Baker. Biopolymers. *Rep. Progr. Phys.*, 47:1133–232, 1984.
10. C. Chothia. Principles that determine the structure of proteins. *Ann. Rev. Biochem.*, 53:537–72, 1984.
11. M. Levitt and C. Chothia. Structural patterns in globular proteins. *Nature*, 261(5561):552–8, 1976.
12. J. S. Richardson . β sheet topology and the relatedness of proteins. *Nature*, 268(5620):495–500, 1977.
13. J. S. Richardson. The anatomy and taxonomy of protein structure. *Adv. Protein Chem.*, 34:167–339, 1981.

10

Myoglobin and Hemoglobin

Hemoglobin is the vital protein that conveys oxygen from the lungs to the tissues and facilitates the return of carbon dioxide from the tissues back to the lungs [1]–[6]. Myoglobin accepts and stores the oxygen released by hemoglobin and transports it to the mitochondria. The pathways are shown in Fig. 10.1. Together, these two proteins show how protein structure determines function and how Nature uses components, in this case myoglobin, to build more complex proteins consisting of functional subunits. In Part III we explore myoglobin dynamics in more depth; this chapter provides a brief overview of a very complex system and points to why Part III will concentrate on myoglobin rather than hemoglobin.

Myoglobin is about as simple a protein as can be found, and we use it as a prototype for studying many of the aspects of protein physics. Hemoglobin is already more complex. While Mb is a monomer, Hb is a tetramer, basically built of 4 Mb monomer, as we show in Fig. 10.2. Hb is the prototype of an allosteric protein that is regulated by specific molecules in the environment. Moreover, Hb shows cooperative behavior: it can bind four oxygen molecules, but the binding free energy is a function of how many oxygen molecules are bound. The first oxygen binds with a much smaller free energy than the last. Mb from different species are similar, whereas Hb shows pronounced and fascinating species adaptation [4, 7].

In this chapter we give some of the basic features and numbers concerning the binding of small ligands, O_2 and CO, to Mb and Hb.

10.1 Ligand Binding to Myoglobin

We use Mb as a prototype of a monometric protein and write a differential equation for the reaction of deoxyMb with a ligand, for instance O_2,

$$\mathrm{Mb} + \mathrm{O_2} \underset{\lambda_{off}}{\overset{\lambda'_{on}}{\rightleftharpoons}} \mathrm{MbO_2}. \tag{10.1}$$

H. Frauenfelder, *The Physics of Proteins*, Biological and Medical Physics, Biomedical Engineering, DOI 10.1007/978-1-4419-1044-8_10,

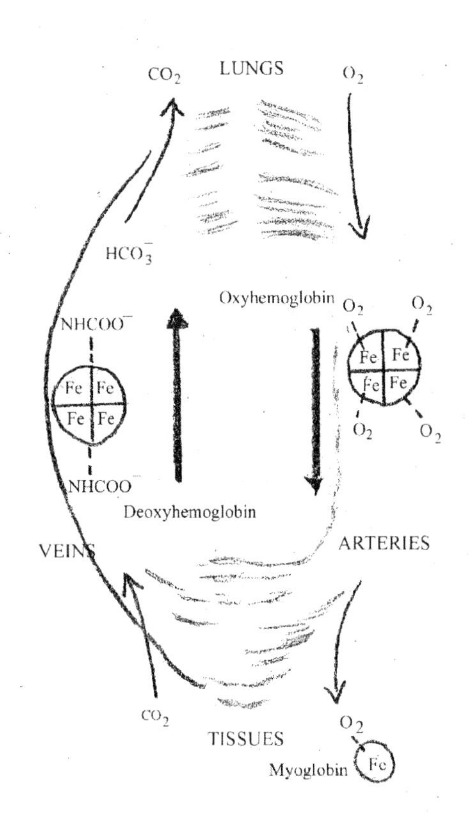

Fig. 10.1. Oxygen is carried from the lungs to the tissues by hemoglobin, and carbon dioxide is carried back to the lungs. Part of the CO_2 transport occurs because the amino terminal of the four chains in hemoglobin can bind CO_2 directly to form carbamino compounds: $R\text{–}NH_2 + CO_2 \rightarrow R\text{–}NH\text{–}COO^- + H^+$. After Dickerson and Geis [6].

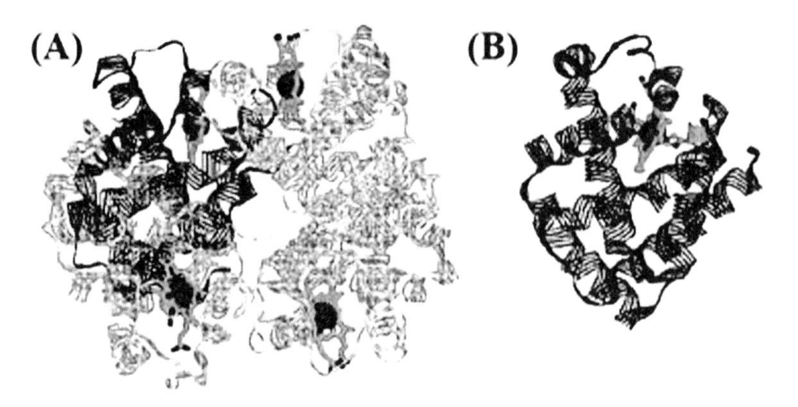

Fig. 10.2. Comparison of the X-ray structure of (A) deoxy Hb and (B) deoxy Mb.

Here λ'_{on} is the rate coefficient for association and λ_{off} the corresponding coefficient for dissociation. Since these coefficients can change over many orders of magnitude depending on temperature, we refuse to call them constants. Moreover, for reasons that will become clear later, we denote them with λ and not with k. We keep k for elementary steps. The prime on λ'_{on} denotes that it is a second-order coefficient, with units different from the first-order coefficient λ_{off}; λ'_{on} depends on the concentration of the reactants, λ_{off} does not. Denoting concentration (mole/liter) by brackets [ ], we write the rate equation for the reaction (10.1) as

$$-\frac{d[\text{Mb}]}{dt} = +\frac{d[\text{MbO}_2]}{dt} = \lambda'_{\text{on}}[\text{Mb}][\text{O}_2] - \lambda_{\text{off}}[\text{MbO}_2]. \tag{10.2}$$

In equilibrium, d[Mb]/dt = 0, so we examine the *steady-state* conditions:

$$\Lambda_a = \frac{\lambda'_{\text{on}}}{\lambda_{\text{off}}} = \frac{[\text{MbO}_2]}{[\text{Mb}][\text{O}_2]}. \tag{10.3}$$

Here Λ_a is the equilibrium coefficient for association, connected to the equilibrium coefficient for dissociation by

$$\Lambda_a = 1/\Lambda_d. \tag{10.4}$$

If the system is not in equilibrium, the general solution of Eq. (10.2) depends on the initial conditions and is usually found by computer. We return to this problem later and continue here with the equilibrium case. Note that Λ_d has units of moles oxygen/liter.

Assume that Mb is in a *solution in equilibrium with an external reservoir of oxygen*. This means that the partial oxygen pressure above the solution, given by P, is also assumed to be in equilibrium with the solvent and is constant, independent of the concentration of Mb in solution or how much oxygen has been bound by the Mb. In equilibrium, some Mb molecules will be in the deoxy form, and some will have an O_2 bound. The dimensionless degree of *fractional saturation*, y, is given by

$$y = \frac{[\text{MbO}_2]}{[\text{MbO}_2] + [\text{Mb}]} \tag{10.5}$$

or, with Eq. (10.3),

$$y = \frac{\Lambda_a[\text{O}_2]}{1 + \Lambda_a[\text{O}_2]} = \frac{[\text{O}_2]}{[\text{O}_2] + \Lambda_d}. \tag{10.6}$$

Since the $[O_2]$, the concentration of dissolved O_2 is proportional to the partial pressure P of the dioxygen above the liquid, we have:

$$y = \frac{P}{P + P_{1/2}}. \tag{10.7}$$

Here $P_{1/2}$ is the partial oxygen pressure at which half of the Mb molecules are in the deoxy state, and it is a measure of the affinity Λ_d of the protein for oxygen if the solubility of O_2 in the solvent is known. Of course, different solvents and different temperatures have different oxygen solubilities so one has to be careful. In any event, typically Λ_d is given by the external O_2 pressure at 20° C, usually given in torr, where 1 torr = 1 mm Hg = 133.21 Pa (pascal), and 10^5 Pa = 1 bar. At that pressure and temperature, the concentration of O_2 is about 0.2 mM. Do not confuse association rates and equilibrium coefficients. Mb binds oxygen more tightly than Hb, but the binding of O_2 is slower to Mb than to Hb.

Before computers made curve fitting easier, it was common to plot using a transform of the variables that turned the plot into a straight line. It is too bad this habit of linearizing data has died out. The Hill plot is the log of the ratio of oxyMb to deoxyMb, $y/(1-y)$, to the log of the ratio of the oxygen pressure P to $P_{1/2}$, or using Eqs. (10.6) and (10.7)

$$\log(\frac{y}{1-y}) = \log(\Lambda_a/O_2) = \log(P) - \log(P_{1/2}). \tag{10.8}$$

Thus, for a protein with a single binding site for O_2 such as Mb, we expect a plot of $\log(y/(1-y)0$ versus $P/P_{1/2}$ should yield a straight line of slope 1, and the value of Λ_d can be read off at the point where $\log(y/(1-y)) = 0$, that is, where half the Mb molecules have bound an O_2. Figure 10.3 shows a Hill plot of (idealized) binding data for M and Hb. Something very strange is seen in the Hb data: while Mb behaves as expected, the Hb Hill binding curve is **not** a straight line with slope 1! While it is true that Hb binds four oxygen molecules and not one; that does not change the slope of the curve from 1, if there is no effect of the occupation number of oxygens in an Hb molecule, no occupation effects simply means that 1 Hb = 4 Mb. You might guess that maybe the four binding sites in Hb are inequivalent: maybe one binds very tightly, and another binds weakly. The trouble with that is that would mean that the tighter binding site would be occupied first as the concentration of oxygen rises, and the weakest last. But the binding curve is different: the weakest bind at **low** oxygen pressure, then the strongest! It's backwards (and for the biological reason stated earlier). That's quite a trick by Nature, isn't it?

10.2 Ligand Binding to Hemoglobin: The Case for Conformational Complexity

Figure 10.2 demonstrated that Hb does not satisfy the simple (hyperbolic) relation Eq. (10.7) but seems to bind oxygen with a constantly changing affinity rather than a single fixed value! In an ad hoc way to attempt to parameterize the strange binding curve of oxygen to Hb, Hill introduced in 1913 a hypo-

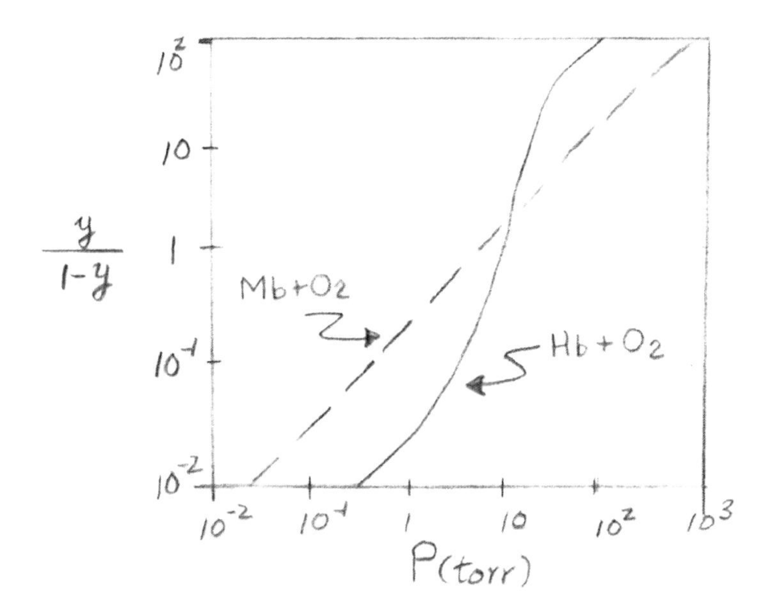

Fig. 10.3. Hill plot of Hb at pH 7 (solid line), compared to Mb (dashed line).

thetical nonlinear equation that is unphysical in that it assumes an n-body interaction [1]:

$$\frac{d[Hb - O_2]}{dt} = \Lambda_a[Hb][O_2]^n - \Lambda_d[Hb - O_2]. \tag{10.9}$$

The unphysical strangeness is in the $[O_2]^n$ term: it implies that n ligands compete simultaneously for one binding site, and this is unphysical. However, if we assume that the concentration of O_2 is proportional to the partial dioxygen pressure we get a new binding isotherm:

$$y = \frac{[\mathrm{Hb(O_2)}_n]}{[\mathrm{Hb(O_2)}_n] + [\mathrm{Hb}]} = \frac{P^n}{P^n + P^n_{1/2}}. \tag{10.10}$$

With Eq. (10.10) we get

$$\log(\frac{y}{1-y}) = n\log(P) - n\log(P_{1/2}). \tag{10.11}$$

Thus, in a Hill plot, $\log[y/(1-y)]$ versus $\log(P/P_{1/2})$, Eq. (10.11) gives a straight line with slope n. Look again at Fig. 10.3, the Hill plot for Hb. At small and large values of P, $n = 1$. This result is reasonable: At very low P, only a small fraction of Hb molecules are occupied, and they have only one of the four sites occupied. The molecule acts like a monomer and the relation deduced for Mb applies. At very large P, nearly all states will be occupied, and again Hb acts like Mb, but with a smaller value of $P_{1/2}$. In the transition region, Eq. (10.11) is a good approximation, with a Hil coefficient $n = 2.8$. The fact that the slope deviates from $n = 1$ is a measure of what is called cooperativity in Hb: the more occupation of the four binding sites of Hb by oxygen, the *stronger* the binding. The possible mechanism seems to be that the *conformation* of the Hb molecule is different if there is one ligand bound or four ligands bound, and the pioneering X-ray crystallography work of Max Perutz confirmed that in a major tour de force [2]. Figure 10.4 shows the conformational changes that occur in Hb between the all-deoxy state with no oxygen bound and the all-oxy state with four oxygen ligands. There is a huge literature on how the conformational changes of Hb achieves the inverted binding curves if weak first and then strong; we won't go into it because we want to stress the importance of conformational changes and not get bogged down in details at this point.

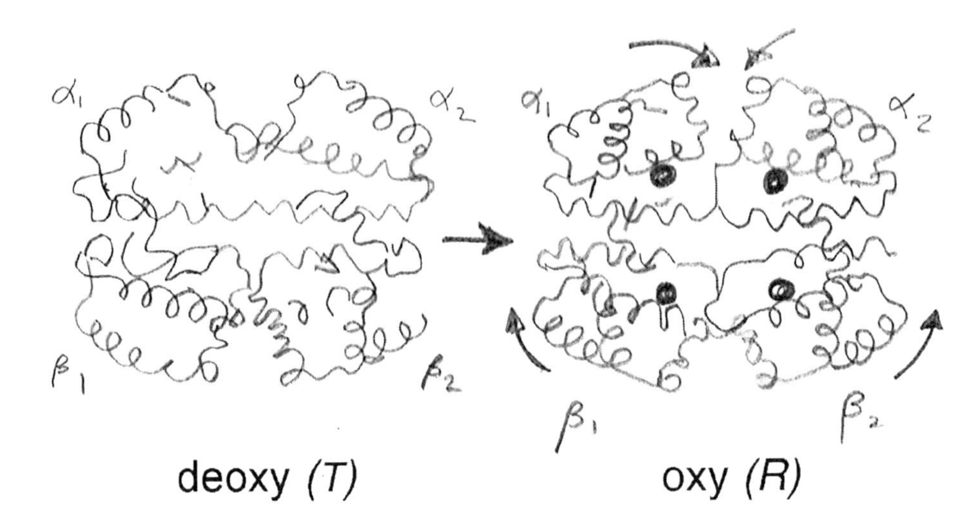

Fig. 10.4. Schematic of the conformational changes that occur in Hb between the dexoy state (T for tense) and the full oxygenated state (R for relaxed). The arrows indicate movement of the $\alpha - \beta$ subunits.

10.3 The Bohr Effect: Further Evidence for Conformational Changes

The Hill equation, as we mentioned, is unphysical, and there is no direct physical interpretation of $n \neq 1$. Rather, it is a way to parameterize the data we don't understand in terms of the basic physics of proteins, and the physics we want to stress is the importance of conformation changes of the protein. Perhaps it isn't too surprising that the four subunits of Hb somehow interact with one another. Linus Pauling first suggested that it was some sort of electric dipole-dipole interaction between the 4 Mb-like subunits of Hb, maintaining the view that protein monomers at least are rigid entities. However, besides the fact that the dipole-dipole interaction energies are far too small to account for the known free energy changes, there are other complexities to Hb of important biological function. Christian Bohr (1855–1911), the father of Niels Bohr, discovered that the affinity of hemoglobin for O_2 depends on acidity. The relevant curves are shown in Figs. 10.5 and 10.6. The cooperative characteristics are not affected by pH. Removal of protons increases the affinity for dioxygen. The biological importance is clear. In the lung, uptake of O_2 is facilitated; in the tissues, O_2 is more easily given off.

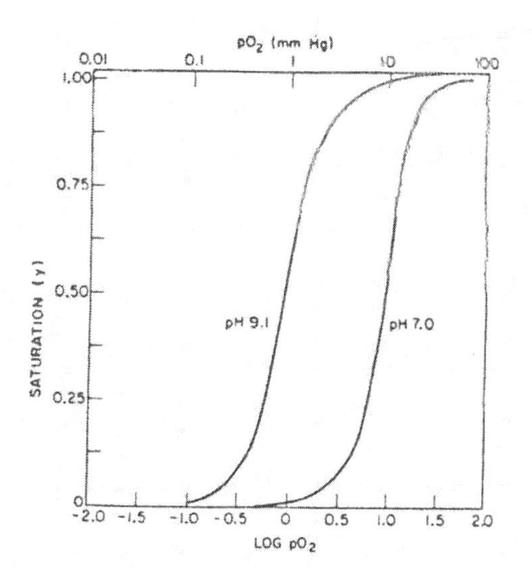

Fig. 10.5. The effect of pH on oxygen equilibrium curves.

These results tell us something extremely important and lead us from the conventional view of proteins that early X-ray crystallography promoted:

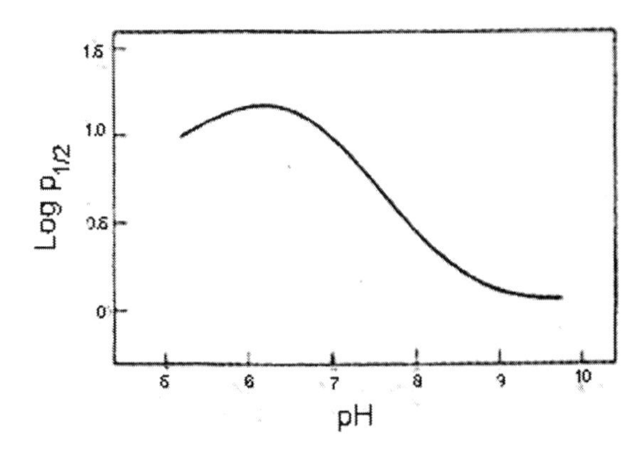

Fig. 10.6. Effect of pH on the affinity of hemoglobin for oxygen on a log-log scale.

proteins as rigid objects that bound ligands in a lock and key mechanism. Instead, it must be true that proteins are conformationally complex; they must thermally access multiple conformational states. The rest of this book will address this conformational complexity at the deepest level we can.

References

1. M. Weissbluth. *Hemoglobin.* Springer, New York, 1974.
2. M. F. Perutz. Hemoglobin structure and respiratory transport. *Sci. American*, 239(6):92–125, 1978.
3. J. M. Berg, J. L. Tymoczko, and L. Stryer. *Biochemistry*, 6th edition. W. H. Freeman, New York, 2006.
4. M. F. Perutz. Species adaptation in a protein molecule. *Mol. Bio. Evol.*, 1:1–28, 1983.
5. M. F. Perutz. Mechanisms of cooperativity and allosteric regulation in proteins. *Q. Rev. Biophys*, 22:139–236, 1989.
6. R. E. Dickerson and I. Geis. *Hemoglobin.* Benjamin/Cumming, Menlo Park, CA, 1983.
7. A. C. T. North and J. E. Lydon. The evolution of biological macromolecules. *Contemp. Phys*, 25:381–93, 1984.

Part III

THE ENERGY LANDSCAPE AND DYNAMICS OF PROTEINS

"...a protein cannot be said to have 'a' secondary structure but exists mainly as group of structures not too different from one another in free energy, but frequently differing considerably in energy and entropy. In fact the molecules must be conceived as trying out every possible structure."

Until now we have treated biomolecules as if they possessed a unique structure, defined by the primary sequence. In reality a protein can assume a very large number of different conformations or *conformational substates*. The substates are organized in a hierarchy and can have the same or different functions. A working protein fluctuates from substate to substate and these thermal fluctuations are essential for function. In Part III, we describe the concept of states and substates and discuss the evidence for the existence of substates and their organization, described by the protein energy landscape. The entire field is still in a primitive state, and physicists can make major contributions.

In the treatment of protein dynamics, two different spaces are used. We are familiar with the ordinary space, in which a protein is seen as its real three-dimensional object. To fully describe the protein at a given instance of time, the conformational space is used. This space has $3N-6$ conformation coordinates. A point ("energy valley") in this space corresponds to a particular arrangement of all atoms in the protein, its hydration shell, and part of the surrounding solvent. A change in the conformation of the protein that involves changes of the 3-D coordinates of many atoms corresponds to a transition from one point to another in the conformation space.

Further reading

1. K. U. Linderstrom-Lang and H. A. Schellman. In P. Boyer, editor, *Enzymes. I,* 2nd edition. Academic Press, New York, 1959. p. 433.

11

Conformational Substates

A characteristic property of a complex system is inhomogeneity; it can assume many different conformations, or as we call them "conformational substates." Glasses are typical examples. Proteins clearly are complex and it is ludicrous to assume that they exist in a unique structure, each atom exactly in a unique position. Here we discuss states and substates and present some of the evidence for conformational substates. Detailed discussion of more experimental and theoretical support for the concept can be found in reviews [1, 2].

11.1 States and Substates

Many proteins are nanomachines and perform some functions. They consequently must possess at least two different *states*, like a switch that can be open or closed. Examples are myoglobin (Mb), which can have a bound dioxygen (MbO_2) or not (deoxy Mb), and cytochrome (c), which can be charged or neutral. In a given state, a protein can assume a large number of *conformational substates (CS)*. This situation is sketched in Fig. 11.1.

Two types of motions are occurrring: Equilibrium fluctuations (EF) lead from one substate to another. Nonequilibrium motions accompany the transition from one state to another, for instance, from Mb to MbO_2. We can call these nonequilibrium transitions FIM, for functionally important motions [1, 2]. The cartoon in Fig. 11.2 explains the equillibrium and nonequilibrium situation clearly.

EF and FIM are related through fluctuation-dissipation theorems. We discuss the simpleset case of this theorem in Section 20.7. The connection is important for the studies of the energy landscape. It means that we can investigate a particular well in the conformational energy surface either by looking at fluctuations or by pushing the system into a nonequilibrium state and watching it relax to equilibrium. Of course, there are limitations to this approach: The extreme fluctuations in equilibrium must reach the region that is involved in the nonequilibrium motion.

H. Frauenfelder, *The Physics of Proteins*, Biological and Medical Physics, Biomedical Engineering, DOI 10.1007/978-1-4419-1044-8_11,

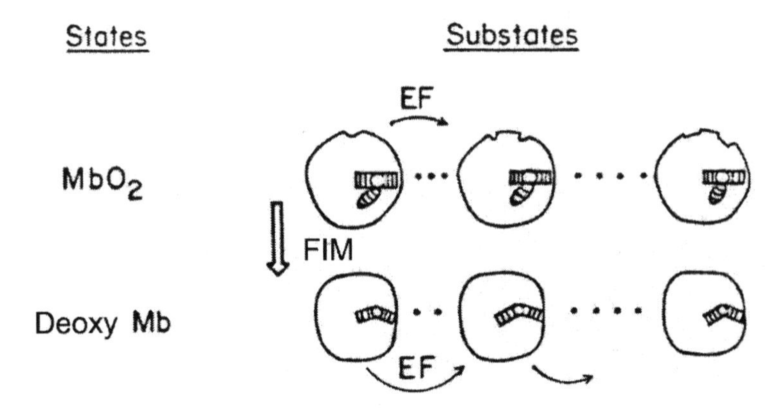

Fig. 11.1. Most proteins can exist in at least two different states, for instance, MbO_2 and deoxy Mb. In each state, the protein can assume a large number of conformational substates.

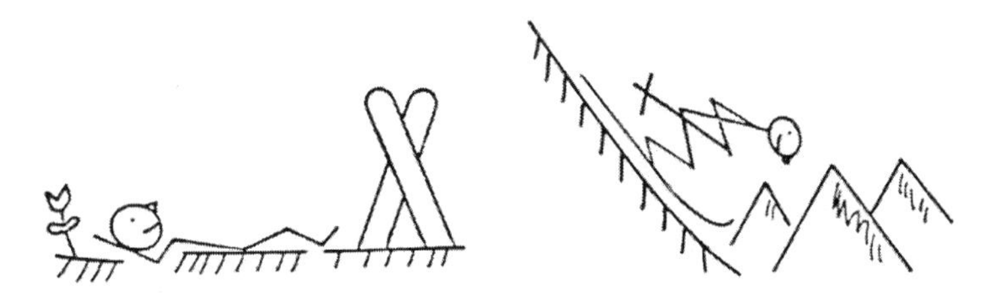

Fig. 11.2. A system at rest and in action.

We can summarize the most basic feature of the conformational energy surface of a protein by drawing a one-dimensional cross section through the hypersurface, and we expect that it will look like the left side of Fig. 11.3. Schematically we can represent the highly degenerate ground state by the drawing on the right. In this picture, only the small circles at the bottom represent conformational substates. The apex is a transition state; a transition from one substate (1) to another substate (n) occurs by a passage through the apex.

Figure 11.3, while giving a first impression of the energy landscape, is misleading in two respects. It appears that a transition from an initial state i to a final state f must follow a unique pathway, and entropy does not play a role. A much better approach is given in Fig. 11.4, where the energy of the protein is shown as a "topographic map" with two conformational coordinates CC1 and CC2. Substates are shown as valleys in this map. The density of valleys in a particular part of the map describes the entropy, and it is clear that many paths lead from i to f.

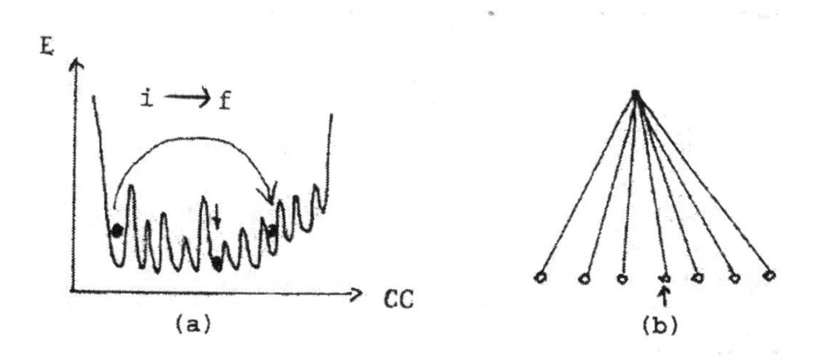

Fig. 11.3. (a) One-dimensional cross section through the conformational energy hypersurface in a particular state. (b) Tree diagram for the energy surface in (a).

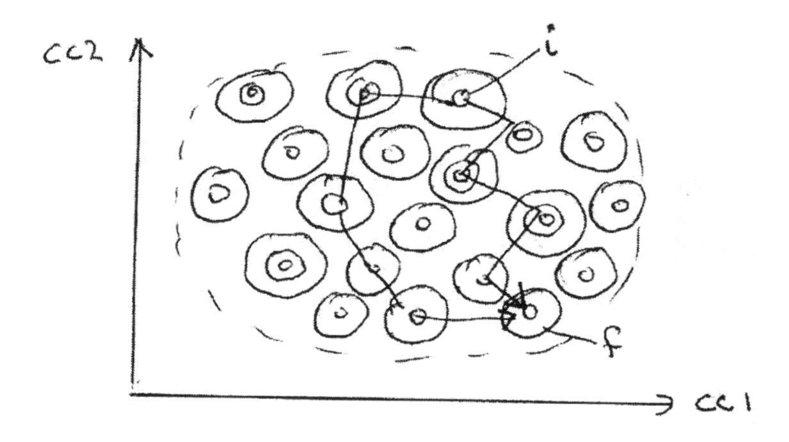

Fig. 11.4. A two-dimensional cross section through the energy landscape. CC1 and CC2 are two of the very large number of conformational coordinates describing a protein. A valley in this map corresponds to a substate.

Figures 11.3 and 11.4 raise a number of questions. Do proteins indeed have conformational substates? If yes, how are they organized and what is their role in protein dynamics and function? In the following sections, some of the experimental and computational evidence for substates will be discussed and the organization of the landscape will be described.

11.2 Nonexponential Kinetics

Dramatic evidence for the complexity of proteins appears in the rebinding kinetics of small molecules such as CO and O_2 to heme proteins at low temperatures [3]. We discuss these experiments in more detail in Part IV and only describe the essential features here.

The "laboratory" for these experiments is the inside of a heme protein, for instance, myoglobin. Figure 11.5(a) shows a schematic cross section through the heme pocket, with a CO molecule covalently bound at the heme iron (state A). The bond between the iron and the CO can be broken by a light pulse at a particular wavelength. The CO then moves into the heme pocket (state B). At low temperatures, where the solvent is solid, the CO cannot escape from the pocket into the solvent, but it rebinds. The potential energy surface for photodissociation and rebinding is depicted in Fig. 11.5(b).

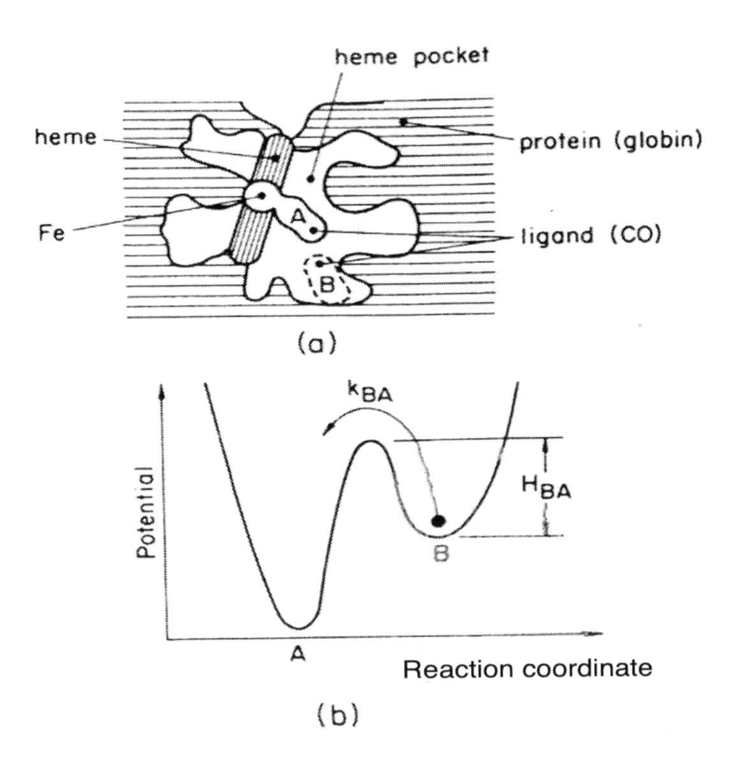

Fig. 11.5. Schematic representation of the heme pocket in heme proteins: (a) Real-space cross section. (b) Potential diagram. A denotes the state where the ligand is bound to the heme iron, B the state where the ligand is in the heme pocket.

The barrier height for the binding process $B \to A$ is H_{BA}. Above about 40 K, where quantum-mechanical tunneling is no longer significant, the rate coefficient, k_{BA}, can be described by an Arrhenius relation of the form

$$k_{BA}(T) = A_{BA}(T/T')\exp[-H_{BA}/RT]. \tag{11.1}$$

T' is a reference temperature, usually set at 100 K, H_{BA} is the barrier height (activation enthalpy), and R (= 8.31 J/mol) is the gas constant. If the pre-exponential factor A_{BA} and the barrier height H_{BA} have unique values, then the rate coefficient has a single value, and rebinding is exponential in time,

$$N(t) = \exp\{-k_{BA}(T)\,t\}. \tag{11.2}$$

Here $N(t)$ is the survival probability, i.e., the fraction of Mb molecules that have not rebound a ligand at time t after photodissociation.

The experimental results differ from the prediction of Eq. (11.2). A typical result is shown in Fig. 11.6 for the rebinding of CO to sperm whale myoglobin. Note that the plot does not display log $N(t)$ versus t, but log $N(t)$ versus log t. In a log-log plot, an exponential appears nearly like a step function, and a power law gives a straight line. The nonexponential time course of $N(t)$ in Fig. 11.6 can indeed be approximated by a power law with N(0) = 1,

$$N(t) \approx (1 + k_0 t)^{-n}, \tag{11.3}$$

where k_0 and n are temperature-dependent parameters.

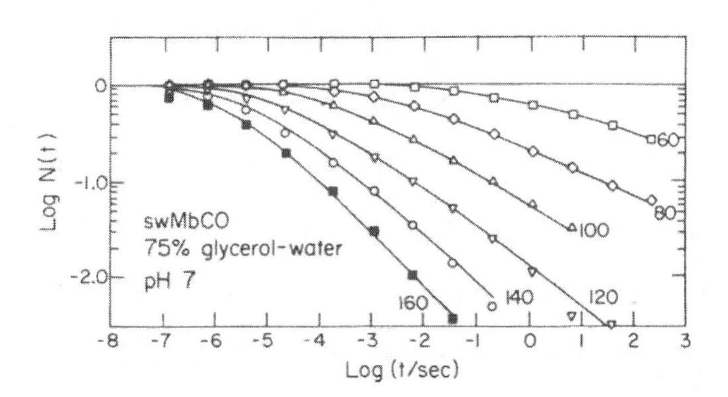

Fig. 11.6. Time dependence of the rebinding after photodissociation of CO to Mb. $N(t)$ is the survival probability, the fraction of Mb molecules that have not bound a CO at time t after the flash. The solid lines correspond to a theoretical fit, based on Eq. (11.3), with the activation enthalpy distribution $g(H)$ as shown in Fig. 11.7. This particular solvent, 75% glycerol–25% water, is solid below 180 K.

The binding of CO to Mb is not an isolated case of nonexponential rebinding kinetics. Nearly every binding process studied at low temperatures yields similar curves [3]. Nonexponential kinetics has also been observed by time-resolved electron paramagnetic resonance (EPR) in the rebinding of nitric oxide (NO) to cytochrome c oxidase and Mb at low temperature. The experimental data for a wide variety of heme protein-ligand combinations thus demonstrates that the binding from the heme pocket in a solid solvent at low temperature is nonexponential in time.

Nonexponential rebinding can be explained in two very different ways, either by assuming inhomogeneous proteins with each protein having a simple pathway as shown in Fig. 11.5(b) or by assuming homogeneous proteins with different pathways in each protein. Experiments unambiguously show that the first alternative is correct [3]. This explanation is just what one expects from conformational substates, as we now show.

Equation (11.2) is valid if the preexpeonential A_{BA} and the barrier height H_{BA} have a unique value. We postulate that proteins in different conformational substates have different barrier heights and denote by $g(H_{BA})dH_{BA}$ the probability of a protein having a barrier height between H_{BA} and $H_{BA} + dH_{BA}$. Equation (11.2) then is generalized to read

$$N(t) = \int dH_{BA}\, g(H_{BA}) \exp\{-k_{BA}(H_{BA}, T)\, t\}, \tag{11.4}$$

where $k_{BA}(H_{BA}, T)$ and H_{BA} are related by Eq. (11.1). If $N(t)$ is measured over a wide temperature and time range, the preexponential A_{BA} and the activation enthalpy distribution $g(H_{BA})$ can be found by Laplace inversion of Eq. (11.4). The preexponentials typically have values near $10^9 s^{-1}$, and some distributions are shown in Fig. 11.7.

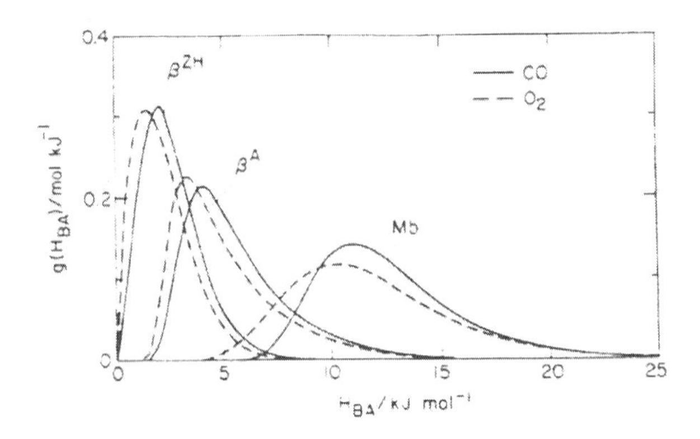

Fig. 11.7. Activation enthalpy distributions. $g(H_{BA})$ for rebinding of CO and O_2 to myoglobin and the separated beta chains of normal (β^{A}) and mutant (β^{ZH}) hemoglobin.

We repeat here the essential definitions and assumptions underlying the discussion. We have defined CS as having the same primary sequence, crudely the same structure, but differing in detail, and performing the same function, but with different rates. The observation of the nonexponential rebinding led to this definition: Eq. (11.4) implies that we can characterize a CS by the barrier height H_{BA}. In this particular substate, rebinding is exponential and given by the rate $k_{BA}(H_{BA}, T)$. The different structures in the different CS result in different barrier heights and hence in different rates. At low T, each protein remains frozen in a particular CS, and rebinding consequently is non-exponential in time. At temperatures well above 200 K, a protein fluctuates rapidly from one CS to another CS and rebinding can become exponential in time [3].

11.3 The Debye-Waller Factor

We now switch to X-ray diffraction. When the nonexponential kinetics discussed in the previous section suggested that a protein can assume a large number of conformations some biochemists reacted by stating that this concept was in disagreement with X-ray diffraction data that gave unique structure [4]–[6]. It turned out, however, that substates exist and that X-ray diffraction helps to prove their existence and yields additional information [7]. To explain this fact, we return to the theory of X-ray diffraction. Rereading the derivation leading to Eq. (25.16) (in Part V), the structure factor equation shows that the derivation implied fixed atoms. We depict the situation in Fig. 11.8(a). Atoms, however, move. The problem of temperature motion was taken up as early as 1914 by Debye, and later by Waller. Since then, the problem has been discussed by many authors. In Fig. 11.8(b), two atoms are shown, each moving within a volume characterized by the mean square radius $\langle x^2 \rangle$.

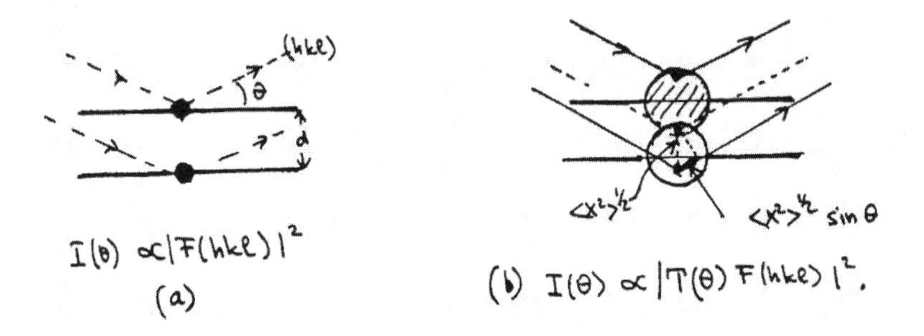

Fig. 11.8. **(a)** Diffraction from fixed atoms. **(b)** Atoms move about their equilibrium position.

The charge density $\rho(\mathbf{r})$ in Eq. (25.4) must be replaced by

$$\rho(\mathbf{r}) \to \int g(\mathbf{r}')\rho(\mathbf{r}-\mathbf{r}')\, d^3\mathbf{r}' \tag{11.5}$$

where $g(\mathbf{r}')$ describes the probability distribution of the atom. Since the Fourier transform of a convolution is equal to the product of the Fourier transforms of the two functions, we obtain the replacement

$$f(\mathbf{q}) \to f(\mathbf{q})T(\mathbf{q}). \tag{11.6}$$

$T(\mathbf{q})$, the Fourier transform of $g(\mathbf{r})$, is called the *temperature* or *Debye-Waller factor*. With Eq. (11.6), Eq. (25.17) becomes

$$I(hk\ell) = \text{const } \mathrm{T}^2(\mathbf{q})|\mathrm{F}(\mathrm{hk}\ell)|^2. \tag{11.7}$$

Assuming isotropic harmonic vibrations about the equilibrium positions, $T(\mathbf{q})$ becomes:

$$T(\theta) = \exp\left[W(\theta)\right] \tag{11.8}$$

$$W(\theta) = 2\langle x^2\rangle(\sin^2\theta)/\lambda^2. \tag{11.9}$$

In the X-ray literature, instead of $\langle x^2\rangle$ one often quotes

$$B = 8\pi^2\langle x^2\rangle. \tag{11.10}$$

In proteins many atoms are spread out about their mean positions in a nonharmonic manner. Equation (11.9) then may be a bad approximation or it may not fit the data at all.

The X-rays scattered from different atoms interfere, and the path difference between two trajectories, for instance, the solid and the dashed ones in Fig. 11.8(b), is given by $2\langle x^2\rangle(\sin^2\theta)$. Equation (11.7) and Fig. 11.8(b) show that $I(\theta)$ decreases with increasing angle because of the decreasing factor F and the decreasing $T(\theta)$. Originally, the factor $T(\mathbf{q})$ was considered to arise entirely from the vibrations about the equilibrium positions. In proteins, $\langle x^2\rangle$ calculated from the measured values of T turned out to be rather large, and the unsatisfactory results were blamed on "bad crystals." However, the actual situation is far more interesting than appeared at first [4, 8]. To discuss the problem in a zeroth approximation, we ask: What phenomena can contribute to $\langle x^2\rangle$? After some thinking, we can write

$$\langle x^2\rangle = \langle x^2\rangle_c + \langle x^2\rangle_d + \langle x^2\rangle_{\ell d} + \langle x^2\rangle_v. \tag{11.11}$$

Here c refers to *c*onformational substates, d to *d*iffusion, ℓd to *l*attice *d*isorder, and v to *v*ibrations. The first term comes from the fact that a given atom in a protein can be in many different conformational substates. To the X-rays, the various substates will appear as a larger $\langle x^2 \rangle$. In some cases, however, when two positions of an atom are far enough apart, X-rays can see them both [8]. The second term can usually be neglected in X-ray work. If diffusion were important, the X-ray pattern would not be sharp. The third term is caused by imperfections in the crystals, and we assume that it is temperature independent and the same for all atoms in the protein. The last term comes from vibrations about the equilibrium position. With Eq. (11.11), $T(\theta)$ in Eq. (11.7) is no longer caused only by temperature motion.

In some favorable cases, X-ray diffraction and Mössbauer effect (Chapter 28) together permit determination of the conformational distribution. From the X-ray intensities, $\langle x^2 \rangle$ is determined for all atoms in a crystal, if possible at many temperatures. Neglecting diffusion, $\langle x^2 \rangle$ is then given by three terms in Eq. (11.11). If the protein contains an iron atom, the Mössbauer effect can be used to determine the recoilless fraction f. At temperatures where conformational relaxation takes place, $\langle x^2 \rangle$ for the Mössbauer effect contains only two terms, v and c. Lattice disorder is not important, because each iron atom acts independently. Thus from f, we can find $\langle x^2 \rangle$, which is given by

$$\langle x^2 \rangle_{Mo} = \langle x^2 \rangle_c + \langle x^2 \rangle_v. \tag{11.12}$$

Thus $\langle x^2 \rangle_x - \langle x^2 \rangle_{Mo} = \langle x^2 \rangle_{\ell d}$; the lattice disorder term can be obtained. Once it is known, we can find $\langle x^2 \rangle_{cv} \equiv \langle x^2 \rangle_c + \langle x^2 \rangle_v$ for all atoms. The separation of the conformational and the vibrational terms requires measurements over a wide temperature range. However, for the function of a protein, the sum of the two terms is important rather than the individual one. We will therefore in the following discuss the behavior of $\langle x^2 \rangle_{cv}$.

The determination of the mean-square deviation (MSD), $\langle x^2 \rangle_{cv}$, of the atoms in a protein was initially a slow and painful process. At present, however, the use of the Debye-Waller factor in protein crystallography has become an industry [9]. Advances in the relevant technology, discussed in Chapter 25, together with automated data evaluation [10], have led to the knowledge of the MSD in many proteins. Here older data will be discussed because they express the relevant concepts clearly.

The initial work on Mb [7] already indicated that protein structural fluctuations are much larger and much more complex than initially anticipated. Figure 11.9 shows the backbone of Mb. The shaded area gives the region reached by fluctuations. It is defined such that fluctuations have a 1% probability of reaching outside the shaded area. However, the evaluation underestimates the motions; the actual displacements may be considerably larger than shown. Figure 11.9 demonstrates that fluctuations are not uniform as they would be in a solid. In some regions, they are very small, in others, for instance, at the corner near 120 (GH corner), they are large.

Figure 11.10 shows the MSD in Mb as a function of residue number [7]. It is evident that the range of substates depends on the position of the atoms in

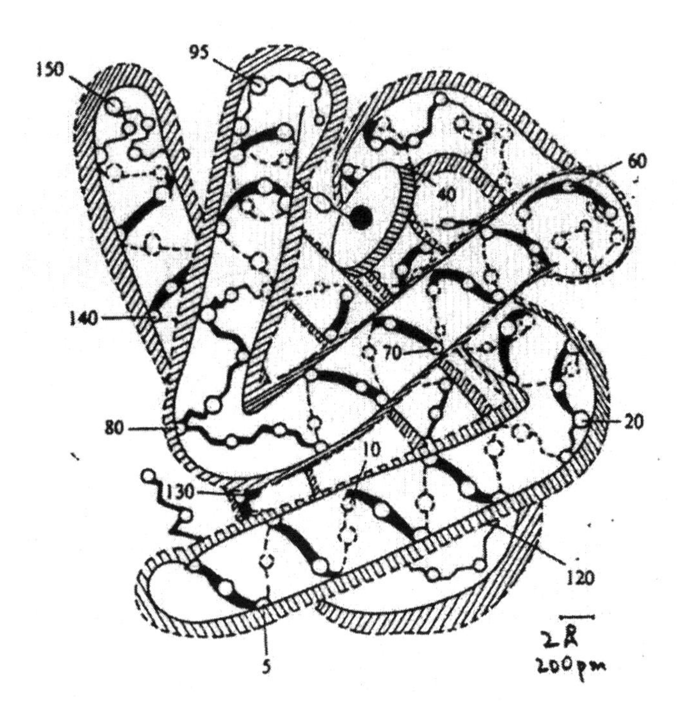

Fig. 11.9. Backbone of myoglobin. The circles indicate the positions of the alpha carbons. Displacements of backbone nonhydrogen atoms have a 1% probability of reaching outside of the shaded region [7].

the protein. The essential point is, however, clear. The Debye-Waller factor supports the concept of conformational substates in proteins. The data in Fig. 11.10 are still crude; the modern sophisticated data evaluation gives a far clearer picture of the range of substates in different protein regions. An example is given in Fig. 11.11 for a different protein, HIV protease. The figure gives the average B factor and the the root-mean-square deviations, evaluated with different models [11].

While the Debye-Waller factor obtained by X-ray crystallography gives valuable information about the existence of substates in proteins, there is a serious limitation. Proteins in cells fold and unfold continuously. The motions involved are much larger than the MSD in Figs. 11.10 and 11.11. The reason is clear. Proteins cannot unfold while embedded in crystals. The lesson is that actual protein substates in working proteins occupy a much larger conformational space than the one obtained from X-ray diffration.

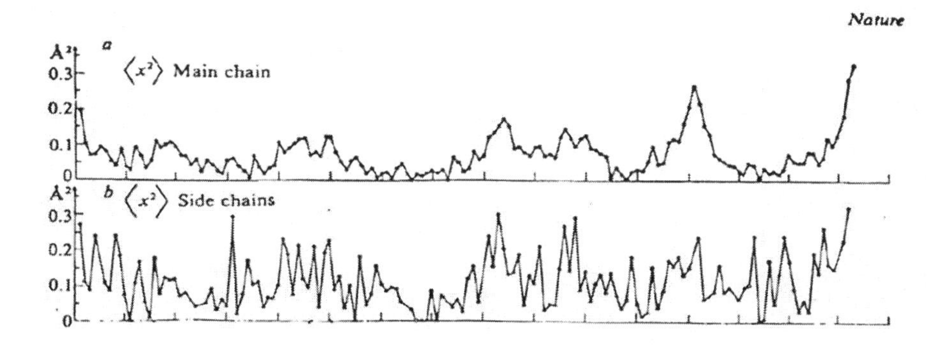

Fig. 11.10. The conformational and vibrational displacement $\langle x^2 \rangle_{cv}$ for myoglobin. The upper part gives the average values for the three main chain (backbone) atoms C_α, C, and N. The lower presents the largest x^2_{cv} in each side chain [7].

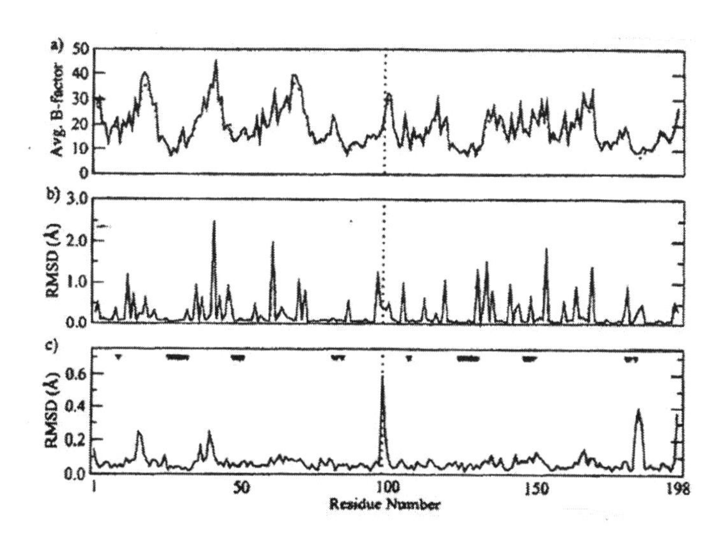

Fig. 11.11. B factors and RMSD per residue, after Blundell and collaborators [11].

11.4 Spectral Hole Burning

If proteins in different substates indeed possess different conformations, we expect that their electronic properties differ also. Spectral lines should not be

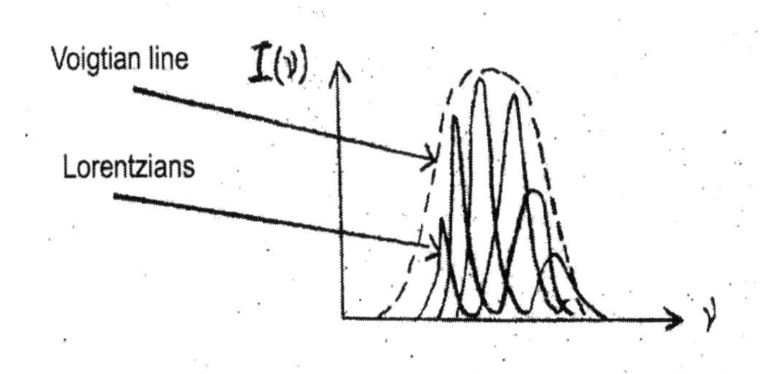

Fig. 11.12. Inhomogeneous spectral lines. The observed line is expected to have a Voigtian shape, namely a Gaussian superposition of Lorentzians.

homogeneous, but should consist of a superposition of homogeneous lines as sketched in Fig. 11.12.

A Gaussian superposition of Lorentzian lines is called a Voigtian. A number of different approaches provide information about the line inhomogeneity. By analyzing the *shape of the lines*, the homogeneous and the inhomogeneous contributions can be separated [12]–[14]. A second approach uses the lifetime of fluorescent residues in proteins.[15] Fluorescence lifetimes are highly sensitive to the environment. If a protein contains only one of a given fluorescent residue, say tryptophan, and if the environment were identical in each protein molecule, the observed decay would have only a single lifetime. In reality, the observed decays can be fit better with a *lifetime distribution*, implying inhomogeneous sites [15].

The most direct evidence for the inhomogeneity comes, however, from *hole-burning* experiments. Because of the power of these experiments, we discuss two different types. *Spectral hole burning* was first introduced to NMR spectroscopy by Bloembergen, Purcell, and Pound [16]. We consider the optical version of hole burning here [17, 18]. Consider first a single homogeneous component in Fig. 11.12. The line consists of a narrow zero-phonon line and a broad phonon sideband. The zero-phonon line represents transitions in which no phonon is emitted or absorbed. The photon sideband occurs because the environment starts vibrating as a result of the chromophore excitation. The relative intensity of the zero-phonon line is given by the Debye-Waller factor (Section 11.3), $\exp\{-S\}$, where S characterizes the strength of the electron phonon coupling. Three different types of hole-burning processes are important. In *transient* hole burning, the incident laser line of wavenumber ν_o excites just one homogeneous line (we neglect the phonon sideband) and the line after excitation shows a transient hole (Fig. 11.13(b)). The hole will usually be refilled quickly by the de-excitation. In *photochemical hole burning* (PHB), the incident laser beam induces a photochemical reaction. The spectrum of the photoproduct is usually well separated from that of the chromophore and

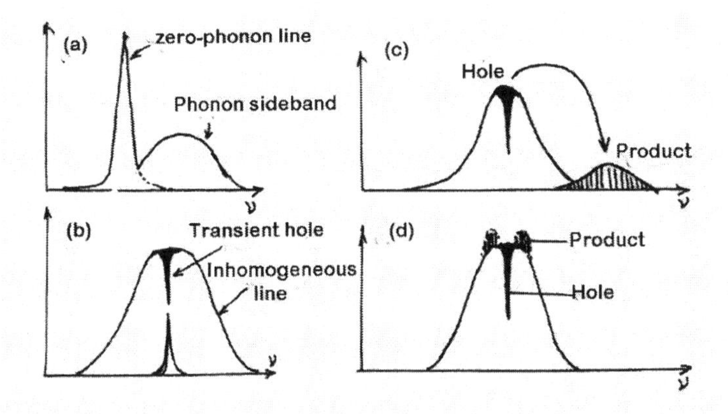

Fig. 11.13. **(a)** Line shape of a chromophore in a perfect lattice. **(b)** Transient hole burning. The phonon sideband is not shown. **(c)** Photochemical hole burning. **(d)** Nonphotochemical hole burning.

the hole consequently remains until the reaction has been reversed. In a *non-photochemical hole burning* (NPHB) process, the incident light modifies the chromophore-protein interaction. The wavenumber of the absorption of the photoproduct is close to that of the photolyzing light. In other words, the incident light transfers the protein from one conformational substate to another. PHB and NPHB are well known and studied in chromophores included in glasses. Hole-burning experiments with proteins have also been performed. Both photochemical and nonphotochemical burnt holes have been observed [19]–[22]. These experiments show unambiguously that the spectral lines of the chromophores are inhomogeneously broadened and add to the evidence for conformational substates.

Hole burning gives a lower limit on the number of substates in a protein. Figure 11.14 shows that the ratio of the width of the broadened line to that of the hole is of the order of 10^4. At present, the ratio is about 10^6 [23]. The number of substates is therefore at least of the order of 10^6. The actual number is much larger because many substates have the same line position.

11.5 Substates Seen in Molecular Dynamics Simulations

If conformational substates exist, they should also show up in molecular dynamics (MD) simulations. In fact, simulations give unambiguous computational evidence and provide additional information that is difficult or impossible to find experimentally [24]–[26]. A discussion of MD is beyond the scope here, but one crucial result related to conformational substates deserves a place. Garcia and collatorators [26] observed a power law walk in conforma-

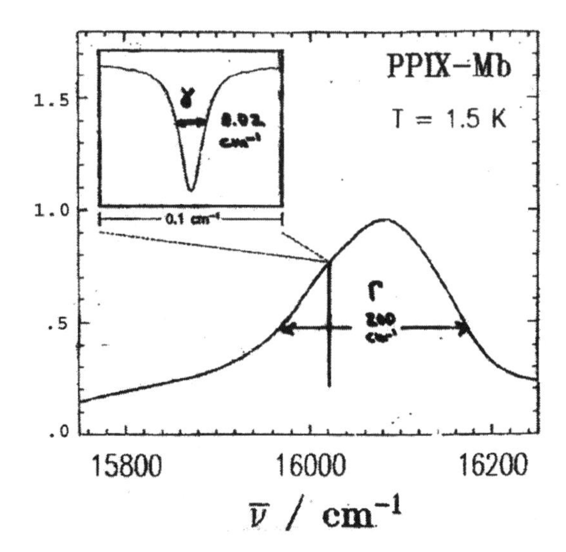

Fig. 11.14. The ratio of the width of a broadened line to the width of a photochemical hole gives a lower limit on the number of substates [23].

tional space and found that the energy valleys in the conformational space are arranged in a hierachy, as shown in Fig. 11.15.

The information given in the present chapter can be summarized by stating that proteins are inhomogeneous and can exist in a very large number of different conformational substates.

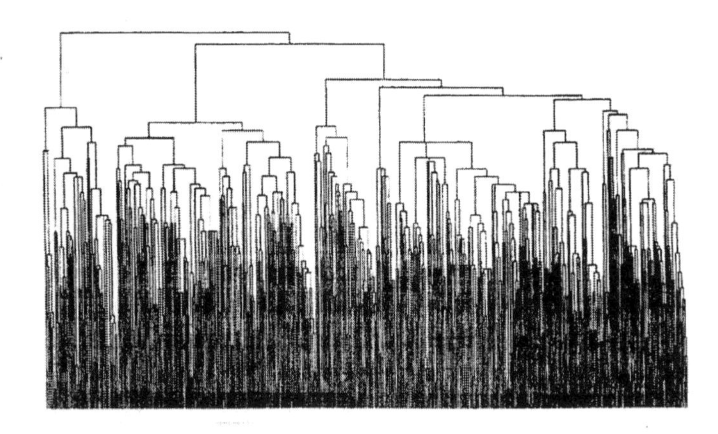

Fig. 11.15. The organization of substates in the small protein crambin as determined by an MD search for conformation minima. From Garcia et al. [26]

References

1. H. Frauenfelder, G. U. Nienhaus, and R. D. Young. Relaxation and disorder in proteins. In *Disorder Effects in Relaxation Processes.* Springer, Heidelberg, 1994. pp. 591–614.
2. G. U. Nienhaus and R. D. Young. Protein dynamics. In *Encyclopedia of Applied Physics. Vol. 15: Power Electronics to Raman Scattering.* VCH Publishers, New York, 1996. pp. 163–84.
3. R. H. Austin, K. W. Beeson, L. Eisenstein, H. Frauenfelder, and I. C. Gunsaluls. Dynamics of ligand binding to myoglobin. *Biochemistry*, 14:5355–73, 1975.
4. H. Frauenfelder. The Debye-Waller factor: From villain to hero in protein crystallography. *Int. J. Quantum Chemistry*, 35:711–5, 1989.
5. B. T. M. Willis and A. W. Pryor. *Thermal Vibration in Crystallography.* Cambridge Univ. Press, London, 1975.
6. G. A. Petsko and D. Ringe. Fluctuations in protein structure from X-ray diffraction. *Ann. Rev. Biophys. Bioeng.*, 13:331–71, 1984.
7. H. Frauenfelder, G. A. Petsko, and D. Tsernoglou. Temperature-dependent X-ray diffraction as a probe of protein structural dynamics. *Nature*, 280:558–63, 1959.
8. A. E. Garcia, J. A. Krumhansl, and H. Frauenfelder. Variations on a theme by Debye and Waller: From simple crystals to proteins. *Proteins*, 29:343–347, 1997.
9. P. A. Rejto and S. T. Freer. Protein conformational substates from X-ray crystallography. *Prog. Biophys. Mol. Bio.*, 66:167–96, 1996.
10. T. D. Romo, J. B. Clarage, D. C. Sorensen, and G. N. Phillips, Jr. Automated identification of discrete substates in proteins: Singular value decomposition analysis of time–average crystallographic refinements. *Proteins*, 22:311–21, 1995.
11. M. A. DePristo, P. I. W. de Bakker, and T. L. Blundell. Heterogeneity and inaccuracy in protein structures solved by X-ray crystallography. *Structure*, 280:831–8, 2004.
12. A. Cooper. Photoselection of conformational substates and the hyposochromic photoproduct of rhodopsin. *Chem. Phys. Lett.*, 99:305–9, 1983.
13. K. T. Schomacker and P. M. Champion. Investigations of spectral broadening mechanisms in biomolecules: Cytochrome-c. *J. Chem. Phys.*, 84:5314–25, 1986.
14. V. Srajer, K. T. Schomacker, and P. M. Champion. Spectral broadening in biomolecules. *Phys. Rev. Lett.*, 57:1267–70, 1986.

15. J. R. Alcala, E. Gratton, and F. G. Prendergast. Interpretation of fluorescence decays in proteins using continuous lifetime distributions. *Biophys. J.*, 51:925–36, 1987.
16. N. Bloembergen, E. M. Purcell, and R. V. Pound. Relaxation effects in nuclear magnetic resonance absorption. *Phys. Rev.*, 73:679–712, 1948.
17. J. Friedrich and D. Haarer. Photochemical hole burning: A spectroscopic study of relaxation processes in polymers and glasses. *Angew. Chem. Int. Ed.*, 23:113–40, 1984.
18. R. Jankowiak and G. J. Small. Hole-burning spectroscopy and relaxation dynamics of amorphous solids at low temperatures. *Science*, 237(4815):618–25, 1987.
19. S. G. Boxer, D. J. Lockhart, and T. R. Middendorf. Photochemical hole-burning in photosynthetic reaction centers. *Chem. Phys. Lett.*, 123:476, 1986.
20. S. R. Meech, A. J. Hoff, and D. A. Wiersma. Evidence for a very early intermediate in bacterial photosynthesis: A photon-echo and hole-burning study of the primary donor band in *Rhodopseudomonas sphaeroides*. *Chem. Phys. Lett.*, 121:287–92, 1985.
21. W. Köhler, J. Friedrich, R. Fischer, and H. Scheer. High resolution frequency selective photochemistry of phycobilisomes at cryogenic temperatures. *J. Chem. Phys.*, 89:871–4, 1988.
22. S. G. Boxer, D. S. Gottfried, D. J. Lockhart, and T. R. Middendorf. Nonphotochemical holeburning in a protein matrix: Chlorophyllide in apomyoglobin. *J. Chem. Phys.*, 86:2439–41, 1987.
23. J. Gafert, H. Pschierer, and J. Friedrich. Proteins and glasses: A relaxation study in the millikelvin range. *Phys. Rev. Lett.*, 74:3704–7, 1995.
24. R. Elber and M. Karplus. Multiple conformation states of proteins: A molecular dynamics analysis of myoglobin. *Science*, 235:318–21, 1987.
25. N. Go and T. Noguti. Structural basis of hierachical multiple substates of a protein. *Chemica Scripta*, 29A:151–64, 1989.
26. A. E. Garcia, R. Blumenfeld, J. A. Hummer, and J. A. Krumhansl. Multi–basin dynamics of a protein in a crystal environment. *Physica D*, 107:225–39, 1997.

12

The Organization of the Energy Landscape

The experimental data and the computational arguments given in Chapter 11 prove the existence of substates: Proteins can exist in a large number of different conformations. How are the substates organized? Here the present understanding will be described. It is clear, however, that we are at a beginning and not at the end, and many of the "facts" and arguments given here may be proven to be incomplete or even wrong in the future.

12.1 The Energy Landscape in One Dimension

In Section 11.2 the nonexponential kinetics of the binding of CO to myoglobin was interpreted in terms of distributed barrier heights, Fig. 11.7. At temperatures above about 250 K, binding of CO becomes exponential in time, implying averaging because of rapid fluctuations among substates. Thus there must be barriers between substates that prevent transitions at low temperature but permit them above about 250 K. These observations led to the simple "landscape" shown in Fig. 12.1 [1]–[3]. In this figure, the barrier height E_{ba} (or H_{BA} in Fig. 11.5) serves as the conformational coordinate. The barrier between substates was implicitly assumed to be due to the protein and its height was only crudely estimated. Later experiments showed that Fig. 12.1 missed in two respects: The landscape is much more complex and the barriers between substates are mainly due to the solvent, not the protein. Nevertheless, Fig. 12.1 gives a glimpse into the complexity of proteins.

12.2 The Hierarchy of Substates

The "energy landscape" in Fig. 12.1 is similar to the landscape already shown in Fig. 11.3. Each valley corresponds to a unique structure. Experiments show, however, that this picture is far too simple [4]. There are valleys within valleys within valleys in the energy landscape, and they can be classified into a

H. Frauenfelder, *The Physics of Proteins*, Biological and Medical Physics, Biomedical Engineering, DOI 10.1007/978-1-4419-1044-8_12,

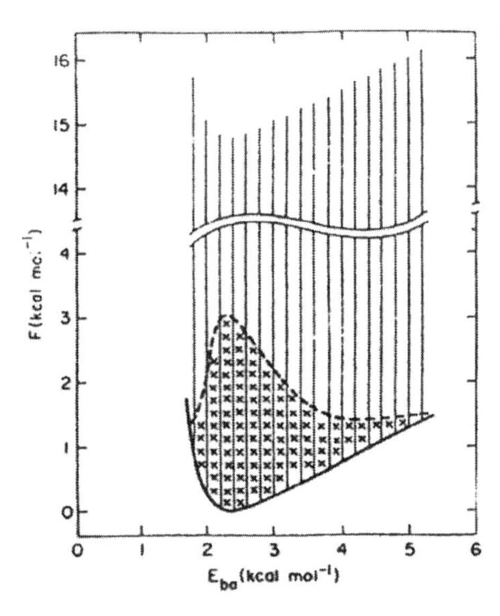

Fig. 12.1. An early version of an energy landscape: The conformational energy F as a function of the activation energy for binding of CO to myoglobin [3]. The vertical lines represent barriers between substates (1 kcal/mol = 418 kJ/mol).

number of tiers as simplified in Fig. 12.2(a). MbCO can assume three different substates in tier 0, denoted by A_0, A_1, and A_3, which we discuss in the next section. A protein in one of these substates can, in turn, assume a very large number of substates in tier 1. In each substate of tier 1 there are substates of tier 2. The ladder of substates continues, but details are still obscure. It appears, however, that the tiers can be characterized as taxonomic, statistical, and few-level, as described later and shown in Fig. 12.2(b).

12.3 Taxonomic Substates (Tier 0)

A description of the taxonomic substates in Mb provides the background for the later discussion of protein dynamics and function. In heme proteins, nature has provided a superb tool, namely the stretch vibration of the CO bound to the heme iron (Fig. 11.5). CO as a gas or in a liquid shows a stretch band centered at about 2140 cm^{-1}. A CO molecule bound covalently to the heme iron is expected to have a somewhat lower frequency, because the Fe–CO bond will weaken the C–O bond. Indeed, the wavenumber of CO bound to the iron atom in myoglobin has a value of about 1950 cm^{-1}. However, instead of just one stretch band, MbCO shows at least three. An example of the transmission spectrum of MbCO, obtained by FTIR spectroscopy, is given in Fig. 12.3. To prove that the prominent lines are caused by CO, a mixture of $C^{12}O^{16}/C^{12}O^{18}$

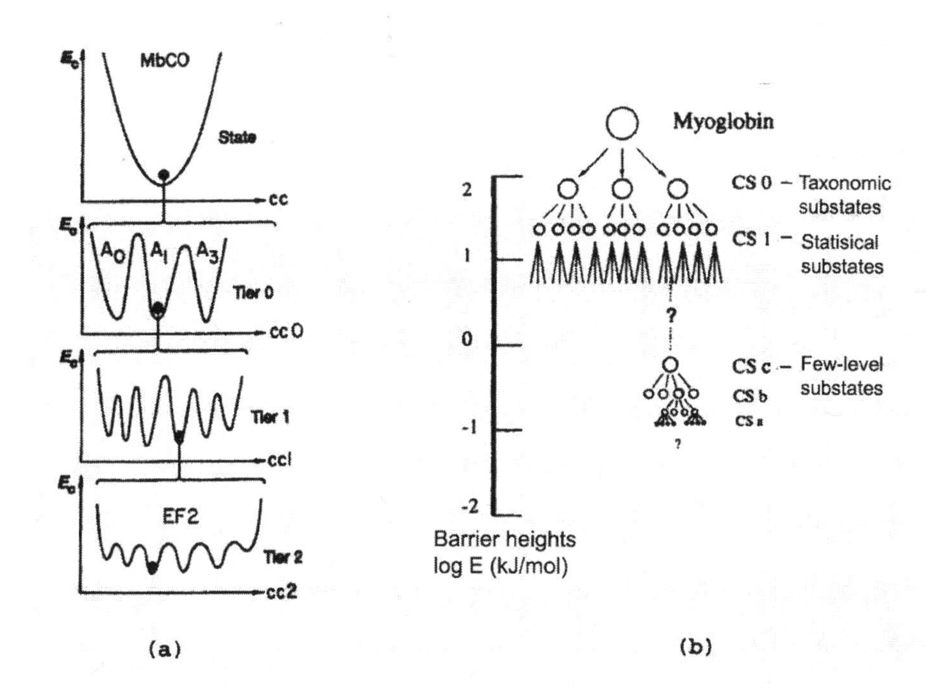

Fig. 12.2. (a) A 1-D cross section through the energy landscape of MbCO. The top three tiers are shown, E_c is the conformational energy, $cc0$ to $cc2$ denote conformational coordinates. (b) Substates can be characterized as "taxonomic," "statistical," and "few-level." The figure gives an approximate indication of the relevant barrier heights separating the substates.

was used. Indeed, two characteristic spectra are obtained that shifted by the expected amount.

While the transmission spectrum in Fig. 12.3 clearly shows CO lines, detailed study is difficult. Improved spectra are obtained with a **difference technique**: The spectrum of MbCO is first measured at low temperature. The CO–Fe bond then is broken by an intense light, and the spectrum of the resulting photolized product Mb* is determined. The difference spectrum of MbCO minus Mb* yields the spectrum of the CO lines both in the bound and the photolyzed state, as shown in Fig. 12.4. The data taken at 5.5 K show three IR bands in the bound and three in the photodissociated state. The bands are labeled as indicated. A weak line may be present between A_1 and A_3. We assume that each IR band characterizes a particular substate of tier 0 and denote these substates with A_0, A_1, and A_3. This assumption is supported by all experimental data. The three taxonomic substates (tier 0) possess properties that make them interesting from the points of view of physics and biology. We discuss three of these properties here.

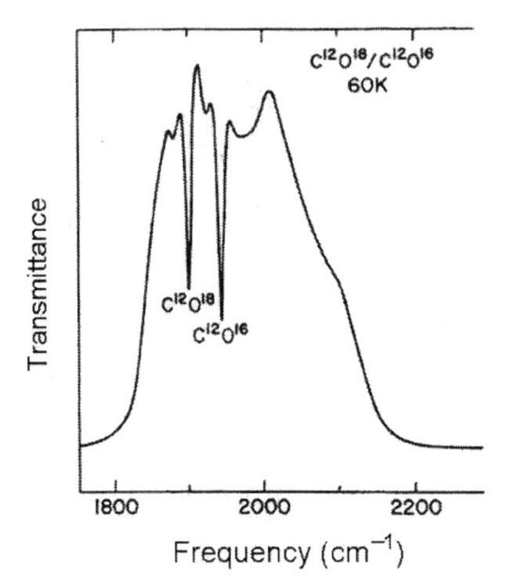

Fig. 12.3. The infrared transmission spectrum of MbCO. Two different isotopes of dioxygen are used.

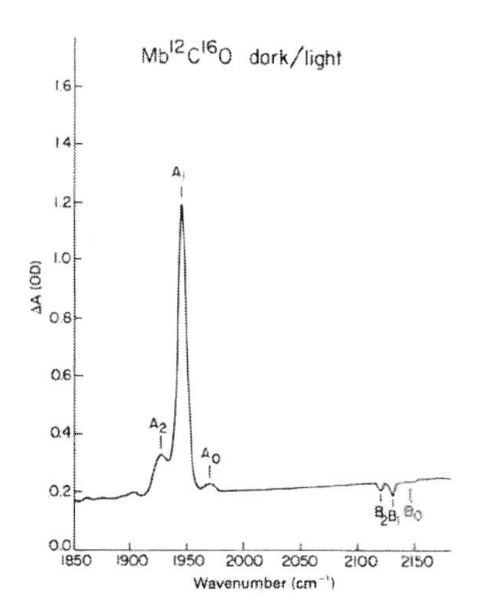

Fig. 12.4. Difference spectrum for MbCO. The difference in absorbance between MbCO and Mb* at 5.5 K is shown as a function of wavenumber. The bound states appear near 1950 cm^{-1} the weak lines of the photodissociated CO are near 2150 cm^{-1}.

1. Structure. The taxonomic substates differ in structure [5]. The X-ray structures of A_0 and A_1 have been determined and a crucial difference is shown in Fig.12.5. In A_1, the distal histidine, His 64, is inside the heme pocket; in A_0, it extends out into the solvent. In addition to this major difference, A_0 and A_1 display a number of smaller differences.

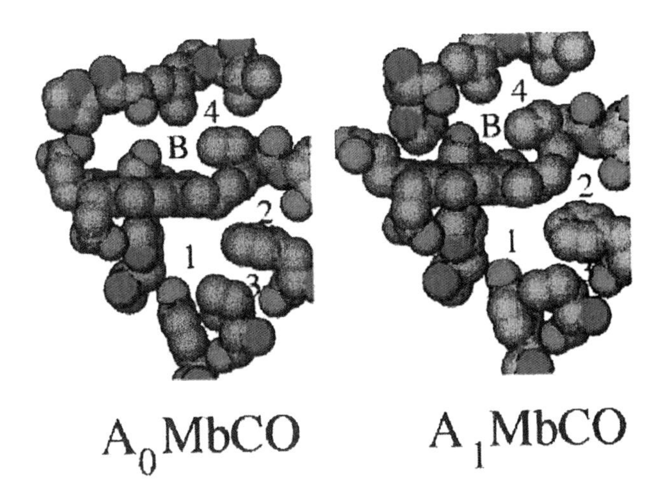

Fig. 12.5. The region around the heme pocket at pH 5, where A_0 dominates, and PH 7, where A_1 dominates (Protein data bank file 1A6G; Yang and Phillips [5]).

2. Dependence on external parameters. While the wavenumbers of the taxonomic substates are nearly independent of external conditions, their relative intensities depend strongly on temperature, pressure, pH, and solvent composition [6]. Of particular importance for function is the dependence on pH. At low pH, A_0 dominates; at high pH, A_1 does. The ratio also depends strongly on pressure and temperature, as shown in Fig.12.6. To discuss the data in Fig.12.6, we assume that the two substates have different energies, entropies, and volumes, as indicated in Fig.12.7. Also indicated in Fig.12.7 are the rate coefficient k_{10} and k_{01} for transitions between A_1 and A_0. Figure 12.7 shows three regions. At low temperatures, below $T_g \approx 200$ K, the raio A_0/A_1 is temperature independent, indicating that the exchange between the substates has ceased. At high temperature ($T > 300$ K), the ratio increasess with increasing temperature. Between the two extremes, the ratio increases with decreasing temperature. The behavior can be understood, and we consider the temperature dependence in more detail.

The Glass Phase ($T < T_g$). One characteristic of a glass is metastability. A glass is a nonequilibrium system, and its properties depend on its history. Metastability implies that the state of a system below its glass temperature depends on its history. History dependence can be proven by a general approach [6]. The properties of the system are measured at a given temperature and pressure below T_g after reaching the point (T, P) by two different pathways, as shown in Fig.12.8. On the solid pathway, the system is first frozen (F) and then pressurized (FP); on the dashed pathway, pressurization (P) is followed by freezing (PF). Figure 12.9 shows the result of such an experiment, where the IR spectrum of MbCO is measured at the four points F, FP, PF,

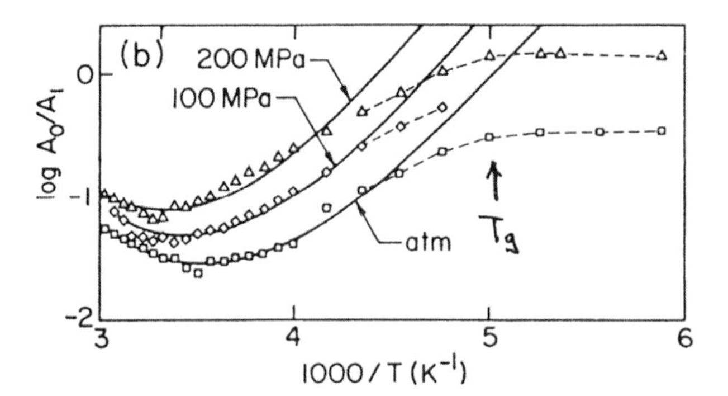

Fig. 12.6. Temperature dependence of ratio A_0/A_1 at pH 6.6 for three pressures [6].

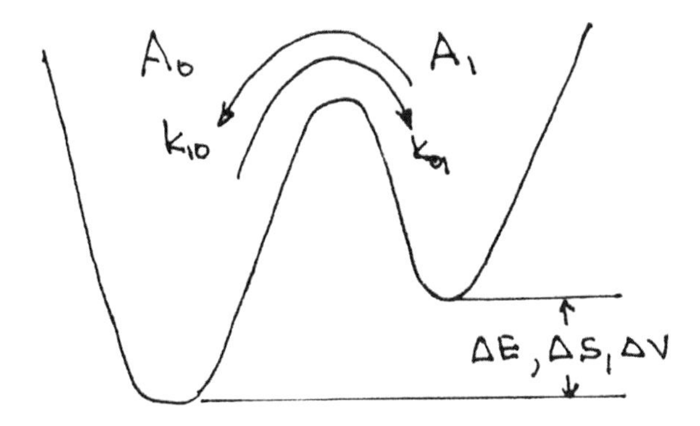

Fig. 12.7. Substates of tier 0.

and PFR in the TP plane, with $T_i = 225\,\mathrm{K}$, $T_f = 100\,\mathrm{K}$, $P_i = 0.1\,\mathrm{M\,Pa}$, and $P_f = 200\,\mathrm{M\,Pa}$. The infrared spectra obtained on the two pathways differ, and metastability is proven.

The Equilibrium Region. Well above T_g, the ratio A_0/A_1 is given by

$$A_0/A_1 = \exp[-\Delta G/RT] = \exp[-(\Delta E + P\Delta V - T\Delta S)/RT] , \qquad (12.1)$$

where ΔE, ΔV, and ΔS are defined in Fig. 12.8.

Equation 12.1, with constatnt values of S, E, and V, predicts a straight line in a plot of $\log(A_0/A_i)$ versus $1/T$ and does not explain the observed change in slope near 300 K. A phase transition near 300 K, with a change from the

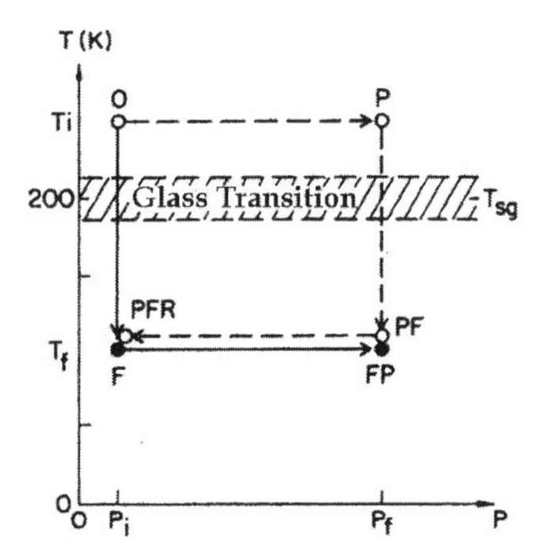

Fig. 12.8. The protein sample is taken from an initial state (T_i, P_i) to a final stsate (T_f, P_f) on different pathways, for instance, $0 \to F \to FP$ (solid path) and $0 \to P \to PF$ (dashed path). In equilibrium, the protin properties in the two states must be identical. Metastability is proven if different properties are found at FP and PF.

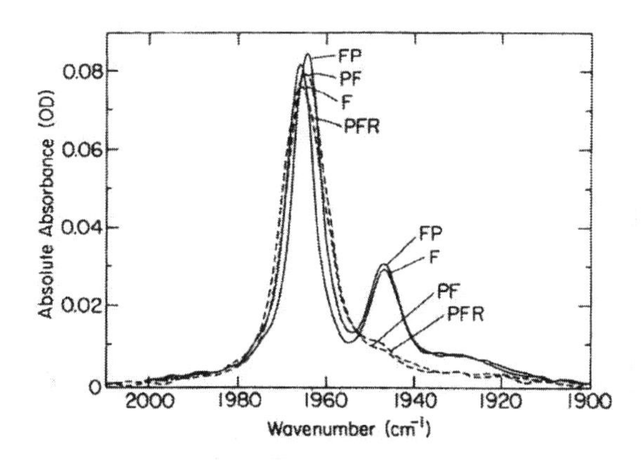

Fig. 12.9. The MbCO stretch spectrum determined at the points $F, FP, PF <$ PFR in the TP plane of Fig.12.8. In equilibrium, the pairs, F and PFR, and FP and PF, should yield identical spectra. The measured spectra differ considerably, proving that proteins at low temperatures are metastable [6].

example, substate A_0 to a new substate A_0' can fit the data. However, no pronounced change in the position or width of either stretch band is observed and this explanation therefore is not attractive. However, the entropy of a glass or of the liquid above the glass temperature is not constant but increases nearly linearly with T. A similar behavior has been observed for proteins. We therefore assume as the simplest possibility that the entropy depends on T and P as

$$S(T,P) = S(0) + sT + vP. \tag{12.2}$$

The term linear in T is justified by experiment, the term linear in P is justified by noting that the Maxwell relation $(\delta S/\delta P)_T = -(\delta V/\delta T)_P$ yields

$$V(T,P) = V(0) - vT. \tag{12.3}$$

The standard relation describing thermal expansion is usually written as

$$V(T) = V(0)[1 + \beta T], \tag{12.4}$$

where β is the coefficient of thermal expansion and where the compressibility has been neglected. Equations (12.3) and (12.4) show that

$$v = -\beta V(0). \tag{12.5}$$

Equation (12.5), with standard thermodynamic relations (Chapter 20), gives

$$E(T,P) = E(0) + vPT + \frac{1}{2}\,sT^2. \tag{12.6}$$

The difference in Gibbs free energy between the substates A_0 and A_1 becomes

$$\Delta G(T,P) = \Delta E(0) + P\Delta V(0) - T\Delta S(0) - PT\Delta v - \frac{1}{2}\,T^2\Delta s\,. \tag{12.7}$$

Fits with Eqs. (12.1) and (12.7) yield the solid lines in Fig.12.6 and at pH 6.6 $\Delta E(0) \approx -70\,\mathrm{kJ/mol}$, $\Delta V \approx -0.1\,\mathrm{cm^3\,mol^{-1}\,K^{-1}}$, and $\Delta S/R \approx -60$. The taxonomic substate A_0 in Mb, with His 64 extended into the solvent, thus has a lower energy, lower entropy, and smaller volume than A_1. This result is given as an example that different taxonomic substates have clear different properties.

3. Function. The results discussed so far would be biologically uninteresting, if they were not also related to function. It turns out, however, that the different substates can differ in function and that they can control the functional rate. In the substate A_1, Mb stores and transports dioxygen, and in A_0, it can act as an enzyme [7]. The result presents a challenge. If a "simple" protein such as myoglobin possesses two different functions, one can hypothesize that many proteins may have multiple taxonomic substates with different functions, to be switched on when needed. Prime examples are prions.

12.4 Statistical Substates (Tier 1)

Statistical substates are so numerous that they cannot be described individually; their properties must be characterized by distributions. An example has already been given in Eq.(11.4). Clear evidence for statistical substates has come from many experiments [4]. Two have already been discussed, the nonexponential rebinding of CO to Mb at low temperature (Fig. 11.6) and spectral hole burning (Fig. 11.14). IR experiments show that rebinding is nonexponential in time in each taxonomic substate. Each taxonomic substate therefore contains a very large number of statistical substates.

Relaxation1 experiments, which will be discussed in Chapter 13, show that the taxonomic substates contain (at least) two tiers, denoted by 1α and 1β. The detailed exploration of these statistical substates is still in its infancy. Moreover, most of the studies have been performed using Mb. It is therefore unclear how the energy landscape of other proteins is organized.

12.5 Few-Level Substates (Tier 2–?)

Naively one expects that all protein motions cease below the "glass temperature" T_g, indicated in Fig.12.6. It turns out, however, that motions in proteins continue below 1 K. Their detailed organization is not yet clear, but some features are emerging. Of course these motions do not disappear at ambient temperatures, but they are difficult to observe, because they are overwhelmed by the motions in the higher tiers. Nevertheless, they may influence processes that are biologically relevant by interacting with motions in the higher tiers [8, 9]. While the biological relevance is doubtful, it is clear that proteins form exceptionally interesting systems to study the physics of amorphous systems at low temperatures [10, 11]. Optical experiments provide surprising information about the energy landscape below T_g (e.g.[12]–[14]). In hole-burning experiments, already discussed in Section 11.4, a narrow hole is burned into an inhomogeneously broadened spectral line [15]. Broadening is caused by the fact that a protein can exist in many substates. For simplicity assume that each line in the broadened spectrum characterizes a substate. The hole in the spectrum then corresponds to a particular substate. If the protein undergoes no conformational fluctuations, the hole will remain unchanged in time. If fluctuations occur, however, the protein moves from substate to substate. Correspondingly, the hole changes frequency with each fluctuation, in a process called spectral diffusion (SD). Special diffusion can be described as a Brownian walk in the conformation space in which barriers have to be overcome. The spectral diffusion leads to line broadening from which a fluctuation rate R(T) can be deduced [10, 13]. Thorn Leeson and collaborators [13] have studied the spectral diffusion in apomyoglobin complexed with Zn-mesoprophyrin IX (Zn-Mb). An Arrhenius plot of log R(T) versus T exhibits three different regions below about 20 K, as shown in Fig. 12.10. At first, this result appears to imply the existence of three different tiers in the energy landscape. A careful evaluation of the data shows, however, that it is more likely that there are three different sites in the same tier with different barrier heights, as indicated in Fig. 12.11 [13]. Moreover, the data show that the hole returns often to the original position. This observation indicates that there are only a small number of substates in this tier—therefore the name "few-level substates." The Arrhenius behavior indicates that the conformational fluctuations occur by over-the-barrier motions. Some motions occur even below 2 K; they can be ascribed to tunneling.

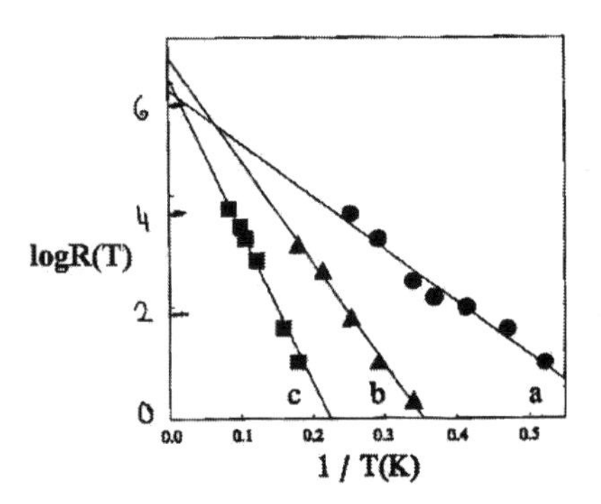

Fig. 12.10. Temperature dependence of the fluctuations rate, R(T), from spectral diffusion measurements in Zn-Mb (after Thorn Leeson et al. [13]).

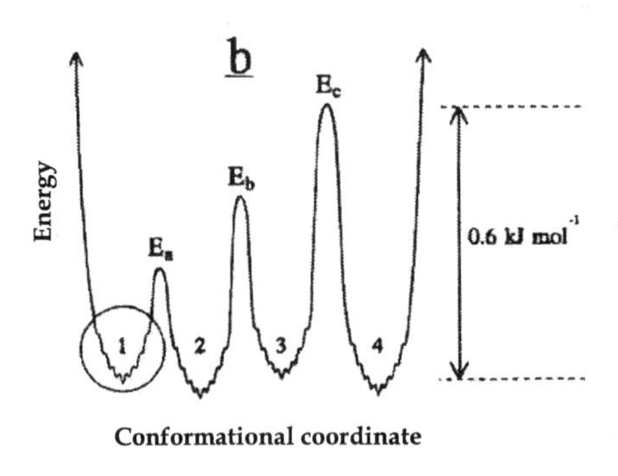

Fig. 12.11. The few-level tier of the energy landscape in Zn-Mb (after Thorn Leeson and collaborators [13]).

References

1. H. Frauenfelder et al., editors. *Landscape Paradigms in Physics and Biology. Concepts, Structures, and Dynamics: Special Issue Originating from the 16th Annual International Conference of the Center for Nonlinear Studies, Los Alamos, NM, USA, 13–17 May 1996.* North Holland, Amsterdam, 1997. Also in *Physica D*, 107:117–435 (1997).
2. H. Frauenfelder, S. G. Sligar, and P. G. Wolynes. The energy landscape and motions of proteins. *Science*, 254:1598–603, 1991.
3. R. H. Austin, K. W. Beeson, L. Eisenstein, H. Frauenfelder, and I. C. Gunsalus. Dynamics of ligand binding to myoglobin. *Biochemistry*, 14:5355–73, 1975.
4. A. Ansari, J. Berendzen, S. F. Bowne, H. Frauenfelder, I. E. T. Iben, T. B. Sauke, E. Shyamsunder, and R. D. Young. Protein states and proteinquakes. *Proc. Natl. Acad. Sci. USA*, 82:5000–4, 1985.
5. F. Yang and G. N. Phillips, Jr. Crystal structures of CO^-, deoxy- and met-myoglobins at various pH values. *J. Mol. Bio.*, 256:762–74, 1996.
6. H. Frauenfelder et al. Proteins and pressure. *J. Phys. Chem.*, 94:1024–37, 1990.
7. H. Frauenfelder, B. H. McMahon, R. H. Austin, K. Chu, and J. T. Groves. The role of structure, energy landscape, dynamics, and allostery in the enzymatic function of myoglobin. *Proc. Natl. Acad. Sci. USA*, 98:2370–4, 2001.
8. R. G. Palmer, D. L. Stein, E. Abrahams, and P. W. Anderson. Models of hierarchically constrained dynamics for glassy relaxation. *Phys. Rev. Lett.*, 53:958–61, 1984.
9. E. Abrahams. Nonexponential relaxation and hierarchically constrained dynamics in a protein. *Phys. Rev.*, E 71:051901, 2005.
10. V. V. Ponkratov, J. Friedrich, J. M. Vanderkooi, A. L. Burin, and Y. A. Berlin. Physics of proteins at low temperature. *J. Low Temp. Phys.*, 137:289–317, 2004.
11. F. Parak. Physical aspects of protein dynamics. *Rep. Progr. Phys.*, 66:103–29, 2003.
12. K. Fritsch and J. Friedrich. Spectral diffusion experiments on a myoglobin-like protein. *Physica D*, 107:218–24, 1997.
13. D. Thorn Leeson, D. A. Wiersma, K. Fritsch, and J. Friedrich. The energy landscape of myoglobin: An optical study. *J. Phys. Chem.*, B 101:6331–40, 1997.
14. J. Schlichter, V. V. Ponkratov, and J. Friedrich. Structural fluctuations and aging processes in deeply frozen proteins. *J. Low Temp. Phys.*, 29:795–800, 2003.

15. P. Schellenberg and J. Friedrich. Optical spectroscopy and disorder phenomena in polymers, proteins and glasses. In A. Blumen and R. Richter, editors, *Disorder Effects on Relaxational Processes.* Springer, Berlin, 1994.

13

Reaction Theory

Reactions and conformational fluctuations govern all aspects of biological processes, from enzyme catalysis to transfer of charge, matter, and information. Any deep understanding of biological reactions must be based on a sound theory of reaction dynamics. Most of the knowledge of reaction dynamics, however, has been deduced from two-body interactions of small molecules in the gas phase [1]. In contrast, biomolecules provide a complex but highly organized environment that can affect the course of the reaction. Fortunately, the complexity implies a richness of phenomena that allows the examination of fundamental aspects of reaction dynamics. Biomolecules, in particular heme proteins, form an excellent laboratory.

In this chapter we describe some of the main features of reaction theory and stress aspects where experiments with proteins have yielded new insight into reaction theory or call for a reevaluation and extension of the theory. These topics include tunneling, friction, gating, adiabaticity versus nonadiabaticity, pressure, and the role of intermediate states. The results indicate that the field is rich. One word concerns the terms kinetics and dynamics. With *kinetics* we mean the description of the time dependence of a reaction in terms of rate coefficients. *Dynamics* is concerned with the explanation of the reaction parameters in terms of the structure of the protein and its environment.

13.1 Arrhenius Transitions and Tunneling

The reaction rates depend on many parameters [2]. The most important is temperature. Empirically many reaction rate coefficients, $k(T)$, depend exponentially on temperature,

$$k(T) = \mathcal{A}\, e^{-E/k_B T}. \tag{13.1}$$

The preexponential or frequency factor $\mathcal{A}$ is essentially temperature independent. The activation energy E controls the temperature dependence; k_B is the

H. Frauenfelder, *The Physics of Proteins*, Biological and Medical Physics, Biomedical Engineering, DOI 10.1007/978-1-4419-1044-8_13,

Boltzman constant. Equation (13.1) was first given by Hood in 1878, then by Arrhenius in 1887 [3]; it is called the Arrhenius equation. From the measured reaction rates, $\mathcal{A}$ and E are found by plotting log k versus $1/T$ (Arrhenius plot).

Let us consider a binding process in more detail by using a simplified case. Assume that the potential between two reacting particles depends only on their distance. We represent the situation by the potential energy map in Fig. 13.1. The map shows two wells, separated by a saddle. Well A is much deeper than B. Two cross sections, as indicated in Fig. 13.1, are given in Fig. 13.2. A particle placed in one well, say B, has a certain probability of moving to the other well. What are the essential aspects of such a process? If well A is much deeper than B, say by about 1 eV (100 kJ/mol), the transition corresponds to a chemical binding.

The energy levels in the two wells are quantized. An enormous amount of energy has been spent by many people to calculate energy levels in detail [4].

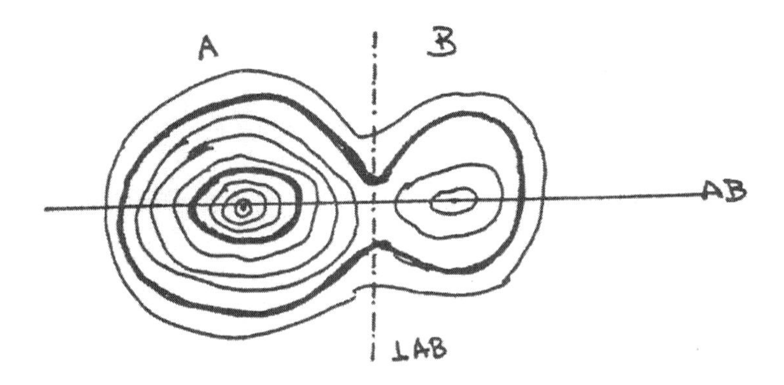

Fig. 13.1. Extremely simplified potential energy map for a transition $B \rightarrow A$.

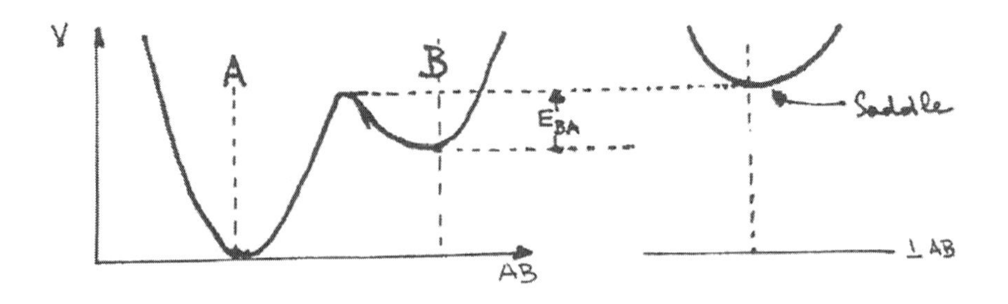

Fig. 13.2. Two cross sections through the energy map of Fig. 13.1.

We sketch in Fig. 13.3 the energy levels in a two-well system as obtained in [4]. Since the two wells form one system, energy levels are always common to both wells, even below the barrier. However, the wave function depends drastically on level, as sketched in Fig. 13.4.

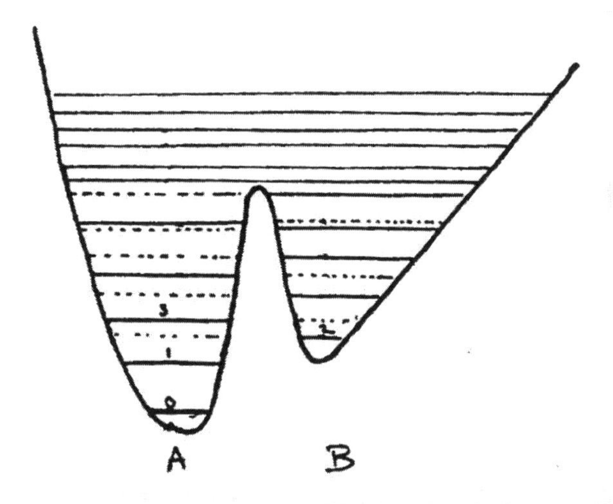

Fig. 13.3. Vibrational eigenvalues in a double well [4]. Levels below the separating barrier are predominantly localized in one or the other well.

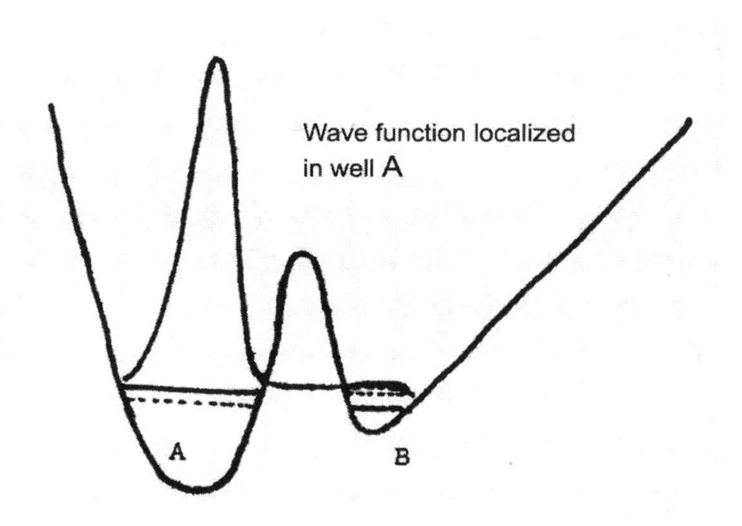

Fig. 13.4. Localized wave function.

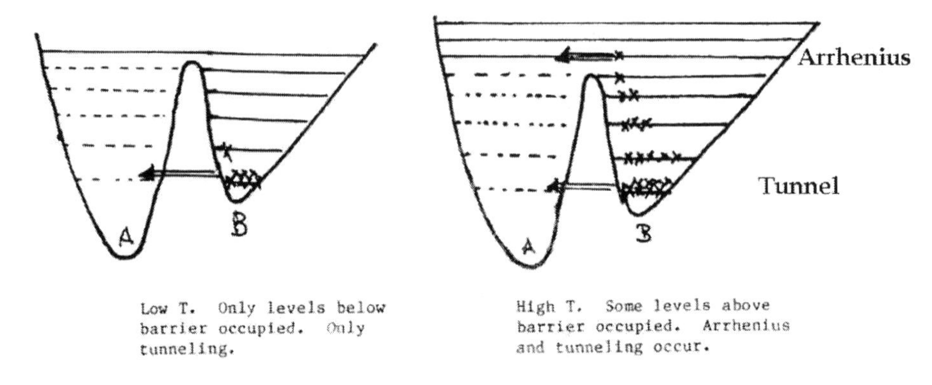

Fig. 13.5. Temperature dependence of transition $B \rightarrow A$.

Assume now that a particle is initially placed in well B and that there are many identical systems so that we can consider averages. At a temperature T, the levels in well B will be occupied according to a Boltzmann distribution, and two situations must be considered as indicated in Fig. 13.5. At very low temperatures, only the lowest levels in B will be occupied. To move to A, particles have to *tunnel* through the barrier. At high temperatures, some levels above the saddle point will be occupied and particles can move over the barrier classically, by the standard Arrhenius motion. We will treat the two cases in more detail in the next sections.

The discussion so far is extremely simplified and corresponds to a single particle potential model. The complex many-body system of a protein is replaced by a potential, and the particle (CO, for instance) moves in a fixed potential. Clearly, such an oversimplified approach misses many of the essential aspects of protein dynamics. Equally clearly, as we know from nearly every field of physics, the single-particle approach is a necessary first step. Improvements come when the most primitive approximation is understood. Figure 13.5 demonstrates that tunneling and classical motion are not two different phenomena, but two aspects of the same process.

13.2 Transition State Theory (TST)

A "derivation" of the Arrhenius relation is given by the transition state theory [1]. Here we give a simple "derivation" of the corresponding rate coefficient k_{TST} for the transition over a barrier of height E. This leads essentially to the Eyring relation. This relation has been used extensively in chemistry in the past 50 years. We will show that its use is not justified in reactions in the condensed phase, in particular in protein reactions. Nevertheless, it forms the starting point for many discussions, and we therefore need to understand its basic features.

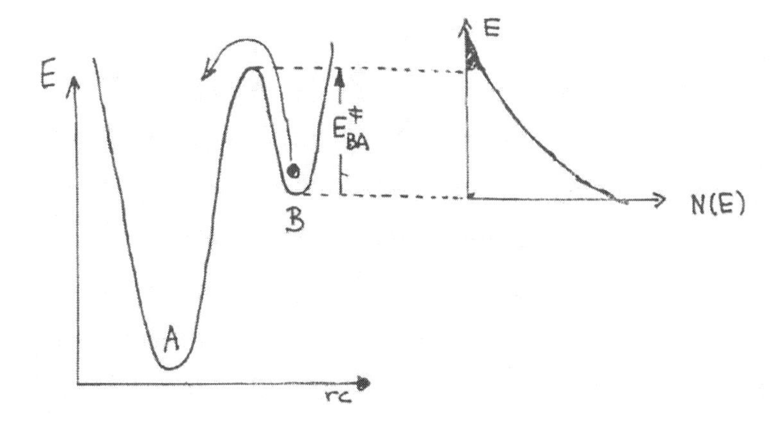

Fig. 13.6. Thermal transitions over the barrier between B and A. The Boltzmann distribution $N(E)$ versus E for the particles in well B are shown at right.

Consider the situation shown in Fig. 13.6 where $E^{\ddagger}_{BA}$ denotes the height of the barrier between the well B and the transition state between B and A. Assume the system to be in equilibrium at temperature T. The fraction of molecules having an energy E greater than E_{BA} is then given by

$$\frac{\int_{E^{\ddagger}_{BA}}^{\infty} \exp[-E/k_BT]dE}{\int_0^{\infty} \exp[-E/k_BT]dE} = \exp[-E^{\ddagger}_{BA}/k_BT] \, . \tag{13.2}$$

The molecules vibrate in well B with a frequency given by $\hbar\omega = k_BT$, or

$$\nu = \frac{k_BT}{h}, \tag{13.3}$$

about $10^{13}\dot{s}^{-1}$ at 300 K. We further denote the number of states (degeneracy) in well B by g_B, on top of the barrier by g_{BA}. The rate coefficient with which a molecule passes from well B to A by overcoming the barrier is given by the product of attack frequency ν, ratio of states g_{BA}/g_B, and probability of having an energy larger than $E^{\ddagger}_{BA}$:

$$k_{TST} = \nu(g_{BA}/g_B)e^{-E^{\ddagger}_{BA}/k_BT}.$$

But

$$g_{BA}/g_B = \exp(S^{\ddagger}_{BA}/k_B), \tag{13.4}$$

where $S^{\ddagger}_{BA}$ is the entropy difference between the bottom of well B and the top of the barrier $B \to A$. The rate thus finally can be written as

$$k_{TST} = \nu \exp(-F^{\ddagger}_{BA}/k_B T), \tag{13.5}$$

where $F^{\ddagger}_{BA} = E^{\ddagger}_{BA} - TS^{\ddagger}_{BA}$ is the difference in Helmholtz energy between well B and the top of the barrier (activation Helmholtz energy).

Assume now in addition that the system is placed under a pressure P and that the volume of the system is V_B in state B and V_{BA} at the transition state on top of the barrier. The particle moving from B to A then has to have energy $E_{BA} + PV^{\ddagger}_{BA}$, with

$$V^{\ddagger}_{BA} = V_{BA} - V_B \ , \tag{13.6}$$

in order to reach the top of the barrier. Here $V^{\ddagger}_{BA}$ is called the *activation volume.* The limit of integration E_{BA} in Eq. (13.2) is replaced by $H^{\ddagger}_{BA} = E^{\ddagger}_{BA} + PV^{\ddagger}_{BA}$, the activation enthalpy. $H^{\ddagger}_{BA}$ then appears instead of $E^{\ddagger}_{BA}$ in all subsequent equations and the final expression can be written as

$$k_{TST} = \nu \exp(-G^{\ddagger}_{BA}/k_B T), \tag{13.7}$$

with

$$G^{\ddagger}_{BA} = H^{\ddagger}_{BA} - T\, S^{\ddagger}_{BA} \tag{13.8}$$

and

$$H^{\ddagger}_{BA} = E^{\ddagger}_{BA} - P\, V^{\ddagger}_{BA}. \tag{13.9}$$

Note that if $V^{\ddagger}_{BA}$ is negative, the reaction becomes faster with increasing pressure.

Comparison of Eqs. (13.7)–(13.9) and Eq. (13.1) shows that

$$\mathcal{A} = \nu \exp(S^{\ddagger}_{BA}/k_B), \tag{13.10}$$

$$E = H^{\ddagger}_{BA}. \tag{13.11}$$

With Eq. (13.3), Eq. (13.7) becomes the standard Eyring expression

$$k_{TST} = (k_B T/h) \exp\{-G^{\ddagger}/k_B T\}. \tag{13.12}$$

This relation, in some form or other, is found in most texts [5]. While it may be justified for gas-phase reactions, its application to biomolecular reactions can lead to serious problems. To discuss improved relations we note that Eq. (13.12) is based on a number of assumptions. The most essential assumptions are:

(i) The system is in thermal equilibrium.
(ii) A trajectory crosses the barrier only once and the system does not return from B to A.
(iii) The system can be treated classically.
(iv) The reaction process is adiabatic and the reaction rate is determined sterically by the potential energy surface (Fig. 13.1).

In the following sections we show that all of these assumptions must be abandoned or changed for biomolecular reactions.

13.3 The Kramers Equation

In 1940, H. A. Kramers [6] treated a simple model of a chemical reaction, namely the escape of a particle from a potential well, by using the Fokker-Planck equation [7]. For a long time, Kramers' work was appreciated only by a few theoreticians [8, 9] and not at all by the experimentalists. Within the last few decades, Kramers' work has become the central point of an improved approach to reaction theory and reactions in solids, and it has also been discovered by the experimentalists. The number of papers has become extremely large so that we refer to reviews for a complete discussion [2, 10]. No really simple discussion of the approach of Kramers exists and his original paper is not easy to read. We therefore follow an earlier, somewhat handwaving, introduction to the Kramers theory [11]. Before doing so, we describe briefly why the approach of Kramers is important for biomolecules.

Biomolecular Reactions. We have pointed out a few times that proteins are dynamic systems and that we expect protein and solvent fluctuations to influence protein reactions. If a protein is embedded in a solvent, the solvent will influence the fluctuations. In particular, we can expect that some protein reactions will decrease with increasing viscosity [12, 13]. Indeed, experiments show that viscosity affects, for instance, enzyme reactions [14] and the binding of ligands to heme proteins [15]. To evaluate such experiments, the following general approach can be used.

Assume that we have measured an observable F as a function of temperature T and viscosity η, keeping other parameters constant. F consequently is a function of the two variables T and η, and $F(T, \eta)$ represents a surface in a T–η plot. For a given solvent, η depends approximately exponentially on temperature. If we plot F versus η, η spans too large a range. Thus we take $\log \eta$ as a second variable. The plot of $F(T, \eta)$, in a T–$\log \eta$ plot is sketched in Fig. 13.7. If the surface is smooth, it can be assumed that the various solvents do not change the protein in an unacceptable manner. *Isothermal* cuts through the $F(T, \eta)$ surface show how the viscosity affects F. *Isoviscosity* cuts provide the temperature dependence of $F(T, \eta)$ at fixed viscosity.

The conclusion of experiments over a wide range of viscosity [15, 16] is clear: Protein reactions are strongly influenced by viscosity. This result implies that protein motions are indeed important for protein function, and it

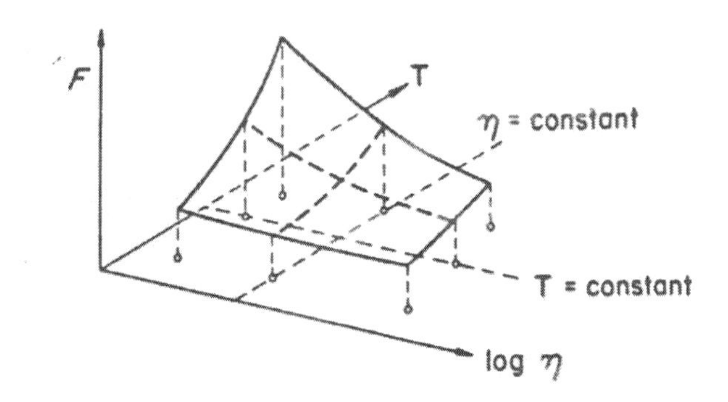

Fig. 13.7. Idealized plot of an observable F as a function of the two independent variables T and $\log\eta$, where T is temperature and η the viscosity. Isothermal ($T =$ constant) and isoviscosity ($\eta =$ constant) sections are indicated.

also suggests that the standard TST relation is not sufficient for the data evaluation.

Reaction Coordinate and Friction. In Fig. 13.6, we show a reaction from state B to state A as a transition along a reaction coordinate rc. Reality, of course, is far more complex. In the actual reaction, the entire protein and its surrounding take part. If we could treat the entire system without approximations, all dynamical effects would be incorporated and there would be no need to introduce viscosity or friction. Indeed, in molecular dynamics calculations such an approach is used. Usually, however, the reaction is treated by separating the coordinates into a reaction coordinate and invisible coordinates. The invisible coordinates exchange energy and momentum with the reaction coordinate. Energy and momentum of the reaction coordinate alone are not conserved; the reaction coordinate can gain or lose energy and momentum. The effect of this exchange on the reaction coordinate is called *friction*. Friction is essential for the trapping of the system in the product state A. The fluctuation-dissipation theorem (Section 20.7) tells us that the friction (dissipation) is necessarily accompanied by fluctuating forces, arising also from the neglected coordinates. When friction is large, the motion along the reaction coordinate looks like a Brownian motion or a random walk, and not the smooth ballistic motion required for the derivation of the TST equation. Friction can be roughly characterized by a velocity autocorrelation time τ_ν given by

$$\tau_\nu = m/\zeta. \tag{13.13}$$

Here m is an effective mass and ζ is the friction coefficient. In a liquid, ζ is approximately related to the viscosity η by Stokes law,

$$\zeta \approx 6\pi\eta a \tag{13.14}$$

where a is a characteristic linear distance.

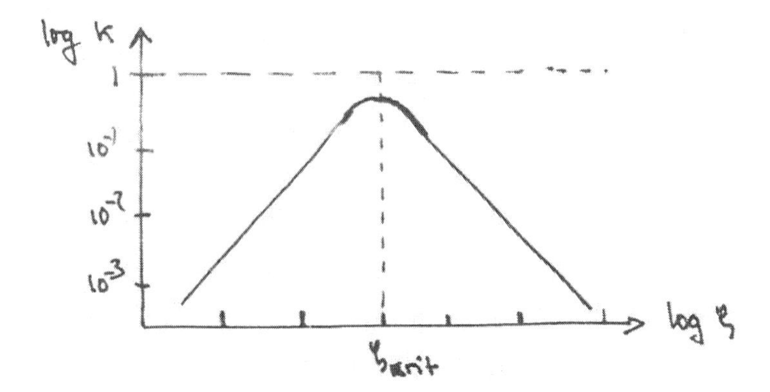

Fig. 13.8. General behavior of the transmission coefficient. κ, as predicted by the Kramers theory.

The Transmission Coefficient. Friction will clearly affect the reaction rate coefficient k. The effect of friction (or the invisible coordinates) can be characterized by a transmission coefficient, κ, by writing

$$k = \kappa \, k_{TST} \tag{13.15}$$

where k_{TST} is given by Eq. (13.7). Kramers' theory shows that κ has the form sketched in Fig. 13.8: At low friction, κ is proportional, and at high friction it is inversely proportional to the friction coefficient. In the intermediate regime, the value predicted for κ by TST, $\kappa = 1$, is an upper limit and can be significantly in error [17].

The general behavior can be easily understood. At low friction, the systems above the barrier in Fig. 13.6 quickly react, but then equilibrium has to be reestablished: Systems with lower energy have to acquire sufficient energy to move over the barrier, and this process is proportional to viscosity. At high viscosity, the system must diffuse over the barrier, and this process is inversely proportional to viscosity.

The rate coefficient k reaches a maximum at a value

$$\zeta_{crit} = dm\omega \tag{13.16}$$

where $\omega = 2\pi\nu$ and d is a numerical coefficient that depends on the reaction energy surface and the damping mechanism but is smaller than 1. With the Stokes relation, Eq. (13.14), and $d = 1$, Eq. (13.16) becomes

$$\eta_{\text{crit}} = (1/3)(m/a)(\omega/2\pi). \tag{13.17}$$

The typical values of $m/a = 1 \times 10^{-15}$ g cm^{-1} and $\omega/2\pi = 10^{13}$ sec^{-1} yield $\eta_{\rm crit} = 7$ millipoise (mP). Water at 300 K has a viscosity of 8.5 mP, so we expect most protein reactions to be overdamped.

Whereas the calculation of the reaction rate over the entire range of friction is difficult [10, 11], the limiting case $\zeta \ll \zeta_{crit}$ and $\zeta \gg \zeta_{crit}$ can be treated by using simple physical arguments based on the two characteristic lengths in the problem. The first length is the *mean free path*, λ, the average distance before the coordinate reverses its direction of motion. From Eq. (13.13), λ is related to the velocity autocorrelation time τ_ν or the friction ζ by

$$\lambda = v_{rms}\tau_\nu \approx (2mk_BT)^{1/2}/\zeta \tag{13.18}$$

where v_{rms} is the root-mean-square velocity. The second relevant length is the *size of the transition region*, ℓ_{TS}, defined by the condition that the energy in this region is within k_BT of the transition state energy. For a parabolic Gibbs energy barrier with curvature $m(\omega^*)^2$ at the col, as drawn in Fig. 13.9, ℓ_{TS} is given by

$$\ell_{TS} = 2[2k_BT/m(\omega^*)^2]^{1/2} \tag{13.19}$$

where ω^* is the undamped frequency of motion in the top of the transition barrier.

The character of the transition is determined by the ratio λ/ℓ_{TS}, given by Eqs. (13.18) and (13.19) as

$$\lambda/\ell_{TS} = m\omega^*/2\zeta\ . \tag{13.20}$$

The system is critically damped if only one collision occurs during the passage so that $\lambda \approx \ell_{TS}$ or $m\omega^* \approx 2\zeta$. This criterion agrees with the general criterion Eq. (13.16). If $\lambda \gg \ell_{TS}$, the system moves ballistically, without collisions,

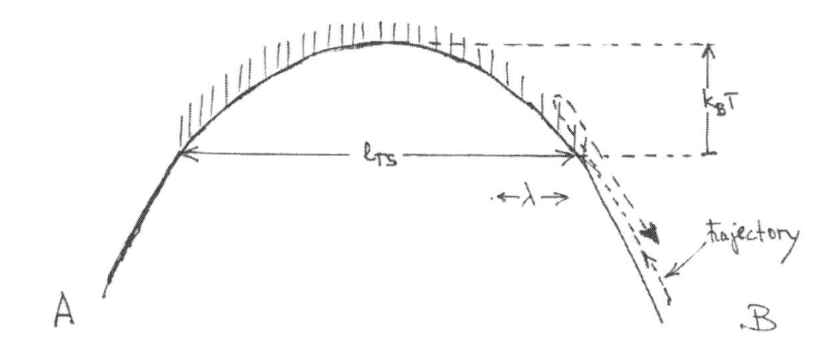

Fig. 13.9. Parabolic barrier, with characteristic lengths. Only the top of the barrier is shown. The transition region is shaded. A typical trajectory in the overdamped regime is shown at the right.

through the transition region; it is underdamped. If $\lambda \ll \ell_{TS}$, the system undergoes many collisions in the transition state and it is overdamped. Both situations can be discussed in simple terms [11]. We first treat the high-friction case.

High Friction. When friction is high, the reacting system is always in equilibrium with the surrounding heat bath, but the particle cannot traverse the transition region in a single attempt. Rather, as sketched in Fig. 13.9, a particle that attempts the crossing will make a random walk over the transition region. Most of the time, it will only penetrate a small distance into the transition region and will then return to the bottom of well B before trying again. The number of attempts required for a successful crossing is given by the ratio $\ell_{TR}/2\lambda$. The transmission coefficient is the inverse of the ratio of attempts and is thus given by Eq. (13.20). Introducing this transmission coefficient into Eq. (13.7) gives, with $\nu = \omega_B/2\pi$,

$$k(T,\zeta) = \frac{m\omega^{*}\omega_B}{2\pi\zeta}e^{-G^{*}/k_BT}. \tag{13.21}$$

Equation (13.21) is the relation Kramers found at the high-damping limit [7]. (This limit is sometimes called the "diffusion limit." *Diffusion* here refers to the Brownian passage over the col and not to the ordinary diffusion that governs "diffusion-controlled" reactions.)

Low Friction. If the friction is low, the system, on leaving the transition state, will not lose sufficient energy to drop into well A (Fig. 13.6) but will bounce off the other side and recross the barrier. If τ_r is the time to traverse the well and τ_E the time to lose approximately the energy k_BT, the system will go back and forth across the barrier roughly $N_c = \tau_E/\tau_r$ times before reacting. In general, τ_E is proportional to the velocity autocorrelation time τ, Eq. (13.13), and consequently proportional to $1/\zeta$. The transmission factor thus is proportional to the friction coefficient ζ, as shown in the left branch of Fig. 13.8. The same result is obtained by considering the time it takes to activate the system.

The result of these considerations can be summarized simply. At low friction, the reaction coefficient is proportional, and at high friction inversely proportional, to the friction coefficient. In the intermediate regime, the reaction coefficient depends somewhat on the potential surface topography and the frictional mechanism, but the value predicted by TST is an upper limit that can be significantly in error.

Experimental Verification of the Kramers Equation. Experiments, in particular by Fleming and collaborators [18], have confirmed the essential aspects of the Kramers theory. Figure 13.10 gives the reaction rate for the photoisomerization of stilbene as a function of the inverse of the diffusion coefficient in gaseous and liquid alkanes. The pattern predicted in Fig. 13.8 is confirmed. The maximum rate coefficient occurs, however, at a viscosity lower than predicted by Eq. (13.17).

Biological Importance. The behavior of the Kramers relation, as shown in Figs. 13.8 and 13.10, can be crucial for the control of biomolecular reactions. If the TST relation were correct, a reaction could be controlled only by changing the Gibbs activation energy, by changing the activation energy, activation volume, or activation entropy. Such changes would usually involve a change within the biomolecule. The Kramers relation shows that control can also be exerted from the outside by changing the friction through changing the viscosity. Such changes can occur, for instance, in membranes where cholesterol is very efficient in affecting the viscosity.

Additional Remarks. Experimental data are usually fitted with an Arrhenius relation, $k = \mathcal{A}\exp[-E/k_BT]$, and activation enthalpy and entropy are obtained with Eqs. (13.10) and (13.11). These approximations are based on the assumption of a temperature-independent transmission coefficient. Figure 13.10 shows, however, that κ depends on friction. Friction is strongly temperature dependent. With

$$\zeta \simeq \zeta_0 \exp(E_\zeta/k_BT), \tag{13.22}$$

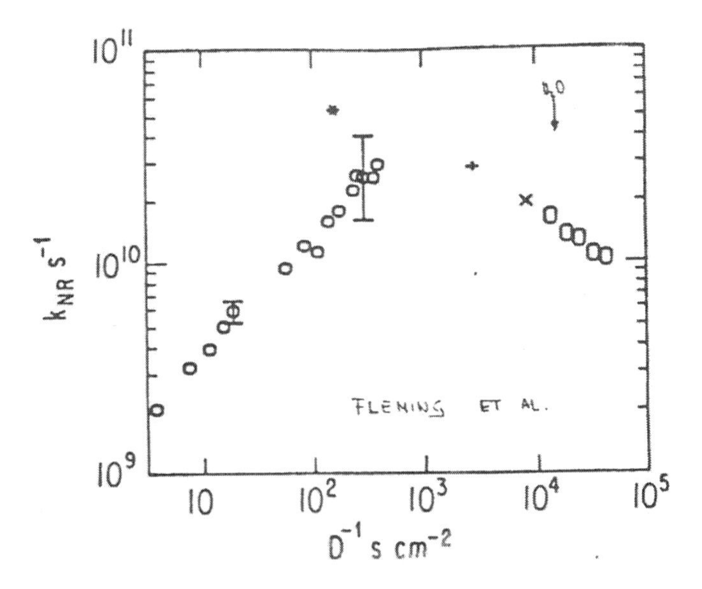

Fig. 13.10. Experimental verification of the Kramers relation. From Fleming et al. [18].

the connections between activation enthalpies and entropies corresponding to the high-damping result Eq. (13.21) are [15]

$$H^{\ddagger} - H^{*} = E_{\zeta}\,, \quad S^{\ddagger} - S^{*} = E_{\zeta}/T_0, \tag{13.23}$$

where T_0 is the average temperature where the data were taken. As before, ‡ denotes data taken in a given solvent and * indicates isoviscosity data. These relations show that the intrinsic activation enthalpies and entropies are smaller than the ones extracted in the customary way from an Arrhenius fit. The relations shown in Eq. (13.23) are valid in the overdamped regime; closer to η_{crit} they must be modified [15]. In very viscous solvents the overdamped motion in the transition state may be more rapid than that used in measuring macroscopic viscosities, and "memory" effects must be taken into account [19]–[23]: Grote and Hynes have shown that one needs only to replace the low-frequency friction coefficient in the Kramers rate by its renormalized value at the overdamped frequency of motion in the transition state. Very viscous solvents have a power-law dependence of viscosity on frequency. Thus at high friction this argument leads to a fractional power-dependence on macroscopic viscosity, as has been observed in reactions of biomolecules [15].

One final remark concerns the time of transition in the overdamped regime. The decrease of the rate coefficient shown in Fig. 13.10 is not caused by the delay of the particle moving over the barrier in a successful trajectory, but by the fact that a particle has to try about ℓ_{TS}/λ times before it succeeds. The actual transition time is, of course, (ℓ_{TS}/λ) times longer than in a ballistic trajectory, but even this time is still very short. The time for a ballistic crossing is of the order of 10 fs. Thus even a considerable lengthening still leads to times less than picoseconds.

13.4 The Tunnel Effect

Quantum-mechanical tunneling is important in nearly every part of physics, from cosmology to condensed matter. We discuss tunneling here from a restricted point of view, namely the binding of ligands to heme proteins. The concepts are, however, valid for many other processes in biological physics and chemistry, and additional information is given in a number of excellent books [24]–[27].

The Tunneling Temperature. In Section 11.2 we briefly discussed the binding of CO to heme proteins at low temperatures and used the Arrhenius equation, Eq. (11.1), to evaluate the data. At first sight it appears that this approach could be used even at very low temperatures, because quantum-mechanical tunneling depends exponentially on the mass. However, the activation barriers, shown in Fig. 11.7, are very small. Intuition obtained from work with standard chemical reactions can therefore fail. To obtain an estimate of the temperature where tunneling should set in, we follow an argument by Goldanskii [28]. The parameters involved are sketched in Fig. 13.11. We first assume that tunneling is given by the well-known Gamow factor, calculated in most texts [29]. Lumping attempt and entropy factors into one coefficient $\mathcal{A}_t$, we write

$$k_t = \mathcal{A}_t \exp\{-\gamma[(2M(H_{BA} - E)]^{1/2}\ell(E)/\hbar\}. \tag{13.24}$$

Here, M is the mass of the tunneling system, H_{BA} the barrier height, E the excitation energy in well B, and $\ell(E)$ the barrier width at the energy E. The value of the numerical factor γ depends on the shape of the barrier: for a triangular barrier, $\gamma = 4/3$, for a parabolic one, $\gamma = \pi/2$, and for a square one, $\gamma = 2$.

Equations (13.1) and (13.24) together imply that Arrhenius transitions will dominate at high, tunneling at low temperatures, just as we explained earlier. To find the temperature T_t at which both have similar rates, we assume equal preexponentials and $H_{BA} \gg E$ (note that E in Eq. (13.1) has the meaning of H_{BA}), we then get

$$T_t = (\hbar/k_B\ \gamma)[H_{BA}/2M]^{1/2}/\ell. \tag{13.25}$$

For a parabolic barrier, with a molecule of $M = 30$, the numerical value of T_t is

$$T_t = 0.9[H_{BA}(\text{kJ/mol})]^{1/2}/\ell\,(\text{nm}). \tag{13.26}$$

For a barrier of 0.1 nm width and 5 kJ/mol height, $T_t \simeq 20\,\text{K}$.

Experimental Results. The value of T_t implied by Eq. (13.26) suggest that it should be easy to observe the molecular tunnel effect in heme proteins. Indeed, the experiments are straightforward and the result is clear [30]–[32]. The

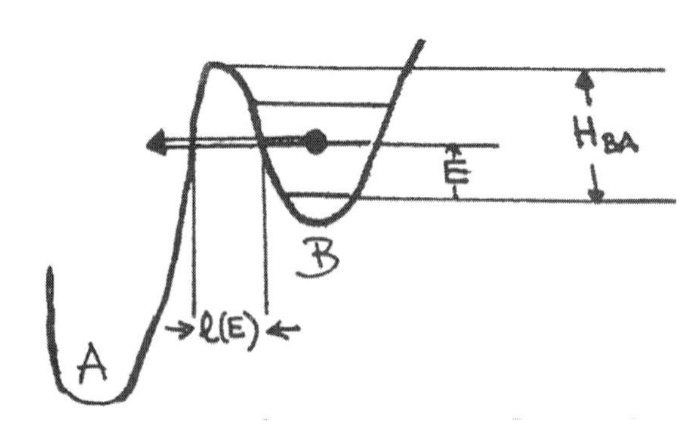

Fig. 13.11. Parameters determining tunneling: height H_{BA}, energy E, and barrier width $\ell(E)$.

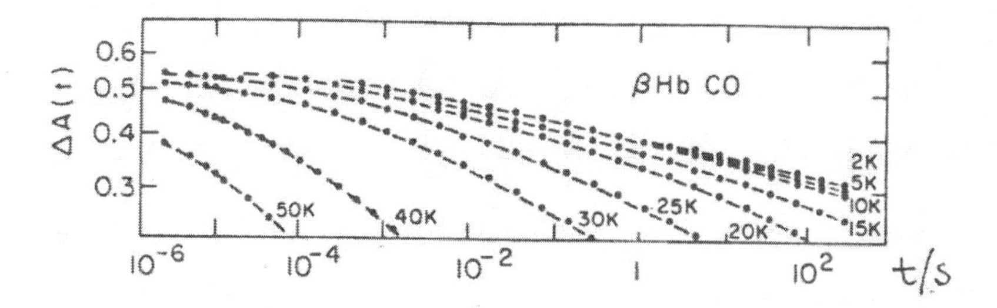

Fig. 13.12. Rebinding of CO to the separated beta chain of human hemoglobin at low temperatures.

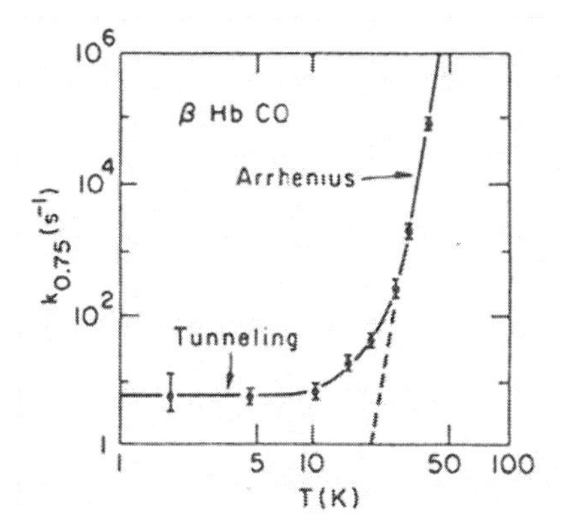

Fig. 13.13. The rate $k_{0.75} = 1/t_{0.75}$ plotted as a function of $\log T$ for binding of CO to β Hb. The steeply dropping part corresponds to classical Arrhenius motion over the barrier. The approximately temperature-independent rate below about 10 K is interpreted as quantum mechanical tunneling through the barrier.

rebinding curves for CO to the β chain of hemoglobin at temperatures below 50 K are given in Fig. 13.12. Rebinding does not slow down as predicted by an Arrhenius behavior; below about 20 K, it becomes essentially temperature independent. To characterize the rate, we define the coefficient $k_{0.75} = 1/t_{0.75}$, where $t_{0.75}$ is the time at which $N(t)$ drops from 1 to 0.75. The coefficient $k_{0.75}$ is shown in Fig. 13.13 as a function of $\log T$. Figure 13.13 demonstrates that tunneling sets in at about 20 K in the rebinding of CO to a *beta* chain of hemoglobin.

The Isotope Effect. Tunneling is characterized by two properties, temperature independence in the limit $T \rightarrow 0$ and a pronounced dependence on the mass of the tunneling system. The first of these characteristics is clearly shown in Figs 13.12 and 13.13. Nevertheless, skeptics remain unconvinced; the temperature independence could be caused by some other phenomenon such as a purely entropic barrier at low temperature. The isotope effect can provide a second independent proof for tunneling. A small mass dependence exists even in a classical Arrhenius rate. The frequency ω_B in Eq. (13.21) depends of course on the mass of the system. The dependence is, however, weak and usually neglected. Equation (13.24), in contrast, shows an exponential mass dependence. With Eq. (13.24) and $\Delta M \ll M$ the mass dependence can be expressed as

$$\ln(k_\ell/k_h) \approx (\Delta M/2M)\ln(\mathcal{A}_t/k_\ell) \tag{13.27}$$

where the subscripts ℓ and h refer to the light and the heavy isotope, $\Delta M = M_h - M_\ell$, and $M = (M_h + M_\ell)/2$. Equation (13.27) tells us that the isotope effect can be observed best at small values of k, hence at long times. If binding were exponential in time, we would have little choice: k would be single-valued at a given temperature and k_ℓ/k_h would be time independent. As Fig. 13.12 shows, however, binding is not exponential in time and the binding rates vary over an enormous range. It is consequently best to measure the rate of rebinding at long times.

While the nonexponential character of rebinding permits us to select a suitably long time, it also causes a problem. Figure 13.12 demonstrates that the curves of $\Delta A(t)$ change very little below about 15 K. If we have to compare two different samples with two different isotopes of CO, it would be exceedingly difficult to measure a small effect of a factor of about 2 or less. Such an experiment would not be believable. Fortunately there exists a method that permits a simultaneous observation of the rebinding rates for two isotopes [33]. Optically, by observation in the Soret region, different isotopes cannot be distinguished. Different CO or dioxygen isotopes have, however, very different infrared spectra as shown in Fig. 13.14 [34]. This fact allows the simultaneous observation of the rebinding of two different isotopes in the same sample. The stretching frequencies of the three isotopes of interest in MbCO are 1945 cm^{-1} for $^{12}C^{16}O$, 1901 cm^{-1} for $^{13}C^{16}O$, and 1902 cm^{-1} for $^{12}C^{18}O$. CO rebinding after photodissociation is monitored by measuring the growth of the absorption spectra at the Mb-bound stretching frequencies. Typical curves are shown in Fig. 13.14. The result is unambiguous: The lighter isotope rebinds faster, as shown in Fig. 13.15 [33]. Values of k_ℓ/k_h at time 1000 s are given in Table 13.1. With some additional assumptions, values of $\Delta M/M$ can be extracted [34].

The data lead to three main conclusions:

(1) The rate coefficient k for binding of CO to Mb shows a mass dependence at 60 K and below.

(2) The most natural explanation for the isotope effect is quantum-mechanical tunneling. The effect at 60 K is about half as big as at 20 K, in agreement with an independent estimate that at 60 K the ratio of tunneling to Arrhenius transitions can be of the order of 0.5.

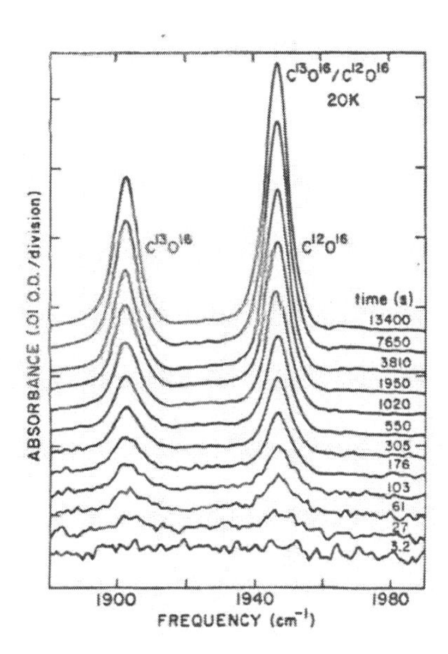

Fig. 13.14. The growth of the CO stretching frequency for two isotopes after photodissociation.

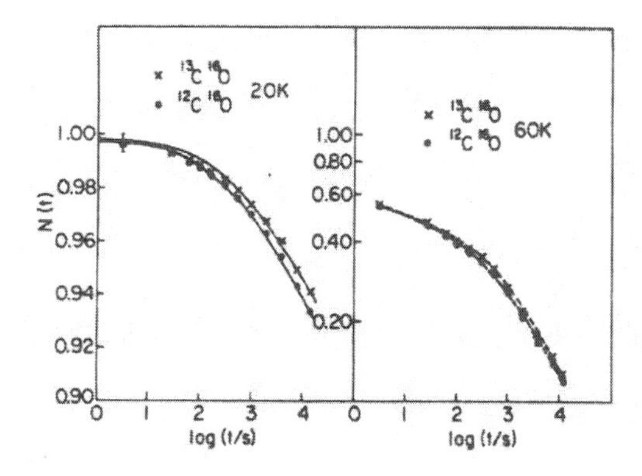

Fig. 13.15. Rebinding of $^{12}C^{16}O$ and $^{13}C^{16}O$ to Mb after photodissociation at 20 and 60 K. $N(t)$ denotes the fraction of Mb molecules that have not rebound CO at time t after the end of the steady-state illumination. The solid curves at 20 K are fits to the data as described in the text. The dashed curves at 60 K are drawn to guide the eye. For 20 K, errors at the earliest times are shown; by 10^3 s, the errors are smaller than the size of the data points. For 60 K the errors are smaller than the data points.

Table 13.1. Values of k_ℓ/k_h at $\approx 10^3$ s and values of $\Delta M/M$, the relative mass change, in the binding of carbon monoxide to myoglobin. The model calculations refer to point particles moving in fixed potentials.

Quantity		$^{12}C^{16}O$ versus $^{13}C^{16}O$	$^{12}C^{16}O$ versus $^{12}C^{18}O$
Experiment:			
k_ℓ/k_h	60 K	1.20 ± 0.05	1.15 ± 0.05
	20 K	1.53 ± 0.05	1.20 ± 0.05
$\Delta M/M$	(20 K)	0.040 ± 0.015	0.019 ± 0.007
Model: $\Delta M/M$, if			
M is the mass of	CO	0.035	0.069
	C	0.080	0
	O	0	0.118
M is the			
reduced mass of	CO–Fe	0.023	0.045
	C–Fe	0.066	0
	O–Fe	0	0.090

(3) The structure of CO affects tunneling. If CO moved as a point particle, the value of $\Delta M/M$ would be about twice as large for the replacement $^{16}O \rightarrow {}^{18}O$ than for $^{12}C \rightarrow {}^{13}C$. The data in Table 13.1 show the opposite behavior. Predictions for some simple models are also given in Table 13.1. It is amusing to note that the value of $\Delta M/M$ obtained in the experiment on $^{12}C^{16}O$ versus $^{13}C^{16}O$ is close to that predicted for a point CO molecule tunneling through a fixed potential. If only one isotope pair would have been studied, excellent agreement with the simplest model would have been noted. Both pairs together exclude any explanation not involving the structure of the CO molecule and the details of the binding process. A quantitative explanation of the tunneling rates may require features such as rotational motion around an axis perpendicular to the C–O vector, excitation of the CO molecule, and participation of other protein constituents near the heme.

Theory. The longer one looks at tunneling in biomolecules, the more complex the theory becomes. We only state the various problems here.

(i) Phonons. Unless the tunneling system can interact with the surrounding heat bath, tunneling may be slow. Two different influences of phonons can be distinguished. As shown in Fig. 13.4, in unequal wells, the wave functions in a given level are highly localized in one well. If the transition occurs with energy conservation, the probability of remaining in the other well will be small. Phonon transition can lead to a level with larger wave function and from there to the bottom of the well. Vibrations can also reduce the distance $d(H)$. Since k_t depends exponentially on $d(H)$, these vibrations can change the tunneling rate [35]. The general behavior expected then is shown in Fig. 13.16. The various contributions can be understood as follows. Quantum theory tells us that the probability of emission of a boson into a state x is proportional to

$$P(x) = \text{const}(1 + n_x) \tag{13.28}$$

where n_x is the number of phonons present in state x. (Phonons are bosons that like each other.) Assume that the energy of state x is E_x. The number of bosons in state x is given by the Planck distribution law,

$$n_x = [\exp(E_x/k_B T) - 1]^{-1}. \tag{13.29}$$

For low temperatures, where $E_x \gg k_B T$, Eq. (13.29) gives

$$n_x \cong k_B T/E_x. \tag{13.30}$$

We can identify a protein with a small solid, E_x with the energy of a phonon or lattice vibration, and the boson with this phonon. Equations (13.28) and (13.30) then explain the linear and quadratic terms in Fig. 13.16. At very low temperatures, only the term 1 in Eq. (13.28) contributes and tunneling becomes temperature independent. This term corresponds to spontaneous emission of a phonon: The energy is removed by single phonon emission. At

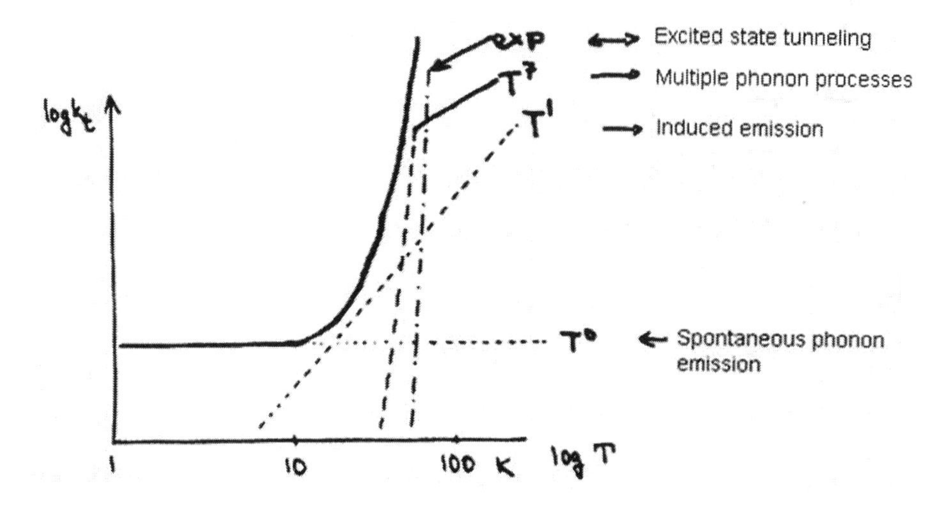

Fig. 13.16. Schematic representation of the temperature dependence of tunneling.

somewhat higher energy, the second term becomes important and tunneling becomes proportional to T. This process is called induced single phonon emission or absorption. At still higher temperatures, multiphonon processes set in and the tunneling rate becomes proportional to a higher power of T, for instance T^7. Such a process corresponds to phonon Raman scattering.

(ii) Friction. In the discussion of the Kramers approach to classical reactions we found that friction plays an essential role. Will friction also be important in tunneling? If we think of friction as a classical effect, the answer to the question is not so clear. We have pointed out, however, that friction is the "fudge factor" that takes the motion of the neglected reaction coordinates into account. Obviously, these coordinates are also neglected even at very low temperatures and friction must play an important role. Friction corresponds to energy dissipation, and thus energy is not conserved. In the ordinary treatment of quantum mechanics, energy is always assumed to be a constant of the motion. It consequently took considerable time before the correct approach was found. The work of Leggett and coworkers [36, 37] Wolynes [38], and Hänggi and collaborators [10, 11] has solved many aspects of this difficult problem, and it is now clear that friction is important and in general reduces the tunneling rate.

(iii) Adiabatic and nonadiabatic tunneling. Tunneling in heme proteins has been treated by Jortner and Ulstrup by using a nonadiabatic approach [39, 40]. We return to the question of adiabaticity later and only note here that it is likely that the tunneling phenomena discussed so far in heme proteins proceed on an adiabatic surface.

(iv) Structure effects. We have already encountered one structure effect when we discussed the isotope effect. A second one is also apparent in Fig. 13.13: Tunneling becomes temperature independent at about 10 K. In paraelastic relaxation in solids, the tunneling rate is proportional to T down to at least 1 K [41]. Why does tunneling become temperature independent at such a high temperature in heme proteins? One possible explanation can be discussed with Fig. 13.17 where we show the heme embedded in the globin, making contact essentially only at the periphery. The longest wavelength, corresponding to the smallest frequency, then is given by the diameter L of the heme. The shortest wavelength is given by the distance between two atoms. The ratio L/d is about 5. Assuming a Debye temperature of about 100 K would thus predict a lower cut-off temperature of about 20 K, in approximate agreement with experiment.

We can summarize the situation by stating that tunneling in heme proteins still offers many possibilities for further research.

13.5 Barriers and Gates

In the reactions described so far we have assumed that the reaction surface, shown for instance in Fig. 13.1 or Fig. 13.2, is fixed and time independent.

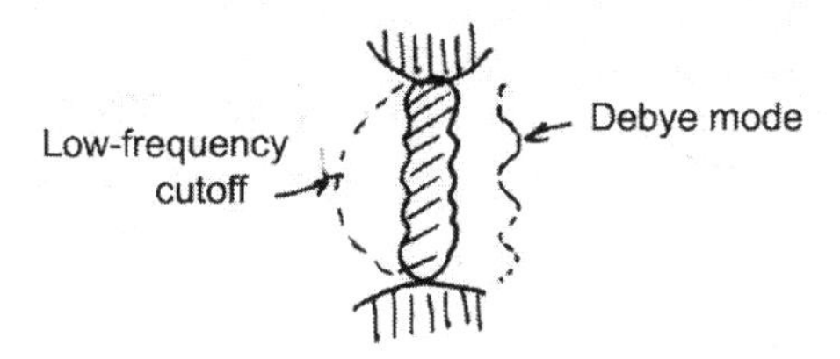

Fig. 13.17. If the heme behaves like a two-dimensional system embedded in a protein, the lowest-frequency mode is, as shown, determined by the size of the heme group. The high-frequency cut-off is given by the Debye mode.

At the same time we have stated many times that proteins are moving and flexible systems. These two statements clearly are contradictory. How can we retain much of the reaction theory described so far while treating the proteins as dynamic systems? One way to take the dynamic aspects into account is by introducing the concept of a gated reaction [15]. The concept of gating is illustrated in Fig. 13.18.

In Fig. 13.18 we consider a specific case, the motion of a CO molecule in the interior of a protein. Case and Karplus have shown that interior groups

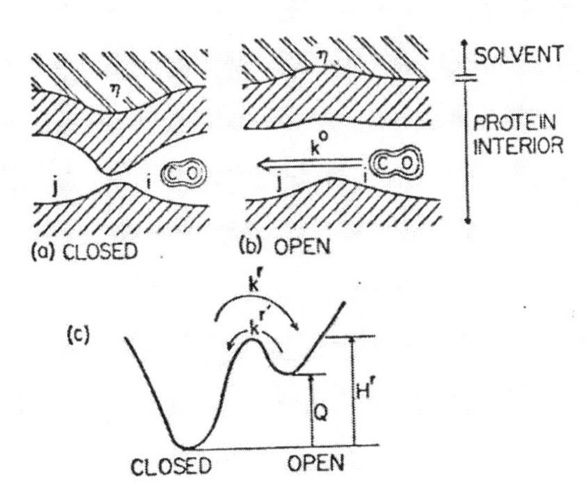

Fig. 13.18. A CO molecule inside Mb moves from state i to j. In **(a)** Mb is in a conformational substate where the channel from i to j is closed, and in **(b)** the gate is open. **(c)** gives the potential for conformational transitions between the open and closed substates [15].

must move in order to let a small ligand escape or enter the heme pocket [42]. We describe this situation by postulating two conformation substates, "closed" and "open." In the open substate, the transition from well i to j occurs with a rate coefficient k_{ij}; in the closed substate the transition cannot occur. With k^r, we denote the relaxation rate coefficient, for the transition from the closed to the open substate; $k^{r'}$ is the coefficient for the reverse step. Parts (a) and (b) of Fig. 13.18 represent the real states (a) and (b) inside the protein. Figure 13.18(c) depicts the open and closed conformational substates. In the simplest case, we assume $k_{ij} = k_{ji}$, $k^r \ll k^{r'}$, and $k_{ji} \gg k^r$. The first assumption implies that the entropies of the states i and j are the same, the second that the probability of finding the system open is small, and the third that the transition $i \leftrightarrow j$ is fast compared to the closing of the passage. With these assumptions, the transitions between i and j are given by $k_{ij} \approx k^r$.

Such a gating model connects the reaction coordinate of Figs. 13.18(a) and (b), with a conformational coordinate, *cc*. Since the motion of the residue or protein can be influenced easily by viscosity, the dependence of a motion inside the protein on external viscosity can be understood [15]. More general treatments of gating can be found in [43]–[45].

13.6 Electronically Controlled Reactions

Most reactions involve motion of the nuclei of the reacting species and changes in their electronic structure. For many reactions the electronic structure adiabatically follows the nuclear motions. The reaction then is controlled by the nuclear motion and is termed *adiabatic* or sometimes *steric* or even *nuclear*. The treatment given so far then is appropriate, and no explicit attention need be paid to the dynamics of the changing electronic structure. When the spins of the reactants change or when long-range electron transfer is involved, however, the changes in the electronic structure may be slower than the nuclear motion. The characteristics of the electronic motion then becomes important and a theory of nonadiabatic transitions from one state to another is needed. In this section, we will first describe our "standard problem," the binding of small ligands to heme protein, then assess the relative importance of nuclear and electronic motions, and finally sketch the adiabatic approach [11].

The Binding of CO and O_2 to the Heme Iron. The inside of the heme pocket in a typical hemoprotein was sketched in Fig. 11.5. The molecular arrangement is shown in Fig. 13.19. At low temperatures, two states are involved in the binding process, as indicated in Fig. 11.5(b). In the initial (deoxy) state B, the ligand is somewhere in the pocket, the heme iron (Fe^{+2}) has spin 2 and lies about 50 pm out of the mean heme plane, and the heme is domed. In the bound state A, the spin is changed to zero by the ligand. The iron has moved closer to the mean heme plane, and the heme is nearly planar. The free CO molecule in state B has closed shells and is in the singlet configuration $^1\Sigma$. Binding hence starts in state B with total spin 2 (quintuplet state). As the

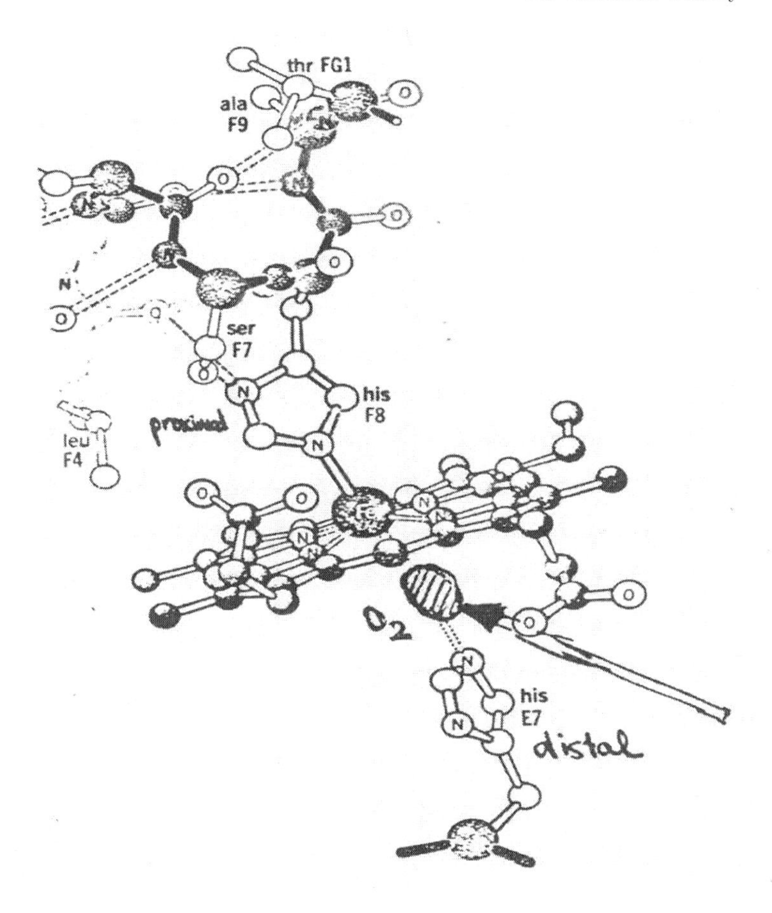

Fig. 13.19. Atomic arrangement at the heme binding site.

CO approaches the iron, it may first encounter a repulsive potential caused by the four nitrogen atoms that surround the iron. A concerted motion then occurs, with the iron atom and the proximal histidine moving closer, the F helix shifting, the heme becoming planar and the spin-0 state moving lower in energy. Finally, upon binding, the spin 0 (singlet) state becomes the ground state. The binding process as described entails both nuclear and electronic motions. The nuclear motions involve changes in the coordinates of most of the atoms shown in Fig. 13.19. The electronic rearrangement consists in the change of the total spin of the system from 2 to 0. We can no longer describe the binding process as an adiabatic motion over a steric barrier, but now must also consider the rate with which the spin change $2 \rightarrow 0$ occurs.

Adiabatic and Nonadiabatic Reaction Surfaces. In chemical reactions, both electrons and nuclei move. Because of their mass difference, we can consider the nuclei at any moment to be fixed and obtain the energy levels for the electronic motion in the fixed field of the nuclei as a function of the distance between the nuclei (the Born-Oppenheimer approximation) [46]. As an example we show in Fig. 13.20 the potential energy curves for the reaction of CO with the heme group in a one-dimensional model in two different approximations. We assume that only two electronic states exist, the bound state s (for singlet, $S = 0$) and the unbound state q (for quintuplet, $S = 2$). If the matrix element V_{sq} connecting the two states is zero, the resulting *diabatic* energy curves (Fig. 13.20(a)) cross each other. If the interaction is not zero, the two curves "repel" each other, as shown in Fig. 13.20(b) [47, 48].

The splitting between the two curves is given by

$$\Delta = 2|V_{sq}| . \tag{13.31}$$

The lower curve starts out as q at large values of r, changes character in the mixing region, and becomes s at small values of r. In the mixing region, both curves are superpositions of s and q. Note that the curves in Fig. 13.20 do not represent potentials to be used in the Schrödinger equation to determine energy levels; they are the energy levels (or energy surfaces) as a function of the distance r.

If the energy splitting Δ is very large compared to k_BT, the upper electronic state is inaccessible and the entire reaction takes place on the lower

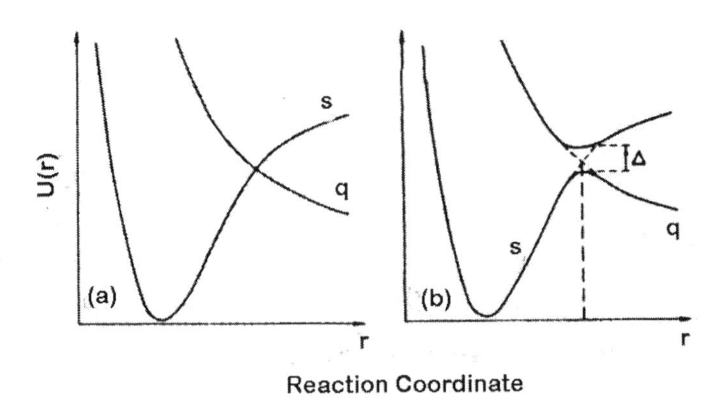

Fig. 13.20. Schematic representation of the potential energy curves for the reaction of CO with the heme iron: (a) diabatic energy levels; (b) adiabatic energy levels. Here s denotes the singlet state, and q the quintuplet state; transitions $s \rightarrow q$ are forbidden in (a).

adiabatic curve. The electronic dynamics can be ignored, except insofar as it determines the adiabatic curve. The reaction is then determined by the nuclear motion and can be described by the Kramers equation. If, however, the splitting is small, the situation becomes more complex, as discussed in the following subsection.

Curve Hopping (The LZS Relation). We now consider the dynamics of the association process in Fig. 13.20 by using a semiclassical model in which the nuclei move classically and the electronic state adjusts to the changing nuclear coordinates. Assume that the system starts out in state B. If $V_{sq} = 0$, the system will remain in state B because the electronic state cannot change even if the nuclear coordinates thermally move to the A configuration. To bind, the system must hop from the q to the s level. Hopping outside of the mixing region can be neglected [49]. The transition $q \to s$ must occur during the passage of the nuclei through the crossing region. If $\Delta \gg k_BT$, the upper state will be inaccessible and the nuclei will move on the adiabatic curve as pointed out earlier. If $\Delta \leq k_BT$, thermodynamic considerations alone do not determine whether the electronic state can change. Depending on the relative time scale of the electronic and nuclear motions, the system can either remain in state q, move along the dashed diabatic curve in Fig. 13.20(b), and reach the upper surface or it can move along the solid adiabatic curve and undergo the transition $q \to s$.

To obtain a criterion characterizing adiabaticity we use the uncertainty relation: If the energy uncertainty of the system in the mixing region is small compared to the splitting Δ, the system will remain on the adiabatic lower surface. The energy uncertainty is given by $\hbar/\tau_{LZ}$, where τ_{LZ} is the time spent in the mixing region, called the Landau-Zener region. If the system moves through the mixing region with a velocity v and if the length of the Landau-Zener region is ℓ_{LZ}, τ_{LZ} is given by

$$\tau_{LZ} = \ell_{LZ}/v. \tag{13.32}$$

Figure 13.20(b) shows that ℓ_{LZ} is approximately given by

$$\ell_{LZ} \cong \Delta/|F_2 - F_1| \tag{13.33}$$

where the forces F_1 and F_2 are the slopes of the diabatic curves at the avoided crossing r_0. The *adiabaticity parameter* γ_{LZ} is defined as the ratio of the splitting to the energy uncertainty,

$$\gamma_{LZ} = \frac{\Delta}{\hbar/\tau_{LZ}} = \frac{\Delta\ell_{LZ}}{\hbar v} = \frac{\Delta^2}{\hbar v|F_2 - F_1|}. \tag{13.34}$$

If $\gamma_{LZ} \gg 1$, the transition is adiabatic. If $\gamma_{LZ} \leq 1$, the system can remain on the upper surface. Not every crossing through the mixing region then results in a reaction $B \rightarrow A$. The probability P for staying on the adiabatic surface in a single crossing has been calculated by Landau [50], Zener [51], Stueckelberg [52] and Ulstrup [53] as

$$P = 1 - \exp\{-\pi\gamma_{LZ}/2\}. \tag{13.35}$$

This expression verifies the hand-waving arguments given earlier; if $\gamma_{LZ} \gg 1$, then P=1 and the transition is adiabatic. If $\gamma_{LZ} \ll 1$, P is then given by

$$P = \frac{1}{2}\pi\gamma_{LZ} = \frac{\pi}{2}\frac{\Delta^2}{\hbar v|F_2 - F_1|}, \tag{13.36}$$

and it is proportional to Δ^2. The dependence on Δ^2 is the hallmark of a fully nonadiabataic reaction. The dependence on Δ^2 is also obtained by deriving the rate coefficient with nonadiabatic perturbation theory (the golden rule) in which Δ is treated as a smaller perturbation that causes the electronic transition.

Friction and the LZS Theory. The treatment of nonadiabatic transition sketched in the previous subsection assumes ballistic motion through the Landau-Zener region. Friction, however, is also present in this process and it is intuitively obvious that it must play a role. The longer the system stays in the mixing region, the more we should expect hopping from one electronic surface to the other to occur. The effect of friction on nonadiabatic transitions has been treated in a number of publications [54]–[56]. The central results can be obtained again by a hand-waving approach, using the three characteristic lengths of the problem; the mean free path λ, the length ℓ_{TS} of the transition region, and the length ℓ_{LZ} of the mixing region [11]. We state only one result here, for $\ell_{TS} \gg \lambda \gg \ell_{LZ}$. On each passage through the transition region, multiple crossings through the Landau-Zener region occur. The number of crossings is approximately given by $N_c = (\ell_{TS}/\lambda)$ and the probability P becomes

$$P = \frac{1}{2}[1 - \exp\{-\pi N_c\gamma_{LZ}\}]\ . \tag{13.37}$$

Even if the Landau-Zener factor, Eq. (13.34), is small, the reaction may appear adiabatic since the parameter $N_c\ \gamma_{LZ}$ can be large. The effective transmission coefficient κ is shown in Fig. 13.21 as a function of the friction coefficient (in units of $m\omega^*$) for three different values of the Landau-Zener coefficient γ_{LZ}. The curves show that the reaction may appear adiabatic for large or small friction coefficients even if it is in reality largely controlled by curve hopping, i.e., electronic factors.

How to Recognize Nonadiabaticity. As shown in the previous sections, entropy, friction, and electronic structure can all affect the preexponential factor $\mathcal{A}$ in an Arrhenius relation. Friction and nonadiabaticity reduce A, whereas

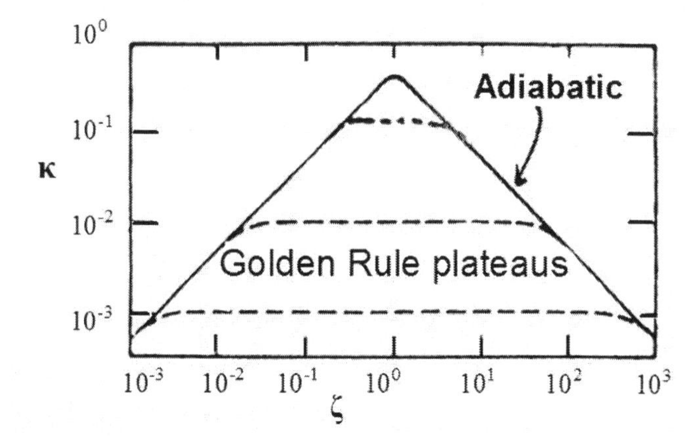

Fig. 13.21. The transmission for κ as a function of the friction coefficient ζ (in units of $m\omega^*$) for three different values of the Landau-Zener parameter. The solid curves give the adiabatic Kramers curve. The dashed lines correspond to nonadiabatic transitions for three different values of the Landau-Zener parameter.

entropy can either decrease or increase A. Some of the effects can be separated if the preexponentials $\mathcal{A}_{AB}$ and $\mathcal{A}_{BA}$ for both reactions $A \rightarrow B$ and $B \rightarrow A$ (Fig. 13.6) can be measured. Friction and nonadiabaticity reduce both coefficients by the same factor; the entropy contribution yields

$$\frac{\mathcal{A}_{BA}}{\mathcal{A}_{AB}} = \frac{\exp(S^*_{BA}/R)}{\exp(S^*_{AB}/R)} = \exp\left[(S_A - S_B)/R\right]. \tag{13.38}$$

The ratio Eq. (13.38) thus provides some information on the role of entropy. A second clue can come from the effect of viscosity on the reaction rate. If the rate coefficients depend strongly on viscosity at constant temperature, the data must be evaluated with the Kramers relation and friction may be responsible for the reduction of a preexponential factor below the TST value. If the rate depends on viscosity, and if the preexponential factor is reduced below the TST value, the adiabaticity factor γ_{LZ} is increased by the amount that the rate coefficient k is decreased. A transition that without friction is nonadiabatic can be made adiabatic by the friction.

13.7 Collective Effects

All reactions discussed so far have been described in terms of fixed and temperature-independent potential energy surfaces. It is easy to see, however, that biomolecular reactions may not always satisfy this assumption. Consider as an example a rotation of an internal group. At high temperatures, the group may be nearly free to rotate; at low temperatures, the protein has contracted and rotation may be hindered. If the energy surface depends on temperature, we can no longer expect that the reaction rate coefficient obeys a simple Arrhenius behavior. Indeed, departures from a standard Arrhenius behavior are found frequently in glasses. This problem will be treated in Chapter 14.

References

1. S. Glasstone, K. J. Laidler, and H. Eyring. *The Theory of Rate Processes.* McGraw-Hill, 1941 (the standard work).
2. P. Hänggi, P. Talkner, and M. Borkovec. Reaction-rate theory: Fifty years after Kramers. *Rev. Mod. Phys.*, 62:251–341, 1990.
3. S. Arrhenius. Über die Reaktionsgeschwindigkeit der Inversion von Rohrzucker durch Säuren. *Z. Physik. Chem.*, 4:226–48, 1889.
4. B. G. Wicke and D. O. Harris. Comparison of three numerical techniques for calculating eigenvalues of an unsymmetrical double minimum oscillator. *J. Chem. Phys.*, 64:5236–42, 1976.
5. See, for instance, R. S. Berry, S. A. Rice, and J. Ross, *Physical Chemistry*, 2nd edition, Oxford Univ. Press, New York, 2000.
6. M. Dresden. *H. A. Kramers: Between Tradition and Revolution.* Springer, New York, 1987.
7. H. A. Kramers. Brownian motion in a field force and the diffusion of chemical reactions. *Physica*, 7:284–304, 1940.
8. H. C. Brinkman. Brownian motion in a field of force and the diffusion theory of chemical reactions. 2. *Physica*, 22:149–55, 1956.
9. R. Landauer and J. A. Swanson. Frequency factors in the thermally activated process. *Phys. Rev.*, 121:1668–74, 1961.
10. P. Hänggi. Escape from a metastable state. *J. Stat. Phys.*, 42:105–48, 1986.
11. H. Frauenfelder and P. G. Wolynes. Rate theories and puzzles of hemeprotein kinetics. *Science*, 229(4711):337–45, 1985.
12. B. Somogyi and S. Damjanovich. Relationship between the lifetime of an enzyme-substrate complex and the properties of the molecular environment. *J. Theor. Bio.*, 48:393–401, 1975.
13. B. Gavish. The role of geometry and elastic strains in dynamic states of proteins. *Biophys. Struct. Mech.*, 4:37–52, 1978.
14. B. Gavish and M. M. Werber. Viscosity-dependent structural fluctuations in enzyme catalysis. *Biochemistry*, 18:1269–75, 1979.
15. D. Beece, L. Eisenstein, H. Frauenfelder, D. Good, M. C. Marden, L. Reinisch, A. H. Reynolds, L. B. Sorensen, and K. T. Yue. Solvent viscosity and protein dynamics. *Biochemistry*, 19:5147–57, 1980.
16. D. Beece, S. F. Bowne, J. Czégé, L. Eisenstein, H. Frauenfelder, D. Good, M. C. Marden, J. Marque, P. Ormos, L. Reinisch, and K. T. Yue. The effect of viscosity on the photocycle of bacteriorhodopsin. *Photochem. Photobiol.*, 33:517–22, 1981.

17. D. G. Truhlar, W. L. Hase, and J. T. Hynes. The current status of transition state theory. *J. Phys. Chem.*, 87:2664–82, 1983.
18. G. R. Fleming, S. H. Courtney, and M. W. Balk. Activated barrier crossing: Comparison of experiment and theory. *J. Stat. Phys.*, 42:83–104, 1986.
19. R. F. Grote and J. T. Hynes. The stable states picture of chemical reactions. II. Rate constants for condensed and gas phase reaction models. *J. Chem. Phys.*, 73:2715–32, 1980.
20. R. F. Grote and J. T. Hynes. Saddle point model for atom transfer reactions in solution. *J. Chem. Phys.*, 75:2791–98, 1981.
21. R. F. Grote and J. T. Hynes. Energy diffusion-controlled reactions in solution. *J. Chem. Phys.*, 77:3736–43, 1982.
22. P. Hänggi and F. Mojtabai. Thermally activated escape rate in presence of long-time memory. *Phys. Rev.*, A26:1168–70, 1982.
23. P. Hänggi. Physics of ligand migration in biomolecules. *J. Stat. Phys.*, 73:401–12, 1983.
24. B. Chance et al., editors. *Tunneling in Biological Systems.* Johnson Research Foundation Colloquia. Academic Press, New York, 1979.
25. R. P. Bell. *The Tunnel Effect in Chemistry.* Chapman and Hall, London, 1980.
26. D. DeVault. *Quantum-Mechanical Tunnelling in Biological Systems*, 2nd edition. Cambridge Univ. Press, Cambridge, 1984.
27. V. I. Goldanskii, L. I. Trakhtenberg, , and V. N. Fleurov. *Tunneling Phenomena in Chemical Physics.* Gordon and Breach, New York, 1989.
28. V. I. Goldanskii. The role of the tunnel effect in the kinetics of chemical reactions at low temperatures. *Dokl. Akad. Nauk SSSR*, 124:1261–4, 1959. See also P. Hänggi, et al., *Phys. Rev. Lett.*, 55:761-4, 1985.
29. See, for instance, L. D. Landau and E. M. Lifshitz, *Quantum Mechanics*, Pergamon Press, London, 1958.
30. N. Alberding, R. H. Austin, K. W. Beeson, S. S. Chan, L. Eisenstein, H. Frauenfelder, and T. M. Nordlund. Tunneling in ligand binding to heme proteins. *Science*, 192(4243):1002–4, 1976.
31. N. Alberding, S. S. Chan, L. Eisenstein, H. Frauenfelder, D. Good, I. C. Gunsalus, T. M. Nordlund, M. F. Perutz, A. H. Reynolds, and L. B. Sorensen. Binding of carbon monoxide to isolated hemoglobin chains. *Biochemistry*, 17:43–51, 1978.
32. H. Frauenfelder. In B. Chance et al., editors, *Tunneling in Biological Systems.* Academic Press, New York, 1979. pp. 627-49.
33. J. O. Alben, D. Beece, S. F. Bowne, L. Eisenstein, H. Frauenfelder, D. Good, M. C. Marden, P. P. Moh, L. Reinisch, A. H. Reynolds, and K. T. Yue. Isotope effect in molecular tunneling. *Phys. Rev. Lett.*, 44:1157–60, 1980.
34. J. O. Alben, D. Beece, S. F. Bowne, W. Doster, L. Eisenstein, H. Frauenfelder, D. Good, J. D. McDonald, M. C. Marden, P. P. Moh, L. Reinisch, A. H. Reynolds, and K. T. Yue. Infrared spectroscopy of photodissociated carboxymyoglobin at low temperatures. *Proc. Natl. Acad. Sci. USA*, 79:3744–8, 1982.
35. J. A. Sussman. A comprehensive quantum theory of diffusion. *Ann. Phys. Paris*, 6:135–56, 1971.
36. A. O. Caldeira and A. J. Leggett. Influence of dissipation on quantum tunneling in macroscopic systems. *Phys. Rev. Lett.*, 46:211–14, 1981.

37. A. J. Leggett, S. Chakravarty, A. T. Dorsey, M. P. A. Fisher, A. Garg, and W. Zerger. Dynamics of the dissipative two-state system. *Rev. Mod. Phys.*, 59:1–85, 1987.
38. P. G. Wolynes. Quantum theory of activated events in condensed phases. *Phys. Rev. Lett.*, 47:968–71, 1981.
39. J. Jortner and J. Ulstrup. Dynamics of nonadiabatic atom transfer in biological systems. Carbon monoxide binding to hemoglobin. *J. Amer. Chem. Soc.*, 101–4:3744, 1979.
40. J. Ulstrup. *Charge Transfer Processes in Condensed Media* (Lecture Notes in Chemistry, 10). Springer, Berlin, 1979.
41. G. Pfister and W. Känzig. Isotopeneffekt in der paraelastischen Relaxation. *Zeitschrift für Physik B Condensed Matter*, 10:231–64, 1969.
42. D. A. Case and M. Karplus. Dynamics of ligand binding to heme proteins. *J. Mol. Bio.*, 132:343–68, 1979.
43. J. A. McCammon and S. H. Northrup. Gated binding of ligands to proteins. *Nature*, 293:316–17, 1981.
44. A. Szabo, D. Shoup, S. H. Northrup, and J. A. McCammon. Stochastically gated diffusion-influenced reactions. *J. Chem. Phys.*, 77:4484–93, 1982.
45. Y. A. Berlin, A. L. Burin, L. D. A. Siebbeles, and M. A. Ratner. Conformationally gated rate processes in biological macromolecules. *J. Phys. Chem. A*, 105:5666–78, 2001.
46. G. Baym, editor. *Lectures on Quantum Mechanics.* W. A. Benjamin, New York, 1969.
47. L. D. Landau and E. M. Lifshitz. *Quantum Mechanics.* Pergamon Press, London, 1958.
48. W. Kauzmann. *Quantum Chemistry.* Academic Press, New York, 1957.
49. E. J. Heller and R. C. Brown. Vibrational relaxation of highly excited diatomics. V. the V-V channel in HF(v)+HF(0) collision. *J. Chem. Phys.*, 79:3336–66, 1983.
50. L. Landau. *Phys. Z. Sow.*, 1:89, 1932. *Z. Phys. Sov.* 2:46 (1932).
51. C. Zener. Non-adiabatic crossing of energy levels. *Proc. Roy. Soc. London*, A137:696–702, 1932.
52. E. C. G. Stueckelberg. *Helv. Phys. Acta*, 5:369–422, 1932.
53. J. Ulstrup. *Charge Transfer Processes in Condensed Media.* Springer, Berlin, 1979.
54. L. D. Zusman. Outer-sphere electron transfer in polar solvents. *Chem. Phys.*, 49:295–304, 1980.
55. R. E. Cline, Jr. and P. G. Wolynes. Stochastic dynamic models of curve crossing phenomena in condensed phases. *J. Chem. Phys.*, 86:3836–44, 1987.
56. I. V. Aleksandrov and V. I. Goldanskii. *Sov. Sci. Rev. B. Chem.*, 11:1–67, 1988.

14

Supercooled Liquids and Glasses

Complex systems range from supercooled liquids to glasses, to proteins, to the brain, to societies. Do these systems share properties? It is likely that they do. A hierarchically organized energy landscape and a range of motions, connecting substates, are candidates. In this chapter we sketch some aspects of the physics of supercooled liquids and glasses. Starting with these materials may appear to be strange, but many properties of glasses and proteins are similar, and it is easier to recognize crucial properties of the dynamics in the less complex system. The information covering the physics of glasses is staggering. A Web search engeine has over 3×10^8 entries for "glass" and 7×10^6 for glass transition! We restrict the treatment to a few salient facts that are useful for understanding related phenomena in proteins. More information can be found in books and selected articles [1]–[6].

If a liquid is cooled, it can solidify either discontinuously to a crystal or continuously to a glass (amorphous solid). The path taken depends on the material and the speed of cooling. If cooled rapidly enough, essentially all liquids form amorphous solids, because the atoms do not have time to move to the proper positions corresponding to the crystal structure. Simplified, a glass is a frozen liquid. Some liquids resist forming a crystal even well below their freezing (melting) temperature; they are supercooled. Despite the technical importance of glasses and the challenging physics problems that they pose, standard condensed matter texts devote only a few pages to them. The biological physicist has to resort to the few available books, to reviews, and to original papers to get the insight into the properties of glasses that help in understanding protein dynamics. The following sections describe characteristic features of ***supercooled liquids and glasses.***

14.1 Glass Structure

The fundamental difference between a glass and a crystal is in the atomic-scale structure. Crystals are periodic and possess long-range order. Glasses

H. Frauenfelder, *The Physics of Proteins*, Biological and Medical Physics, Biomedical Engineering, DOI 10.1007/978-1-4419-1044-8_14,

are aperiodic and have no long-range order. In a crystal, every atom (apart from dislocations) knows where its place is and has no competition for its equilibrium position. Glasses are frustrated [7, 8]: Atoms compete for the same position, and a given atom may have two or more quasi-equilibrium positions. The structure in the glassy state has been studied by many techniques [2], but we are not interested in this aspect here apart from the fact that the arrangement is aperiodic. Crystals are in equilibrium at all temperatures, while glasses are metastable. The metastability of glasses leads to aging [9, 10].

14.2 Energy Landscape

Supercooled liquids and glasses can be described by energy landscapes [11, 12]. The energy landscape of a glass, shown schematically in Fig. 14.1, displays similarities to the protein energy landscape in Fig. 12.2. (i) The landscape is organized in a number of tiers. (ii) The top tier contains a small number of supercooled liquids that form glasses with different overall structures, called *polymorphs* [12, 13]. These polymorphs are similar to the taxonomic substates in proteins. (iii) Each of the polymorphs can exist in a vast number of structures that are locally different, similar to the statistical substates in proteins. (iv) The organization continues to substates with smaller barriers.

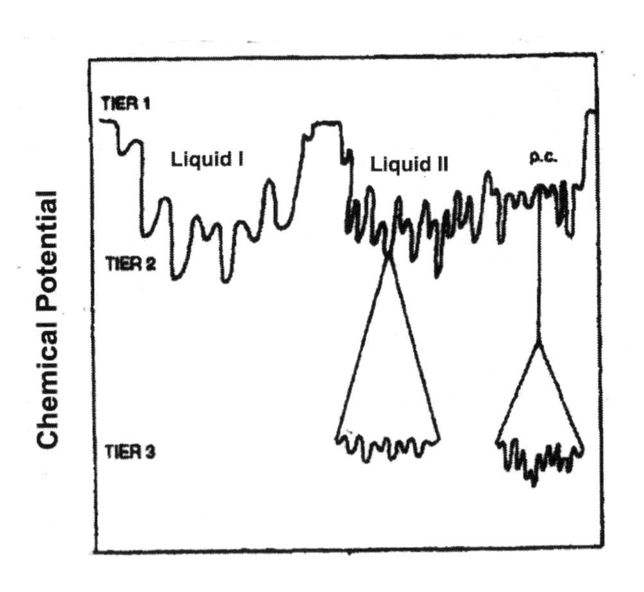

Fig. 14.1. A two-dimensional cross section through the potential energy hypersurface of a glass (from Fan, Cooper, and Angell [12]).

There are, however, substantial differences between the glass and the protein landscapes: (i) Glasses have some deep and very narrow wells, not shown in Fig. 14.1, representing crystals formed under the proper circumstances. Proteins cannot form similar crystals. What is called a protein crystal is a crystal in which the individual components are proteins. Each individual protein can assume a large number of different structures. (ii) Supercooled liquids are not structured to perform specific functions, proteins are.

14.3 Specific Heat (Heat Capacity)

The specific heat of a system is given by

$$C(T) = T\ \partial S(T)/\partial T \tag{14.1}$$

where S is the entropy. Albert Einstein, who knew thermodynamics very well, developed the first theory of the specific heat of crystals to prove that not only radiation, but also matter, is quantified [14, 15]. The modern theory of the specific heat of solids can be found in most condensed matter texts. Here we avoid all details and only describe what will be relevant later for proteins. With the Boltzmann relation

$$S(T) = k_{\mathrm{B}}\ \ln W, \tag{14.2}$$

we rewrite Eq. (14.1) as

$$\partial(\ln W(T))/\partial T = C(T)/k_{\mathrm{B}}T. \tag{14.3}$$

Here $W(T)$ is the number of accessible states at temperature T. Not surprisingly, experiments show that $W(T)$ is larger in a supercooled liquid or glass than in the corresponding crystal. The excess is due to the fact that the atoms in a crystal vibrate around fixed positions, while in a supercooled liquid they cannot only vibrate, but also make conformational motions.

14.4 Viscosity, Glass Transition, and Fragility

The viscosity, $\eta(T)$, of a supercooled liquid increases very rapidly with decreasing temperature, and the glass transition temperature T_g is often defined as the temperature where $\eta(T_g) = 10^{13}$P. (Viscosity is measured in poise (P) or pascal-sec (Pa s), where 1 P = 0.1 Pa s). The story goes that when a glass blower had finished a piece of art, he let it cool until it reached T_g, when it was solid enough to be put aside. In any case, to fully describe the temperature dependence of a supercooled liquid, more is needed than just T_g. Figure 14.2

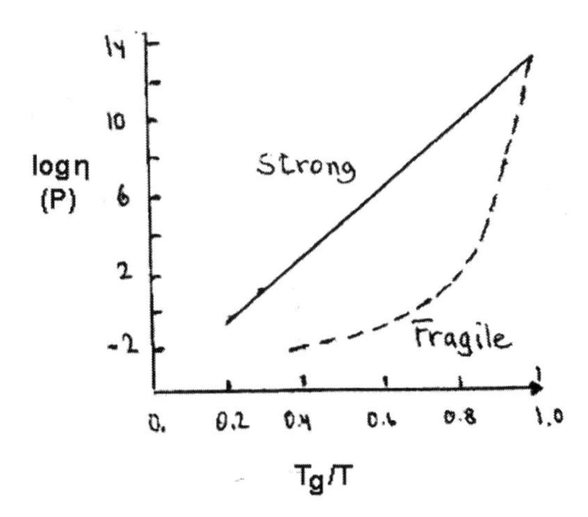

Fig. 14.2. Strong and fragile glass formers (Angell plot) [16].

schematically shows $\eta(T)$ as a function of T_g/T for two extreme types of glass formers [16]. The temperature dependence of a "strong" glass former is approximately given by an Arrhenius law (Eq. (13.12)), while that of a "fragile" glass former [17] can be approximated by the Vogel-Fulcher-Tammann (VFT) expression

$$\eta(T) = \eta_0 \exp\left[\frac{DT_0}{T - T_0}\right]. \tag{14.4}$$

Here D is called the fragility, and T_0 is the temperature where the viscosity would diverge if the relation Eq.(14.4) were still valid at T_0.

14.5 Fluctuation and Relaxation Processes

14.5.1 Fluctuation

Supercooled liquids and glasses undergo thermal fluctuations at all temperatures. If pushed out of thermal equilibrium, for instance, by a temperature or pressure jump, they relax to the new equilibrium. Fluctuations and relaxations are connected by the celebrated fluctuation-dissipation theorem [18]. The literature covering these processes in glasses and supercooled liquids is immense, but a number of reviews and papers summarize the essential results [5, 12, 19]–[23]. Here we sketch the features that are relevant for protein dynamics.

Two main approaches are used to study relaxations, often called time-domain and frequency-domain techniques. In the time-domain technique, a rapid perturbation forces the system out of equilibrium to a new state. The relaxation to the new state is observed, for instance, optically, by following the position of a spectral line. In simple systems, the relaxation usually has the form

$$\phi(t) = \phi(0)\ \mathrm{e}^{-kt} \tag{14.5}$$

where $\phi(t)$ describes a quantity characterizing the measurement, and k is the relaxation rate coefficient. In most complex systems, however, Eq. (14.5) is not adequate, but a stretched exponential,

$$\phi(t) = \phi(0)\ \exp\{-(kt)^{\beta}\} \tag{14.6}$$

with $\beta < 1$, often approximates the observed $\phi(t)$. A remark about relaxation and reaction experiments is on order here. The standard approach of the past, still in vogue in many laboratories, is to fit $\phi(t)$ with two exponentials if one does not fit. This technique is related to the fact that kinetic data in these laboratories are traditionally plotted as $\phi(t)$ versus t. In the same laboratories, data would never be plotted versus H, where H is the hydrogen ion concentration. Instead pH is used, where p denotes -log. Similarly kinetic data should **always** be plotted versus $-pt$, or log t. Of course, if data are only obtained over a small time range, the form of plotting is irrelevant, but such data are useless anyway. A figure of log $\phi(t)$ versus log t helps to distinguish between two or more exponential processes and a single nonexponential process.

The frequency-domain approach, promoted by Peter Debye [24], permits the study of relaxation processes over more than 18 orders of magnitude in frequency, from about 10^{-6} to more than $10^{12}\mathrm{s}^{-1}$ [25, 26]. In the broadband dielectric spectrometry, the sample is placed in a capacitor and exposed to a monochromatic electric field $\mathbf{E}(\omega) = \mathbf{E_0}\exp(-i\ \omega t)$. The resultant polarization $\mathbf{P}(\omega)$ is measured as a function of the radial frequency ω that is connected to the "technical" frequency ν by $\omega = 2\pi\nu$. The radial frequency ω used in the frequency domain is identical to the rate coefficient k in the time domain. The sample properties are expressed in terms of the complex electric permittivity

$$\varepsilon(\omega) = \varepsilon'(\omega) - i\ \varepsilon''(\omega) = 1 + P(\omega)/\varepsilon_0\ E(\omega). \tag{14.7}$$

Here ε_0 is the vacuum dielectric permittivity. The real part $\varepsilon'(\omega)$ represents the component of $\mathbf{P}(\omega)$ that is in phase with $\mathbf{E}(\omega)$, while $\varepsilon''(\omega)$ has a $\pi/2$ phase shift. The frequency dependence of the complex permittivity is caused by the fact that the polarization of a material does not respond instantaneously to the applied electric field. In Fig. 14.3, a dielectric permittivity spectrum shows a number of dispersion phenomena. The fluctuation-dissipation theorem then implies that the same phenomena should also occur in the equilibrium fluctuations. This fact will be important in the discussion of protein dynamics.

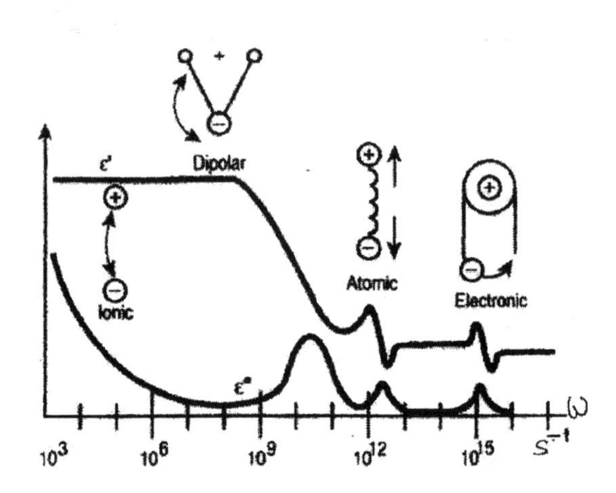

Fig. 14.3. An outline of a typical dielectric permittivity spectrum, taken from Wikipedia. The real and imaginary parts of ε are shown, and some processes are sketched. The declining slope at lower frequencies is a conductivity contribution. The appearance of resonances is important for protein dynamics.

14.5.2 Primary (α) and Secondary (β) Relaxations

A very large number of experiments in many different supercooled liquids and glasses demonstrate that two main classes of relaxation processes occur, distinguished by their time, temperature, and viscosity dependence. Figure 14.4 is a sketch of a dielectric spectrum at three different temperatures [27]. At the lowest temperature, only one peak appears. At a somewhat higher temperature, two peaks are clearly recognizable. Finally, at the highest temperature only one peak is observable. The observations are interpreted in terms of two different relaxation processes, α and β. The slower one is called the primary or α relaxation, or sometimes Debye relaxation. The faster one is the secondary or β relation [28]. At the higher temperature, the two are difficult to separate cleanly. The two relaxations can be described by the temperature dependence of their rate coefficients $k_\alpha(T)$ and $k_\beta(T)$, obtained from the peak positions of the respective peaks in $\varepsilon''(\omega)$. A typical result, given in Arrhenius plot in Fig. 14.5, shows one α and two β relaxations [29]. The α relaxation is curved similarly to the viscosity in Fig. 14.2 and thus cannot be described by a standard Arrhenius law. A VFT expression, Eq. (14.4), approximates the data over a broad range of temperatures. The curves for the β relaxations appear to be straight and thus suggest Arrhenius processes. Three observations imply, however, that the β relaxation is not simply a transition over a unique energy barrier. The relaxation spectrum is inhomogeneously broadened, the time dependence is given by a stretched exponential, and the Arrhenius factor is often larger than the expected 10^{12}s^{-1}.

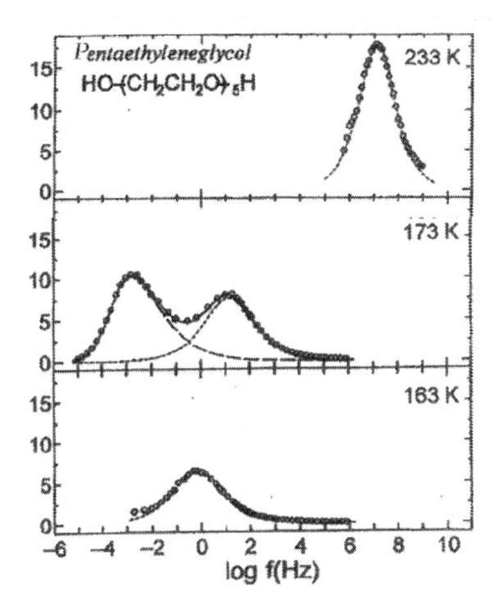

Fig. 14.4. The primary (α) and secondary (β) relaxation in a typical glass-forming liquid at two temperatures as seen by dielectric spectrometry.

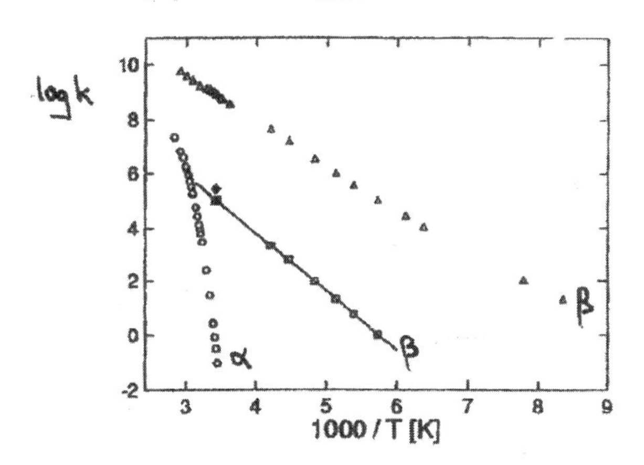

Fig. 14.5. Arrhenius plot of the relaxations in a triepoxy glass former (TPMTGE). From Ngai and Paluch [29]. The α and two β relaxation processes are evident. The α relaxation can be approximated by a VTF relation, Eq.(14.4). The two β relaxation processes can be approximated by Arrhenius relations (Eq. (13.1)), with respective activation enthalpies of 20 kJ/mol and 40 kJ/mol and preexponential factors 10^{10}s^{-1} and 10^{12}s^{-1}.

A transition over an inhomogeneously broadened energy barrier can lead to the so-called Ferry relation,

$$k(T) = A \exp\{-(T/T_F)^2\} . \tag{14.8}$$

T_F is different from T_0 in the VFT relation. The Ferry relation has been derived for a random walk of an excitation within a Gaussian density of states [30]–[32]. It is not clear if the Ferry relaxation fits β relaxations better than the standard Arrhenius law.

How can the two (or more) relaxation processes be interpreted? The description in terms of the hierarchically organized landscape described in Fig. 14.1 is obvious. The α relaxation corresponds to motion in tier 1, the β relaxations to motions in the lower tiers. But what is the structural origin? The α relaxation involves large-scale motions in the supercooled liquid, to be described later in the section on Theory. This fact is expressed by the Maxwell relation that connects $k_\alpha(T)$ and the viscosity $\eta(T)$,

$$k_\alpha(T) = G \ / \ \eta(T). \tag{14.9}$$

Here G is a shear modulus that depends only weakly on temperature and on the material. At 300 K in glycerol-water mixtures G is approximately $\approx 4 \times 10^{11}$ cP s^{-1}. The structural origin of the β relaxations is less clear. Some involve only part of the molecules, such as side chains. Johari and Goldstein have pointed out, however, that β relaxations exist even in glasses composed of rigid molecules [28]. These processes must originate from motions involving the entire molecule. The name Johari-Goldstein (JG) relaxations has been suggested for these β processes.

14.6 Low-Temperature Phenomena in Glasses

The phenomena of glasses at low temperatures are particularly fascinating; they differ fundamentally from those of crystals. This fact had been known since 1911 [33], but it took sixty years before the importance of low-temperature phenomena was appreciated. Zeller and Pohl [34] carefully measured the thermal conductivity and the specific heat of glasses down to 50 μK and found that they behave differently from crystals The difference can be explained by the existence of excitations that are not present in crystals. Zeller and Pohl's work triggered a wide range of experimental and theoretical studies [35, 36]. Examples where glasses and crystals differ are the specific heat and the thermal conductivity. While the specific heat of crystals is proportional to T^3 as $T \to 0$, it is proportional to T in glasses and it is larger than that of crystals. The thermal conductivity has a plateau around 10 K and is proportional to T^2 at lower temperature. A broad peak in the low-frequency region of, e.g., neutron and Raman scattering, called the Boson peak, appears also in dielectric relaxation experiments and the specific heat, but does not occur in crystals [37, 38].

Naively, one would expect that conformational motions stop at very low temperatures, say below 1 K, but experiments demonstrate otherwise. Motions in glasses occur even well below 1 K, as can be seen, for instance, by using hole-burning spectroscopy, already discussed in Section 11.4. The concept of such experiments can be explained in terms of the energy landscape (Fig. 14.1). Figure 14.6 is a sketch of a two-dimensional cross section through one of the substates in a low tier, say tier 3, of the high-dimensional energy landscape. This substate contains substates of the next lower tier. Each of the substates, represented by the small circles, corresponds to a different structure of the glass with a different wavelength of the selected inhomogeneously broadened spectral line. A laser can burn a "hole" at a particular wavelength, corresponding, for instance, to the black circle in Fig. 14.6. If no conformational motions take place, the selected frequency does not change and the black circle remains in its place. If, however, the conformation changes, the frequency changes and the glass moves to another position in the energy landscape as indicated in Fig. 14.6. Every change in the structure leads to another jump in the energy landscape. The process thus corresponds to a Brownian walk in the energy landscape, called spectral diffusion [39]–[41]. The spectral diffusion leads to a broadening of the spectral hole, which can be observed as a function of temperature and time.

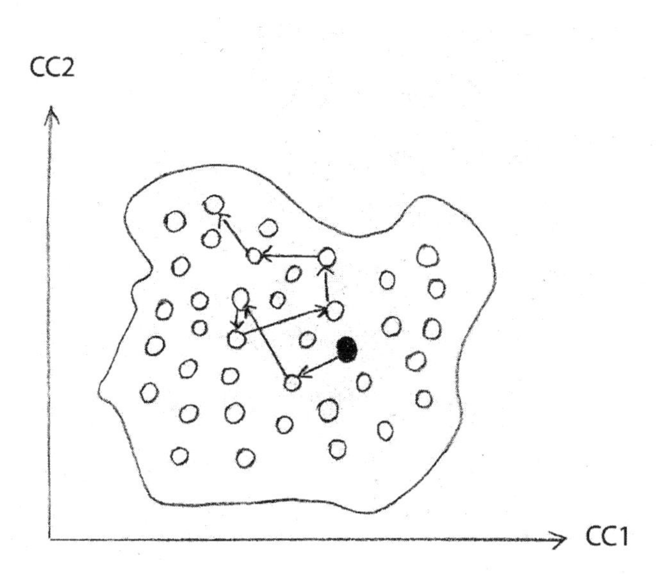

Fig. 14.6. Spectral diffusion. A spectral hole burned initially at the position characterized by the black hole moves in conformational space if the conformation changes.

14.7 Theory

Many attempts have been made to formulate a theory that explains the distinctive features of supercooled liquids and glasses [42]. None of the proposed theories have yet been accepted universally. Here we sketch the concept underlying the random first-order transition theory (RFOT theory) [6, 43]. The theory postulates a mosaic structure in the supercooled liquid. The liquid molecules form cooperatively rearranging regions or cluster ("icebergs"). The size of these regions increases with decreasing temperature. While they are stringlike (fractal) at higher temperatures, they become compact as the glass transition is approached [44]. Different domains undergo first-order transitions at different temperatures, thus leading to a continuous transition to the glassy state. The RFOT theory formalizes this concept and yields a quantitative explanation of the properties of supercooled liquids and glasses.

References

1. E. Donth. *The Glass Transition.* Springer, New York, 2001.
2. R. Zalle. *The Physics of Amorphous Solids.* Wiley, New York, 1983.
3. S. Brawer. *Relaxation in Viscous Liquids and Glasses.* American Ceramic Society, Columbus, Ohio, 1985.
4. C. A. Angell. Formation of glasses from liquids and biopolymers. *Science*, 267:1924–35, 1995.
5. M. D. Ediger, C. A. Angell, and S. R. Nagel. Supercooled liquids and glasses. *J. Phys. Chem.*, 100:13200–12, 1996.
6. V. Lubchenko and P. G. Wolynes. Theory of structural glasses and supercooled liquids. *Ann. Rev. Phys. Chem.*, 58:235–66, 2006.
7. G. Toulouse. Theory of the frustration effect in spin glasses. *Comm. Physics*, 2:115–9, 1977.
8. K. H. Fischer and J. A. Hertz. *Spin Glasses.* Cambridge University Press, Cambridge, 1991.
9. V. Lubchenko and P. G. Wolynes. Theory of aging in structural glasses. *J. Chem. Phys.*, 121:2852–65, 2004.
10. M. Henke, M. Pleimling, and R. Sanctuary. *Ageing and the Glass Transition.* Springer, Berlin, 2007.
11. M. Goldstein. Viscous liquids and the glass transition: A potential energy barrier picture. *J. Chem. Phys.*, 51:3728–39, 1969.
12. J. Fan, E. I. Cooper, and C. A. Angell. Glasses with strong calorimetric β-glass transitions and the relation to the protein glass transition problem. *J. Phys. Chem.*, 98:9345–49, 1994.
13. P. H. Poole, T. Grande, C. A. Angell, and P. F. McMillan. Polymorphism in liquids and glasses. *Science*, 275:322–3, 1997.
14. A. Pais. *Subtle Is the Lord ....; The Science and the Life of Albert Einstein.* Oxford Univ. Press, Oxford, 1982.
15. A. Einstein. Die Plancksche Theorie der Strahlung und die Theorie der spezifischen Wärme. *Annalen der Physik.*, 22:180–90, 1907.
16. C. A. Angelll. The old problem of glass and the glass transition, and the many new twists. *Proc. Natl. Acad. Sci. USA*, 92:6675–82, 1995.
17. C. A. Angell. Liquid fragility and the glass transition in water and aqueous solutions. *Chem. Rev.*, 102:2627–50, 2002.

18. D. Chandler. *Introduction to Modern Statistical Mechanics.* Oxford Univ. Press, Oxford, 1987.
19. R. Richert and A. Blumen, editors. *Disorder Effects on Relaxational Processes.* Springer, Berlin, 1994.
20. P. Lunkenheimer, U. Schneider, R. Brand, and A. Loidl. Glassy dynamics. *Contemp. Phys.*, 41:15–36, 2000.
21. C. A. Angell, K. L. Ngai, G. B. McKenna, P. F. McMillan, and S. W. Martin. Relaxation in glassforming liquids and amorphous solids. *J. Applied Phys.*, 88:3113–50, 2000.
22. R. Richert. Heterogeneous dynamics in liquids: Fluctuations in space and time. *J. Phys. Condens. Matter*, 14:R703–38, 2002.
23. F. Affouard, M. Descamps, and K. L. Ngai, editors. *Relaxations in Complex Systems.* Elsevier, Netherlands, 2006. The papers have also been published in *J. Non-Crystalline Solids*, 352: 4371–5227, 2006.
24. P. Debye. *Polar Molecules.* Chemical Catalog Company, New York, 1929.
25. F. Kremer and A. Schönhals. *Broadband Dielectric Spectroscopy.* Springer, Berlin, 2003.
26. U. Kaatze and Y. Feldman. Broadband dielectric spectrometry of liquids and biosysytems. *Meas. Sci. Tech.*, 17:R17–35, 2006.
27. S. Sudo, S. Tsubotani, M. Shimomura, N. Shinyashiki, and S. Yagihara. Dielectric study of the α and β processes in supercooled ethylene glycol oligomer-water mixtures. *J. Chem. Phys.*, 121:7332–40, 2004.
28. G. P. Johari and M. Goldstein. Viscous liquids and the glass transition. II: Secondary relaxations in glasses of rigid molecules. *J. Phys. Chem.*, 53:2372–88, 1996.
29. K. L. Ngai and M. Paluch. Classification of secondary relaxation in glass-formers based on dynamic properties. *J. Phys. Chem.*, 120:857–73, 2004.
30. H. Bässler, G. Schönherr, M. Abkowitz, and D. Pai. Hopping transport in prototypical organic glasses. *Phys. Rev. B*, 26:3105–13, 1982.
31. R. Zwanzig. Diffusion in a rough potential. *Proc. Natl. Acad. Sci. USA*, 85:2029–30, 1988.
32. J. D. Bryngelson and P. G. Wolynes. Intermediates and barrier crossing in a random energy model (with applications to protein folding). *J. Phys. Chem.*, 93:6902–15, 1989.
33. A. Eucken. Temperature variation of heat conductivity of non-metals. *Annalen der Physik*, 34:185–221, 1911.
34. R. C. Zeller and R. O. Pohl. Thermal conductivity and specific heat of noncrystalline solids. *Phys. Rev. B*, 4:2029–41, 1971.
35. W. A. Phillips, editor. *Amorphous Solids: Low-Temperature Properties.* Springer, Berlin, 1981.
36. V. Lubchenko and P. G. Wolynes. The microscopy theory of low temperature amorphous solids. *Adv. Chem. Phys.*, 136:95–206, 2007.
37. C. A. Angel, et al. Potential energy, relaxation, vibrational dynamics and the Boson peak, of hyperquenched glasses. *J. Phys. Condens. Matter*, 15:S1051–68, 2003.
38. V. Lubchenko and P. G. Wolynes. The original of the Boson peak and thermal conductivity plateau in low-temperature glasses. *Proc. Natl. Acad. Sci. USA*, 100:1515-18, 2003.

39. P. Schellenberg and J. Friedrich. Optical spectroscopy and disorder phenomena in polymers, proteins, and glasses. In R. Richert and A. Blumen, editors, *Disorder Effects on Relaxational Processes*. Springer, Berlin, 1994.
40. R. J. Silbey, J. M. A. Koedijk, and S. Voelker. Time and temperature dependence of optical linewidths in glasses at low temperature: Spectral diffusion. *J. Phys. Chem.*, 105:901–9, 1996.
41. J. M. A. Koedijk, R. Wannemacher, R. J. Silbey, and S. Voelker. Spectral diffusion in organic glasses: Time dependence of spectral holes. *J. Phys. Chem.*, 100:11945–53, 1996.
42. J. C. Dyre. Colloquium: The glass transition and elastic models of glass-forming liquids. *Rev. Mod. Phys.*, 78:953–72, 2006.
43. T. R. Kirkpatrick, D. Thirumalai, and P. G. Wolynes. Scaling concepts for their dynamics of viscous liquids near an ideal glassy state. *Phys. Rev. A*, 40:1045–54, 1989.
44. J. D. Stevenson, J. Schmalian, and P. G. Wolynes. The shapes of cooperatively rearranging regions in glass-forming liquids. *Nature Physics*, 2:268–74, 2006.

Part IV

FUNCTION AND DYNAMICS

In the first three parts, we discussed the "primitive" aspects of proteins, namely their structure and energy landscape. The object of the discussions could just as well have been glasses instead of proteins. We did not use the fact that proteins are parts of living systems and have well-defined functions. In this part we turn to the function of proteins and the connection between structure, dynamics, and function. While we already had to be selective in the first three parts, here we have to select even more. An enormous amount is already known about the function of proteins, and in many cases, the esssential features of protein reactions are well known and well characterized. We are interested here in a more fundamental approach: Can we connect structure, dynamics, and function? Even in very simple cases, these relations are not fully understood and certainly not yet described quantitatively. We will therefore take some very simple protein reactions, namely the binding of small ligands such as dioxygen or carbon monoxide, to heme proteins. The biological reaction here appears to be extremely simple, namely the association and dissociation of, for instance, O_2 to myoglobin (Mb) or hemoglobin (Hb). Nevertheless, the detailed study of such a simple biological reaction demonstrated that it is exceedingly complex and can already teach us a great deal about biological reactions. Much is not known, and many problems remain to be solved. We consequently take these binding reactions as prototypes that will teach us the underlying principles.

The approach taken here involves a number of areas:

(a) The phenomenological description of the function.
(b) The molecular description of the same function, assuming essentially rigid proteins ("single particle model").
(c) The exploration of the motions of the protein (dynamics).
(d) The dynamic picture of the function: connect motions and function.
(e) In order to treat even (a), we require a knowledge of reaction theory. Because reaction theory is involved in all aspects, we treat it after (a).

15

Protein Dynamics

Proteins are not static as shown in texts; they fluctuate continuously. Moreover, the "working protein" consists not only of the folded chain of amino acids. Proteins are surrounded by their hydration shell and are embedded in a bulk solvent. The protein proper, the hydration shell, and the bulk solvent are all involved in the protein motions and all three are necessary for the functions. Protein motions are transitions between different conformational substates that are described by the energy landscape (EL). The energy landscape has already been introduced and will be described in more detail in the present chapter. The experimental exploration of the EL is largely done by studying protein motions. We have therefore put the cart before the horse by treating the EL first. The reason is logic. Once the concept of a hierarchical EL is accepted, the existence of various types of motions is a logical consequence.

The concepts introduced for supercooled liquids and glasses, namely aperiodicity, non-Arrhenius temperature dependence, nonexponential time dependence, frustration, and metastability at low temperature, also apply to proteins [1, 2]. However, there are significant differences. The structure of a glass is random both locally and at large distances; there is both local and global "disorder." Proteins, however, have local random "disorder," but the global structure appears to the untrained eye to be unique as shown by the standard X-ray diffraction structure. (The word "disorder" is not really proper. The substates that simulate disorder are needed for function and they are the results of the design of the primary sequence.)

Einstein reportedly said that if he could have only one property to understand a physical system he would choose specific heat. Equation (14.1) tells why: $C(T)$ gives the number, $W(T)$, of accessible states at temperature T. The functions of proteins demand a large number of states. The measurement and interpretation of $C(T)$ for proteins will not be discussed here, except to note that $C(T)$ deviates from that of crystals below about 150 K and that $C(T)$ implies an exceptionally broad range of relaxation times [3]–[5]. In this chapter, aspects of the protein dynamics are described. However, protein dy-

H. Frauenfelder, *The Physics of Proteins*, Biological and Medical Physics, Biomedical Engineering, DOI 10.1007/978-1-4419-1044-8_15,

namics is far from fully known and understood. Some arguments given in the following sections will likely turn out to be misleading or wrong.

15.1 Equilibrium and Nonequilibrium Motions

Equilibrium fluctuations and nonequilibrium relaxations are involved in protein dynamics and function, as was described in Section 11.1. Consider a double well as shown in Fig. 13.6. In equilibrium, the ratio of the populations in the two wells is given by

$$N_B/N_A = \exp(-\Delta G/RT). \tag{15.1}$$

Here ΔG is the free energy difference between the bottoms of the two wells. The number of transition from A to B and B to A are equal,

$$N_A k_{AB}(T) = N_B k_{BA}(T). \tag{15.2}$$

Equations (15.1) and (15.2) together give

$$k_{AB}(T)/k_{BA}(T) = \exp(-\Delta G/RT). \tag{15.3}$$

Experimentally one desires to determine the rate coefficients $k_{AB}(T)$ and $k_{BA}(T)$ over a wide range of temperatures. If the barrier in Fig. 13.6 has a unique height, the rate coefficient for relaxation is given by

$$k_r(T) = k_{AB}(T) + k_{BA}(T) \ , \tag{15.4}$$

and the relaxation function is exponential in time. With Eqs. (15.3) and (15.4), the individual rate coefficients can be found.

If the populations in the two wells are not very different, the rate coefficients can be found by using either fluctuations [6]–[8] or relaxations [9, 10]. Fluctuations are discussed in Section 20.7. The fluctuation of frequency f in the number of particles in one well, say A, is given by

$$|\delta N_A(f)|^2 = \frac{4\langle N_A\rangle[1-\langle N_A\rangle]}{k_r[1+(2\pi f/k_r)^2]} \ . \tag{15.5}$$

The fluctuations permit the determination of k_r. The autocorrelation gives another approach to find k_r:

$$\langle \delta N_A(t)\ \delta N_A(t+t')\rangle = \text{const}\ \exp(-k_r t) \ . \tag{15.6}$$

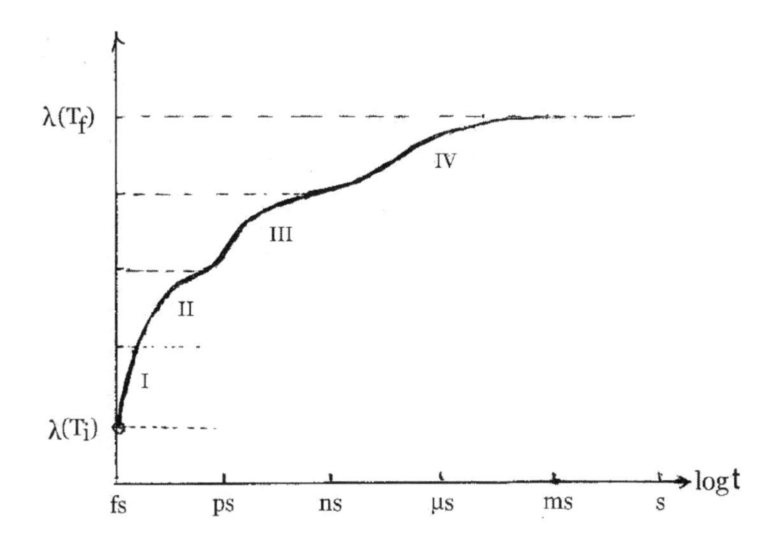

Fig. 15.1. Sketch of the relaxation of a protein. The conformation of the protein is characterized, for instance, by the wavelength $\lambda(T)$ of a spectral line that depends on the protein conformation. Immediately after a jump $T_i \rightarrow T_f$, the conformation and therefore the wavelength have not changed. The wavelength will then relax to the position at the final temperature T_f. In proteins, this relaxation is complex, as described in the text. Moreover, the individual steps are usually nonexponential in time.

In the relaxation method, a sudden perturbation is applied to the system, for instance, by a jump in temperature or pressure or by a laser flash. The perturbation changes the conformation of the proteins. The system then relaxes to the new equilibrium state, as sketched in Fig. 15.1. Assume that the state of the system can be characterized by the wavelength $\lambda(T)$ of a suitably chosen spectral line. Before the temperature jump, the line is at $\lambda(T_i)$. Immediately after the jump, it will still be at this position, but it relaxes toward the final position $\lambda(T_f)$. In proteins, the relaxation process is complex and depends on the protein, the hydration shell, the bulk solvent, and temperature and pressure. Schematically we distinguish four processes:

(I). The protein relaxes elastically, without concomitant change in its topology. This relaxation is caused, for instance, by the thermal expansion after a T jump.

(II). The protein relaxes vibrationally with femtoseconds (fs), also without a conformation change [11]–[13] (Section 15.2).

(III). The α relaxation, already encountered in supercooled liquids, sets in within picoseconds (ps) (Section 15.5).

(IV). The fourth phase of the relaxation process is caused by the β relaxation (Section 15.6). There may be more than one β relaxation. Additional relaxation processes exist that are studied at cryogenic temperatures (Section 15.7). Figure 15.1 is oversimplified (and may be wrong) because the time order of the relaxations can depend on temperature. Ideally, the various relaxation processes are described by separate relaxation functions. In reality, the separation can be extremely difficult.

The functions of proteins involve equilibrium fluctuations and nonequilibrium relaxations. If the excursion from equilibrium is small, the fluctuation-dissipation theorem is valid and can be used to relate nonequilibrium experiments to functions that are based on equilibrium fluctuations. In some cases, for instance, reactions started by the absorption of light that lead to major conformational changes, the fluctuation-dissipation theorem may not hold.

15.2 Vibrations and Normal Modes

In 1907 Einstein removed a major difficulty in understanding the low-temperature behavior of the specific heat of crystals by assuming that all atoms vibrate with the same frequency, thus quantizing the system [14]. Debye later improved the model by considering the crystal to be an elastic continuum. The models are treated in most condensed matter textbooks [15]. For proteins, these models are too simple. Each protein atom can vibrate with its own frequency. What is of interest for dynamics are not these individual frequencies, but the normal modes. These are time-independent constants of the motion and are described by the vibrational density-of-state spectrum, the number of normal modes per frequency interval [16, 17]. The normal modes are obtained as the solutions of the relevant eigenvalue equation. Modern computers can solve this equation even for proteins. Figure 15.2 gives an example of a vibrational density-of-state spectrum, calculated by Melchers and collaborators [17]. Experimentally, information about the normal-mode spectrum has been obtained, for instance, by using the phonon-assisted Mössbauer effect [18] or by high-resolution X-ray measurements [19].

Normal mode analysis has limitations, as implied by the energy landscape shown in Fig. 12.2. Barriers between substates can be so high that vibrations cannot overcome them. Molecular dynamics calculations complement the normal mode approach, and they can explore jumps between wells [20]–[22].

15.3 Protein Quakes

An earthquake is the result of a sudden release of stored energy in the Earth's crust. Tectonic earthquakes, caused by a sudden slip of plates, result in two observables, the change in "conformation" along the fault and seismic waves. Quakes occur on the moon, the planets, the sun, stars, and in proteins. Figure

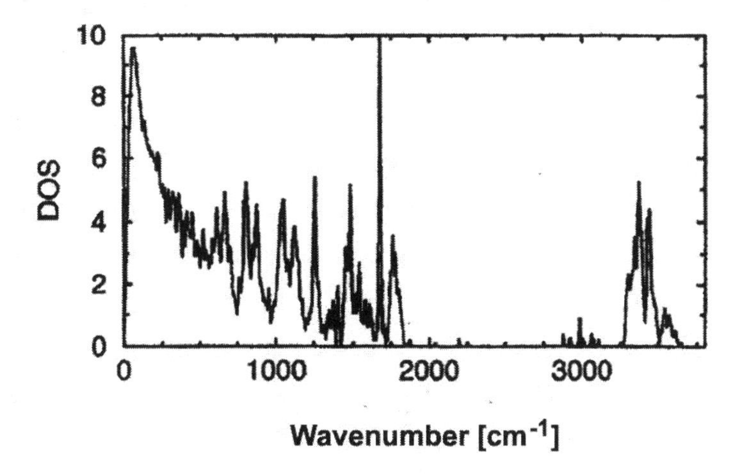

Fig. 15.2. The calculated vibrational density of state spectrum of MbCO with 170 crystal water molecules. From Melchers et al. [17].

15.3, a cartoon of a proteinquake, shows myoglobin in the bound-state structure [23]. A laser flash breaks the CO–Fe bond and the protein relaxes to the deoxy structure. Part of the released energy is emitted as waves. Compared to the total binding energy of the system, proteinquakes are more powerful than earthquakes or starquakes, because the released bond energy is of the same order of magnitude as the binding energy of the protein, about 100 kJ/mol. Proteinquakes show the two characteristic features of quakes, changes in the conformation and the emission of waves.

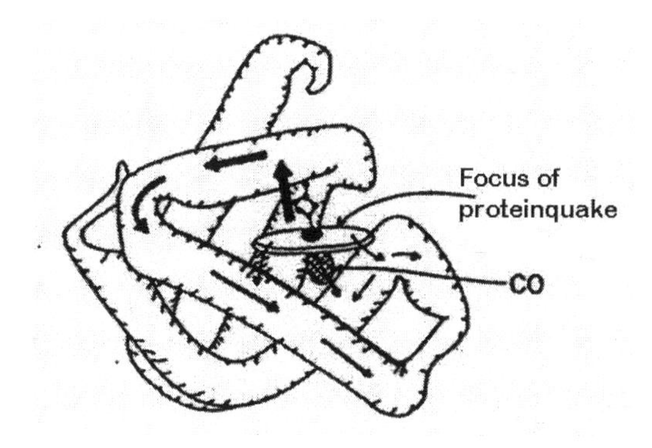

Fig. 15.3. A proteinquake. The figure shows the backbone of myoglobin, with a CO molecule bound to the heme iron. A laser pulse breaks the CO–Fe bond. The myoglobin, initially in the bound-state structure, relaxes to the deoxy structure, and some of the relased energy is emitted in the form of waves. From [23].

Evidence for a structural change during a proteinquake came from the fact that the X-ray structures of the MbCO and deoxyMb are different. Synchrotron radiation gives far more information; it permits even the study of the time dependence of the structure during the quake [24, 25]. Very fast volume changes, which are the equivalent of the waves in earthquakes and are shown as (I) in Fig.15.1, have been observed by Miller and colaborators [26, 27]. The primary features of proteinquakes thus are established. Details of the motions in a proteinquake are, however, not fully explored. The complexity, which causes the diffculties for a complete experimental study, is apparent in Fig. 12.2: The energy landscape is organized hierarchically in more than four tiers. The relaxation from the bound- to the deoxy-state starts, after the elastic relaxation, presumably in the lowest tier. After some transitions in this tier, the protein jumps to another well in the next higher tier and finally winds up in a substate of the fully relaxed protein. A one-dimensional picture of the relaxation [23] does not do justice to this process that takes place in the high-dimensional energy landscape. A more elaborate description of the energy landscape is described in the next section. The properties of the structural changes in the quakes in Mb and in other proteins, for instance, in the photoactive yellow protein (PYP) [28, 29], remain to be explored in depth.

The fluctuation-dissipation theorem implies that the motions seen in proteinquakes also appear in thermal fluctuations. Indeed, entrance of a ligand such as dioxygen into a protein and diffusion through the protein involve fluctuations. Experiments, sketched in Chapter 17, and computational and theoretical treatments [30]–[34] show the crucial roles of proteinquakes and the related fluctuations in protein function.

15.4 Another Look at Protein Landscapes: Conformation, Structure, and Reaction

Protein functions depend on the protein conformations. To describe the connection between function and conformation, consider the case of a protein with two conformational substates, 0 and 1, and two reaction states, A and B. A and B can be the states, described in Fig. 11.5: In A, the CO molecule is bound to the heme iron inside myoglobin; in B, the photodissociated CO is still inside the heme pocket but has moved away from the binding site. Rebinding, $B \to A$, is assumed to be different in the substates 0 and 1 and given by k_0 and k_1. Assume further that k_r is the rate coefficient for the transitions between the substates 0 and 1. Two limiting cases are easy to describe. In the static case, $k_r \ll k_0, k_1$, rebinding takes place with the two different rate coefficients, k_0 and k_1. Generalization to many conformations leads to the nonexponential rebinding discussed in Section 11.2. If the relaxation is much faster than the rebinding, $k_r \gg k_0, k_1$, binding is exponential in time, with an average rate coefficient.

In the general case there are more states along the reaction coordinate and a very large number of conformational substates in the conformation space. Motion along the reaction coordinate takes place under the influence of conformational fluctuations. This situation has been treated by Agmon and Hopfield [35, 36]. They assign a rate coefficient $k(x)$ for each conformational substate x. At low temperatures, the coordinate x is frozen. At higher temperatures, x diffuses in the high-dimensional conformation space and the diffusion is described by a Smoluchowski equation. The coordinate x is different from the reaction coordinate and Agmon and Hopfield call their approach "diffusion perpendicular to the reaction coordinate."

The full complexity of protein dynamics becomes apparent if we take another look at the EL, sketched in Fig. 12.1. This look is also motivated by the observation that the EL is often misinterpreted. To describe protein functions completely, three "landscapes" are actually needed, as shown in Figs. 15.4 and 15.5. The top of Fig. 15.4 represents the *energy landscape*. As discussed earlier, the EL is a construct in a hyperspace of $\approx 3N$ coordinates, where N is of the order of 10^3 or 10^4. A point in the EL characterizes a protein structure fully; it

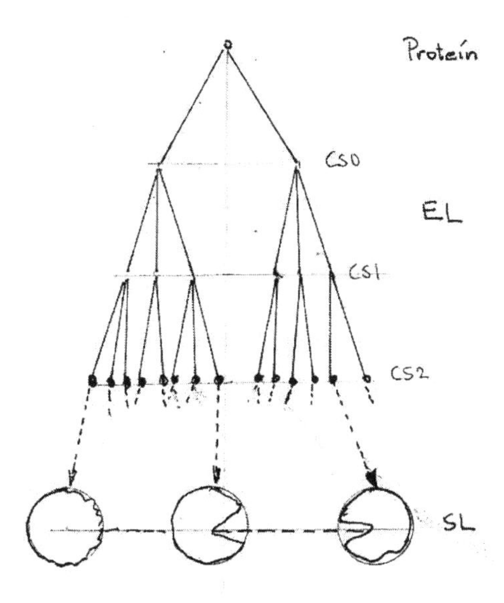

Fig. 15.4. The energy landscape (EL) exists in a hyperspace of 3N dimensions. The figure shows a simplified tree diagram of the EL of myoglobin. Tier 0 contains a small number of taxonomic substates; two are shown. Each of the $CS0$ contains a very large number of substates in tier 1. Each of these substates again contains a large number of substates of tier 2. Lower tiers are not shown. The EL is complemented by the structure landscape (SL). Each point in the $3N$-dimensional EL corresponds to a 3D protein structure. There are as many different structures as points in the EL.

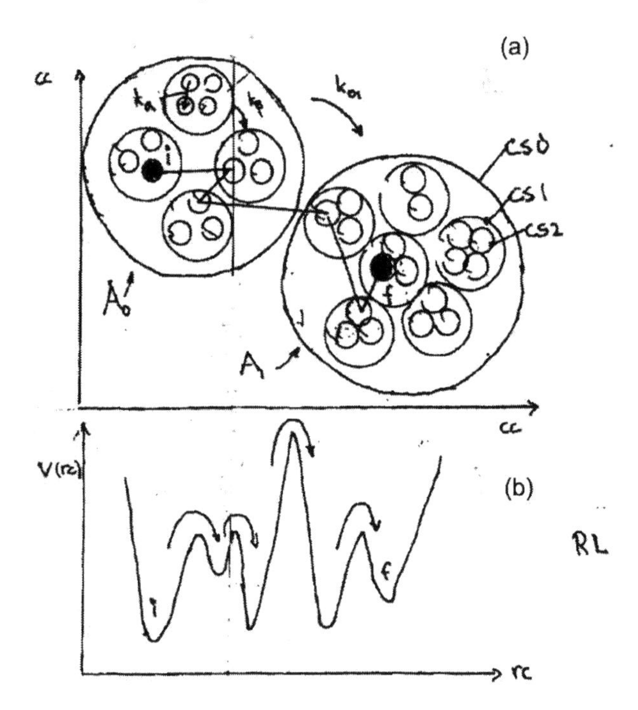

Fig. 15.5. (a) A 3D cross section through the EL at the $CS2$ level. Two dimensions are given by conformational coordinates. The third dimension is the potential energy of the substates, indicated as craters in the EL. Transitions between substates in tier 1 are given by $k_\beta(T)$, between substates in tier 2 by $k_\alpha(T)$. (b) Part (a) implies that the number of paths going from an initial state i to a final state f is extremely large. Fig. 15.5(b) shows a hypothetical potential along one of these paths in the reaction landscapes (RL).

gives the coordinates of all atoms of the protein, its hydration shell, and part of the bulk solvent. This connection is indicated in Fig. 15.4 by the arrows relating the points in the EL, with structures pictured schematically in the *structure landscape* (SL). The number of protein substates is extremely large. The number of corresponding structures is therefore also extremely large. This fact is easily understood; for instance, the rotation of a water molecule in the hydration shell already changes the structure and the corresponding point in the EL. Thus the number of substates is much larger than "astronomic." Figure 15.4 still does not expose the complex dynamics of protein motions fully; it presents only a one-dimensional cross section through the EL. A two-dimensional cross sections at the $CS2$ tier, given in Fig. 15.5(a), provides more insight; it shows the various substates as craters in a topographic map. Only a very small number of substates in the tiers 1 and 2 are shown. The different tiers are characterized by different rate coefficients. Transitions in tier 2 are postulated to be given by $k_\alpha(T)$, in tier 1 by $k_\beta(T)$.

Figure 15.5 shows the need for a third protein landscape, the *reaction landscape* (RL). The RL describes the potential energy along a given path in the EL from i to f as sketched in Fig. 15.5(b). This path is not unique; the transition from i to f can take place in a nearly infinite number of ways. Moreover, while the protein transits, the conformation can change. Thus even a simple process in a protein is truly complex, and there is a need for simplified concepts and descriptions. Two such approaches are described in the following sections. A word of caution: Protein dynamics is still in its infancy, and a review in a few years may look very different.

15.5 The α Fluctuations and Slaving

The complexity of the energy landscape as described in Section 15.4 implies that proteins can undergo a broad range of thermal fluctuations and nonequilibrium relaxations. These motions can differ from protein to protein. Rather than studying the individual motions in many proteins, as is often done in computational approaches, different insight comes from looking at the conspicuous features and describing their properties and roles. A guide comes from the supercooled liquids and glasses described in Chapter 14. Three types of motions have been studied in some detail, the primary or α fluctuations, the secondary or β fluctuations, and the low-temperature TLS tunneling motions. This knowledge helps to classify the motions in proteins. This section treats the α fluctuations. Their outstanding characteristics in glasses are the nonexponential time dependence and the non-Arrhenius temperature dependence. These characteristics also appear in proteins. The nonexponential time dependence has been observed even at room temperature by Anfinrud and collaborators [37]. It will not be further discussed except to note that, with modern equipment, more such measurements should be performed over a wide range of temperatures and pressure, and for more proteins. (We note again that the observables should be plotted as a function of $\log t \equiv pt$.)

Experiments over a broad range of temperatures and using myoglobin embedded in cryosolvents provide information about protein dynamics. A number of experiments have studied the effect of solvent viscosity on the dynamics of myoglobin [38]–[42]. A problem arises in the interpretation of such experiments. While the viscosity in the bulk solvent can be measured, the viscosity in the interior of a protein must be calculated [43]. It is therefore better to use the dielectric relaxation rate coefficient $k_\alpha(T)$, which is related to the viscosity by the Maxwell relation Eq. (14.9) and which can be approximated by the VFT relation, Eq. (14.4),

$$k_\alpha(T) = k_0 \exp\{-D\, T_0/(T - T_0)\}. \tag{15.7}$$

Measuring the rate coefficient $k(T)$ of protein motions as a function of viscosity leads to the realization that large-scale motions depend strongly on viscosity

[38]–[41, 44]. Examples of such large-scale motions are the exit of ligands from myoglobin, the exchange between the substates A_0 and A_1 shown in Fig. 12.5, and fluctuations of the entire protein. Comparing the respective rate coefficients $k_{\text{exit}}(T)$, $k_{01}(T)$, and $k_c(T)$ to $k_\alpha(T)$ shows that the protein motions follow $k_\alpha(T)$ over many orders of magnitude in rate, but are slower by large factors (Fig. 15.6(a)). This observation leads to the concept of slaving: Large-scale protein motions follow the large-scale α fluctuations in the bulk solvent but are often much slower.

Why are the large-scale protein motions much slower than the α fluctuations in the bulk solvent? The answer leads to more insight into protein dynamics. A large-scale protein motion is not like opening and closing a rigid door; it takes place in a large number of steps, and it can be understood in terms of the hierarchical energy landscape, Fig. 15.4. Assume that the initial state i represents a closed conformation, with CO inside the protein, and f the substate where a gate to the solvent is open. Assume further that each step occurs with a rate coefficient $k_\alpha(T)$ and that the protein makes $n(T)$ steps to reach for the open conformation. The exit rate is then given by

$$k_{\text{exit}}(T) = k_\alpha(T)/n(T). \tag{15.8}$$

Doster and collaborators measured $k_{\text{exit}}(T)$ for a number of solvents with different values of $k_\alpha(T)$ and found that $k_{\text{exit}}(T)$ follows $k_\alpha(T)$ over many orders of magitude in rate [41]. Shibata, Kurita, and Kushida determined the rate coefficient $k_c(T)$ for the large-scale fluctuations in myoglobin also as a function of solvent viscosity [42]. Their result is shown in Fig. 15.6(b): The

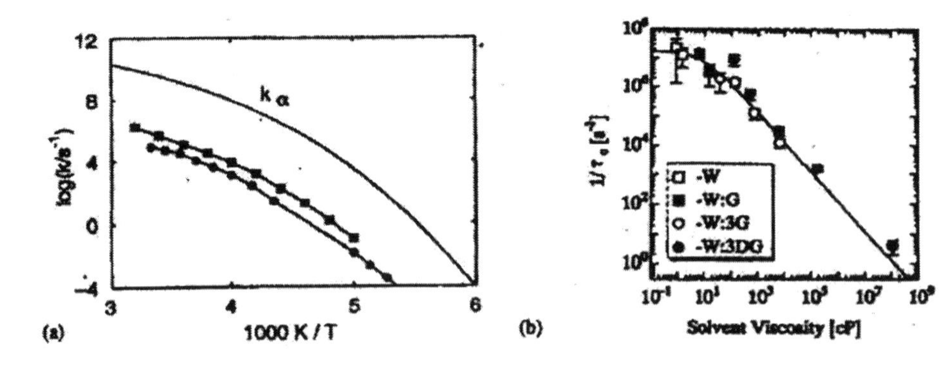

Fig. 15.6. (a) An Arrhenius plot of the rate coefficient $k_\alpha(T)$ for a 3/1 glycerol/water cryosolvents. The rate coefficient $k_{\text{exit}}(T)$ for the exit of CO from Mb is shown by filled squares. The solid circles denote the exchange between the substates A_0 and A_1. (b) The rate coefficient $k_{\text{exit}}(T)$ versus the solvent viscosity for Mb embedded in different cryosolvents. Inverted triangles: 60% ethylene glycol/water, circles 75% glycerol/water, squares 90% glycerol/water, diamonds 80% sucrose/water. From [42].

flucuations in the protein follow the fluctuations in the solvent over about eight orders of magnitude, justifying the concept of slaving: The protein follows the solvent and not the other way around [45]. Moreover, the protein follows the solvent up to the smallest viscosity, about 1 cP. The solvent drives the protein motions! The protein behaves more like a liquid than a solid [46, 47]. The internal protein viscosity is about 1 cP, or even smaller, in agreement with Ansari's calculation that shows that the internal friction is similar to that of water [43]. The bulk solvent contributes the enthalpic barrier, the protein the entropic barrier for the control of large-scale protein motions. Incidentally, slaving is also important for protein folding [48]. Lubchenko and Wolynes have formulated a theory of slaving [49]. It is based on the RFOT theory, mentioned in Chapter 14. The theory shows how the molecular motions of the glass-forming solvent distort the boundary of the protein and thereby slave large-scale conformational motions of the protein.

15.6 Internal Motions, Hydration, and β Fluctuations

Most protein functions stop below 200 K and it has often been claimed that below this temperature proteins undergo only harmonic motions. However, internal motions, for instance, the transit of small molecules from one cavity to another, shown schematically in Fig. 15.7, occur well below 200 K [41].

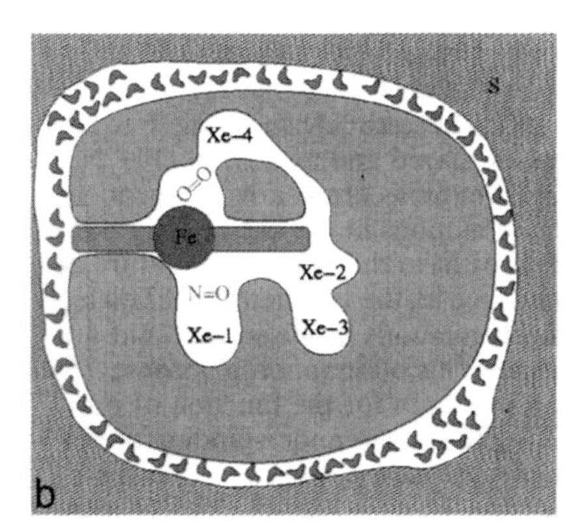

Fig. 15.7. A simplified sketch of a cross section through myoglobin, its hydration shell, and bulk solvent. CO or O_2 bind at the heme iron. After photodissociation, the CO molecule moves through the protein, visiting some or all of the cavities labeled $Xe-1$ to $Xe-4$.

Enzymatic reactions have also been observed to occur below 200 K. The study of protein dynamics at low temperatures is important not only for learning about the physics of these processes, but also because the results are relevant if extrapolated to biological temperatures. The α fluctuations alone cannot be responsible for the internal motions, because they become very slow below 200 K as shown in Fig. 15.6. Different fluctuations must be responsible. The β fluctuations, briefly discussed in Chapter 14, are candidates [50]. Experimental observations support this suggestion:

1. The internal motions continue even if the protein is encapsulated in a solid environment, such as poly-(vinyl) alcohol [51].
2. Internal motions disappear if the protein is dehydrated.
3. The activation enthalpies of the internal ligand transit are of the order of 40 kJ/mol [41]. The activation enthalpies of the β fluctuations in thin water films on a wide range of substances cluster around 45 kJ/mol.

These three observations together suggest that the hydration shell [52]–[55], about two layers of water surrounding the protein as displayed in Fig. 15.7, is the source of the β fluctuations. Experiments to test this conjecture lead to a remarkable conclusion: The internal motions as seen by the heme iron in myoglobin and the protons, mainly in the side chains of the backbone, follow the β fluctuations in the hydration shell [56]. The fluctuations in the hydration shell are measured using dielectric spectroscopy [57]; the internal motions are observed using the Mössbauer effect (e.g., [58]) and neutron scattering [59].

As an example of the connection between the β fluctuations in the hydration shell and the internal motions, we describe an experiment using the Mössbauer effect [60]. This effect involves a nuclear energy level with energy E_o that emits a gamma ray with mean life τ_{Mo}. Usually, such gamma rays are Doppler broadened and their energies are shifted to lower values. Mössbauer discovered that at low temperatures a fraction $f(T)$ of gamma rays is emitted without energy loss and has the natural line shape. $f(T)$ is called the Lamb-Mössbauer factor and is determined from the area of the elastic Mössbauer line. The favorite nuclide is ^{57}Fe, with a lifetime τ_{Mo} =140 ns and E_0 =14.4 keV. The experimental data are unfortunately not given as the measured area $f(T)$, but by quoting the mean-square displacement (msd), $\langle x^2(T)\rangle$, of the iron atom. $f(T)$ and $\langle x^2(T)\rangle$ are related by the Lamb–Mössbauer factor

$$f(T) = \exp\{-{k_o}^2\langle x^2(T)\rangle\} , \tag{15.9}$$

where $k_0 = E_0/2\pi\hbar c$ and for ^{57}Fe, ${k_0}^2 = 53$ Å^{-2}. Figure 15.8 gives $\langle x^2(T)\rangle$ for ^{57}Fe in a myoglobin crystal [61]. Two outstanding features in this figure are the fact that the $\langle x^2(T)\rangle$ can be observed even above 300 K and that it deviates markedly from the linear low-temperature behavior at about 200 K. The first feature is explained by noting that the α fluctuations are extremely slow or absent in a crystal. The steep increase requires more discussion. It is also seen in neutron scattering, where it has been dubbed a "dynamical transition" and given birth to an industry. Actually no new physics is needed to explain the "dynamical transition"; the increase is due to the β fluctuations in the hydration shell. Consider a Mössbauer emitter in a fluctuating environment [62]. As long as the conformational fluctuations are slower than τ_{Mo} only

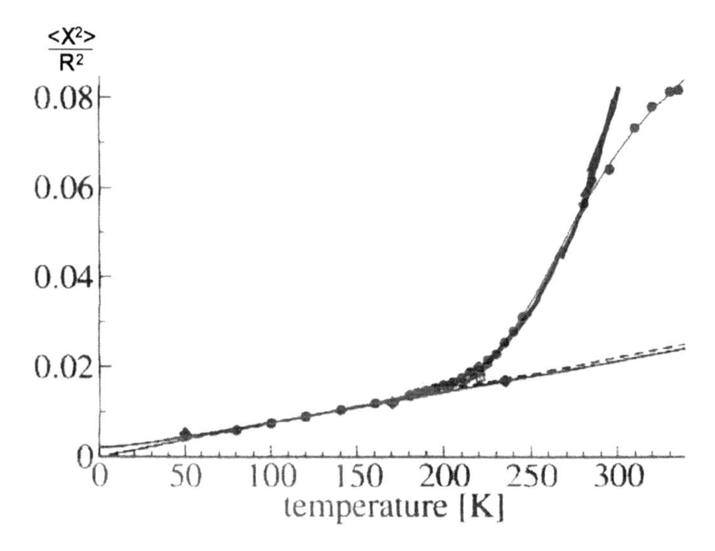

Fig. 15.8. The temperature dependence of the mean-square displacement (msd) of the iron atom in myoglobin, observed with the Mössbauer effect. After Parak and coworkers [58]. The red curve is the msd predicted using the β relaxation in the hydration shell as input.

vibrations decrease $f(T)$. The vibrations lead to the linear increase of $\langle x^2(T)\rangle$ as seen below 200 K. If the fluctuations are faster than τ_{Mo}, an additional fraction of the gamma rays becomes inelastic. To analyze the data in Fig. 15.8, we assume that the total msd is composed of a vibrational component and a conformational component (Section 11.3),

$$\langle x^2\rangle_t = \langle x^2\rangle_c + \langle x^2\rangle_v. \tag{15.10}$$

$\langle x^2\rangle_v$ is found by linearly extrapolating the low-temperature data and $\langle x^2\rangle_c$ is then given by $\langle x^2\rangle_t - \langle x^2\rangle_v$. Inserting $\langle x^2\rangle_c$ into Eq. (15.9) yields $f_c(T)$.

The key step for comparing the fluctuations seen by the iron atom inside the protein with the exterior fluctuations is the quantitative evaluation of the β fluctuations in the hydration shell. Figure 15.9(a) is the dielectric relaxation spectrum recorded at 220 K in a water/myoglobin solution embedded in solid PVA, where the α fluctuations are absent. The vertical line at $\log\,(k/s^{-1}) = 6.85$ represents $k_{Mo} = 1/\tau_{Mo}$. The area $a_\beta(T)$ to the left of k_{Mo} gives the fraction of fluctuations that are too slow to reduce the Lamb-Mössbauer factor. Relaxation spectra measured at many temperatures, shown in Fig. 15.9(b), yield $a_\beta(T)$. Taking $f(T) = a_\beta(T)$ and using Eq. (15.9) yields the msd shown as the red line in Fig. 15.8. The predicted msd agrees with the measured msd up to 270 K. The data above 270 K have large errors, because it is difficult to measure both the dielectric spectrum and the Mössbauer msd.

The result is clear: the fluctuations seen by the heme iron follow the fraction $a_\beta(T)$ in the hydration shell. Support for this conclusion comes from elastic neutron scattering experiments, where the evaluation also shows that it follows the β fluctuations in the hydration shell. The conclusion is surprising. Naively one would expect that different parts of the protein's interior fluctuate with different rates. The observation that the heme iron and interior protons fluctuate like the hydration shell suggests a different picture: Major internal motions are governed by the β fluctuations in the hydration shell.

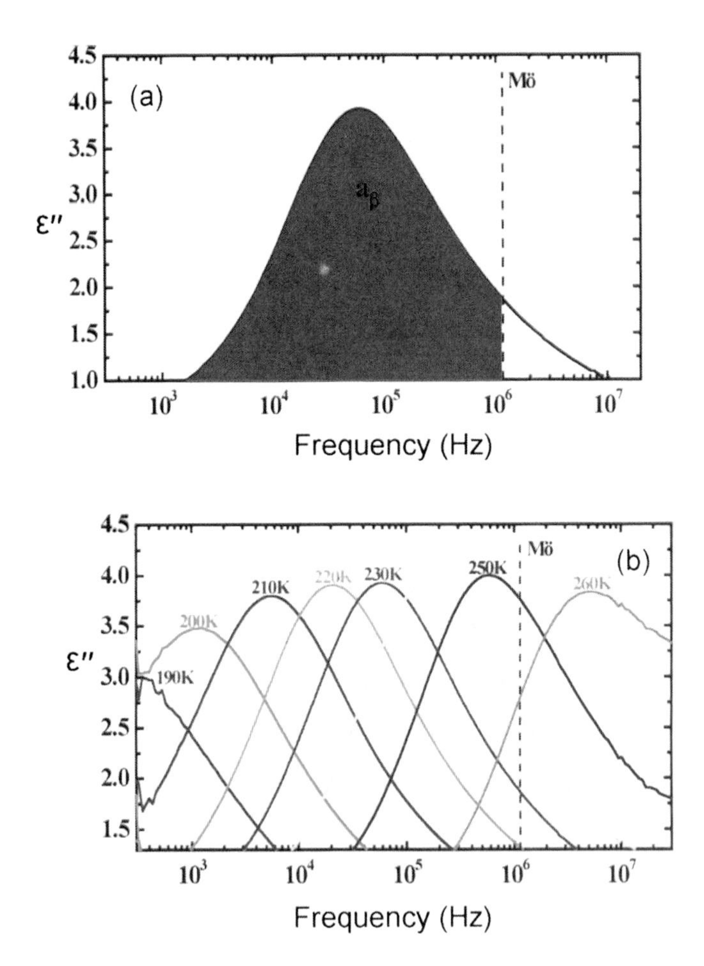

Fig. 15.9. (a) The dielectric spectrum of Mb embedded in a glycerol/water solvent at 225 K. The fluctuations in the red area are too slow to affect the Mössbauer effect in ^{57}Fe. The fraction $a_\beta(T)$ is therefore equal to the Lamb-Mössbauer factor $f(T)$. (b) The dielectric spectrum measured as a function of temperature.

15.7 Low-Temperature Dynamics, or Lost in the Energy Landscape

Protein motions occur even at temperatures well below 1 K [63]. It is not clear if these motions play a role in biological processes, but they are of considerable interest for the physics of complex systems. Spectral hole-burning is the technique that has given much insight into the low-temperature motions [64]–[68]. The technique, already described in Section 11.4, uses a structure-sensitive probe, for instance, a native prosthetic group in a protein. Spectral lines of such probes are inhomogeneously broadened. A laser burns a hole in such a spectrum. Broadening of the hole provides information about the protein motions. To comprehend the technique, the cooling process must be understood. Assume that the envelope of the energy landscape is harmonic. In equilibrium, the substates are occupied according to a Boltzmann distribution that is essentially Gaussian, with a width proportional to $T^{\frac{1}{2}}$. Further assume that this distribution is reflected in the spectral line whose inhomogenerous width is then also proportional to $T^{\frac{1}{2}}$. Two extreme cases of hole-burning are easy to discuss. If cooling from 300 to 4 K is faster than any transition among substates, the 300 K distribution is frozen and motions at 4 K start from this broad nonequilibrium distribution. If cooling is infinitely slow, the equilibrium distribution at 4 K is very narrow. Reality lies between these two extremes and the distribution at 4 K depends on the cooling rate and the solvent. The properties of the α and the β relaxations discussed earlier help to predict features of the low-temperature distribution. If the protein environment is rigid at 300 K, for instance, if the protein is embedded in trehalose or poly-vinyl alcohol, the α relaxation is absent and no transitions in tier 2 occur. If, furthermore, there is not hydration, the β relaxation is also absent and transitions in tier 1 are also absent or very slow. The distribution at 4 K then is essentially the same as the one at 300 K. If, on the other hand, the protein is embedded in a glycerol/water solvent, the distribution freezes at a temperature that lies between 100 and 200 K, depending on the solvent and the cooling rate. The distribution at 4 K in the glycerol/water solvent is then expected to be narrower than in trehalose. Indeed Friedrich and collaborators have measured widths of 100 cm^{-1} and 249 cm^{-1}, respectively, for cytochrome-*c* embedded in glycerol/water and in trehalose. Hole-burning experiments at low temperatures thus depend on the cooling history and the solvent. In general, regardless of the solvent, the initial state at 4 K will not be in equilibrium.

Section 12.5 already gives an idea of the complexity of the energy landscape. Studies of the dynamics at low temperatures give more insight into the complexity. Figure 15.10 suggests what happens after a very narrow laser has bleached a single chromophore in the protein. The resulting hole is indicated at the lowest tier by the heavy arrow, denoted by i. Fluctuations change the environment of the active chromophore and thereby change its frequency. In the energy-landscape picture, this change is described as a step in the energy

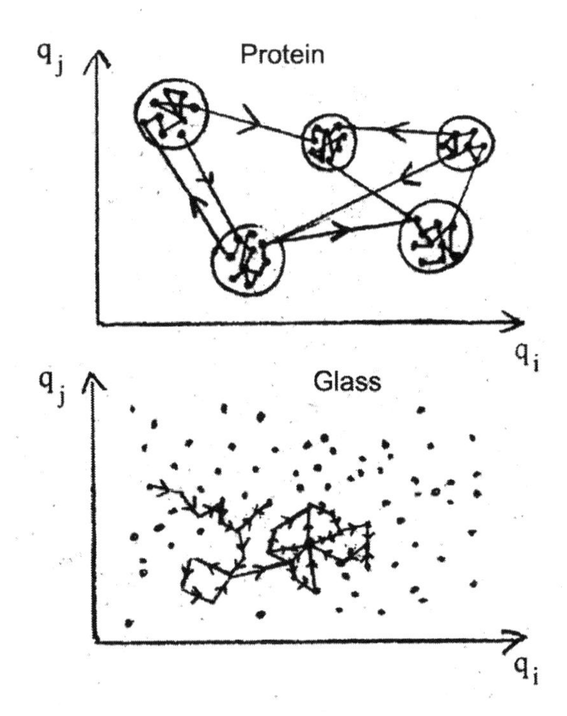

Fig. 15.10. Two-dimensional cross section of a low tier in the energy landscape of a protein and a glass. The protein displays the hierarchical arrangement, demonstrating that proteins have a structural organization that is not random. In contrast, glasses show no such organizations. After Fritsch and Friedrich [71].

landscape, as shown in Fig. 15.10. Successive fluctuations appear as a Brownian walk in the energy landscape. The line width of the hole remains roughly unchanged during this walk. Single-molecule experiments permit a direct examination of the walk [69]. In a protein ensemble, however, each protein takes a different path, and the hole broadens. The process is called *spectral diffusion,* and the broadening is characterized by the spectral diffusion width σ. At very low temperatures, say a few mK, the random walk explores only the basin of the lowest tier. As the temperature is increased, larger and larger areas in the energy landscape are reached. In a protein sample embedded in a glycerol/water solvent, tier 2 is explored above about 150 K. Finally, tier 1 becomes acessible above about 200 K, as shown beautifully by Shibata, Kruita, and Kushida [42]. Spectral hole-burning thus can provide detailed insight into the protein energy landscape from mK to 300 K.

Three types of hole-burning experiments are used. The experiments start from the protein sample cooled to T_o, for instance, 4.2 K. In a waiting time experiment, the hole is burned at $t = 0$ and read after the waiting time t_w.

In an aging experiment, t_a is the time after the sample has reached T_0 and before the hole is burned. In a temperature cycling experiment, the hole is burned at T_0. The temperature is then raised to the excursion temperature T_{ex}, remains there for a fixed time, and then lowers again to T_0, where the hole is remeasured. The energy landscape suggests what can be expected in these experiments. As an example, consider a waiting time experiment. As described earlier, unless the cooling from 300 K has been infinitely slow, the protein will not be in equilibrium at 4 K. The waiting time experiment thus starts from a nonequilibrium state, and the result depends on the cooling procedure. Two features stand out in a waiting time experiment. First, σ follows a power law, $\sigma = \text{const}\ t_w{}^{\alpha}$, with $\alpha \approx 0.25$. Second, σ is larger for the protein embedded in trehalose than in glycerol/water, as expeced from its dependence on the cooling process [70]. The first feature reflects that complexity of the protein; the second implies that the position of the protein in the energy landscape at $t_w = 0$ depends on the cooling protocol.

In Chapter 14, the low-temperature dynamics of glasses was sketched. Glasses and proteins share similarities, in particular the existence of an energy landscape. A closer look indicates differences that appear even in the cryogenic experiments. While σ follows a power law in proteins, it is given by a logarithmic dependence in glasses. The walks in the energy landscape differ as sketched in Fig. 15.10, because the energy landscape of proteins has a hierarchic organization. The result discussed here and in the earlier chapters shows that the energy ladscape and the dynamics of proteins are indeed very complex. Some of the fundamental features are known, but there is as yet no complete picture. We are still lost in the energy landscape.

References

1. I. E. T. Iben et al. Glassy behavior of a protein. *Phys. Rev. Lett.*, 62:1916–19, 1989.
2. J. L. Green, J. Fan, and C. A. Angell. The protein-glass analogy: Some insights from homopeptide comparisons. *J. Phys. Chem.*, 98:13780–90, 1994.
3. G. I. Makhatadze and P. L. Privalov. Energetics of protein structure. *Advan. Protein Chem.*, 47:307–425, 1994.
4. G. Sartor, A. Hallbrucker, and E. Mayer. Characterizing the secondary hydration shell on hydrated myoglobin, hemoglobin, and lysozyme powders by its vitrification behavior on cooling and its calorimetric glass→liquid transition and crystallization behavior on reheating. *Biophys. J.*, 69:2679–94, 1995.
5. D. Idiyatullin, I. Nesmelova, V. A. Dargan, and K H. Mayo. Heat capacities and a snapshot of the energy landscape in protein GB1 from the pre-denaturation temperature dependence of backbone NH nanosecond fluctuations. *J. Mol. Bio.*, 325:149–62, 2003.
6. G. Feher and M. Weissman. Fluctuation spectroscopy: Determination of chemical reaction kinetics from the frequency spectrum of fluctuations. *Proc. Natl. Acad. Sci. USA*, 70:870–5, 1973.
7. M. B. Weissman. Fluctuation spectroscopy. *Ann. Rev. Phys. Chem.*, 32:205–32, 1981.
8. G. U. Nienhaus, editor. *Protein-Ligand Interactions.* Humana Press, Totowa, N.J., 2005.
9. M. Eigen. New looks and outlooks on physical enzymology. *Q. Rev. Biophys.*, 1:3–33, 1968.
10. R. Richert and A. Blumen, editors. *Disorder Effects on Relaxational Processes.* Springer, Berlin, 1994.
11. M. D. Fayer. Fast protein dynamics probed with infrared vibrational echo experiments. *Ann. Rev. Phys. Chem.*, 52:315–56, 2001.
12. H. Fujisaki and J. E. Straub. Vibrational energy relaxation in proteins. *Proc. Natl. Acad. Sci. USA*, 102:6726–31, 2005.
13. M. Gruebele and P. G. Wolynes. Vibrational energy flow and chemical reactions. *Acc. Chem. Res.*, 37:261–7, 2004.
14. Q. Ciu and I. Bahar, editors. *Normal Mode Analysis. Theory and Applications to Biological and Chemical Systems.* Chapman and Hall/CRC, New York, 2006.

15. C. Kittel. *Introduction to Solid State Physics*, 4th edition. Wiley, New York, 1971.
16. K. Hinsen. Normal mode theory and harmonic potential approximations. In Q. Ciu and I. Bahar, editors, *Normal Mode Analysis. Theory and Applications to Biological and Chemical Systems.* Chapman and Hall/CRC, New York, 2006.
17. B. Melchers, E. W. Knapp, F. Parak, L. Cordone, A. Cupane, and M. Leone. Structural fluctuations of myoglobin from normal-modes, Mössbauer, Raman, and absorption spectroscopy. *Biophys. J.*, 70:2092–9, 1996.
18. K. Achterhold, C. Keppler, A. Ostermann, U. van Buerck, W. Sturhan, E. E. Alp, and F. G. Parak. Vibrational dynamics of myoglobin determined by the phonon-assisted Mössbauer effect. *Phys. Rev. E.*, 65:061916, 2002.
19. B. M. Leu et al. Vibrational dynamics of biological molecules: Multi-quantum contributions. *J. Phys. Chem. Solids*, 66:2250–6, 2005.
20. A. E. Garcia. Large-amplitude nonlinear motions in proteins. *Phys. Rv. Lett.*, 68:2696–9, 1992.
21. A. Kitao, S. Hayward, and N. Go. Energy landscape of a native protein: Jumping-among-minima model. *PROTEINS: Structure, Function, and Genetics*, 33:496–517, 1998.
22. O. M. Becker, A. D. MacKerell, Jr., B. Roux, and M. Watanabe, editors. *Computational Biochemistry and Biophysics.* Marcel Decker, Inc., New York, 2001.
23. A. Ansari, J. Berendzen, S. F. Bowne, H. Frauenfelder, I. E. T. Iben, T. B. Sauke, E. Shyamsunder, and R. D. Young. Protein states and protein quakes. *Proc. Natl. Acad. Sci. USA*, 82:5000–4, 1985.
24. T.-Y. Teng, V. Srajer, and K. Moffat. Photolysis-induced structural changes in single crystals of carbonmonoxy myoglobin at 40 K. *Nature Structural Biology*, 1:701–5, 1994.
25. D. Bourgeois, B Vallone, A. Arcovito, G. Sciara, F. Schotte, P. A. Anfinrud, and M. Brunori. Extended subnanosecond structural dynamics of myoglobin revealed by Laue crystallography. *Proc. Natl. Acad. Sci. USA*, 103:4924–9, 2006.
26. R. J. Miller. Energetics and dynamics of deterministic protein motion. *Acc. Chem. Res.*, 27:145–50, 1994.
27. A. M. Nagy, V. Raicu, and R. J. D. Miller. Nonlinear optical studies of heme protein dynamics: Implications for proteins as hybrid states of matter. *Biochimica et Biophysica Acta*, 1749:148–72, 2005.
28. A. Xie, L. Kelemen, J. Hendriks, B. J. White, K. J. Hellingwerf, and W. D. Hoff. Formation of a new buried charge drives a large-amplitude protein quake in photoreceptor activation. *Biochemistry*, 40:1510–7, 2001.
29. K. Ito and M. Sasai. Dynamical transition and protein quake in photoactive yellow protein. *Proc. Natl. Acad. Sci. USA*, 101:14736–41, 2004.
30. J. N. Onuchic and P. G. Wolynes. Energy landscapes, glass transitions, and chemical reaction dynamics in biomolecular or solvent environment. *J. Chem. Phys.*, 93:2218–24, 1993.
31. S. Dellerue, A.-J. Petrescu, J. C. Smith, and M.-C. Bellissent-Funel. Radially softening diffusive motions in globular protein. *Biophys. J.*, 81:1666–76, 2001.
32. N. Miyashita, J. N. Onuchic, and P. G. Wolynes. Nonlinear elasticity, proteinquakes, and the energy landscapes of functional transitions in proteins. *Proc. Natl. Acad. Sci. USA*, 100:12570–5, 2003.
33. O. Miyashita, P. G. Wolynes, and J. N. Onuchic. Simple energy landscape model for the kinetics of functional transitions in proteins. *J. Phys. Chem. B*, 109:1959–69, 2005.

34. K. Okazaki, N. Koga, S. Takada, J. N. Onuchic, and P. G. Wolynes. Multiple-basin energy landscapes for large-amplitude conformational motions of proteins: Structure-based molecular dynamics silmulations. *Proc. Natl. Acad. Sci. USA*, 103:11844–9, 2006.
35. N. Agmon and J. J. Hopfield. Transient kinetics of chemical reactions with bounded diffusion perpendicular to the reaction coordinate: Intramolecular processes with slow conformational changes. *J. Chem. Phys.*, 78:6947–59, 1983.
36. N. Agmon and J. J. Hopfield. CO binding to heme proteins: A model for barrier height distributions and slow conformational changes. *J. Chem. Phys.*, 79:2042–53, 1983.
37. M. Lim, T. A. Jackson, and P. A. Anfinrud. Nonexponential protein relaxation: Dynamics of conformational change in myoglobin. *Proc. Natl. Acad. Sci. USA*, 90:5801–4, 1993.
38. D. Beece, L. Eisenstein, H. Frauenfelder, D. Good, M. C. Marden, L. Reinisch, A. H. Reynolds, L. B. Sorenson, and K. T. Yue. Solvent viscosity and protein dynamics. *Biochemistry*, 19:5147, 1980.
39. W. Doster. Viscosity scaling and protein dynamics. *Biophys. Chem.*, 17:97–103, 1983.
40. J. T. Sage, K. T. Shoemaker, and P. M. Champion. Solvent-dependent structure and dynamics in myoglobin. *J. Phys. Chem.*, 99:3394–3405, 1995.
41. T. Kleinert, W. Doster, H. Leyser, W. Petry, V. Schwarz, and M. Settles. Solvent composition and viscosity effects on the kinetics of CO binding to horse myoglobin. *Biochemistry*, 37:717–33, 1998.
42. Y. Shibata, A. Kurita, and T. Kushida. Solvent effects on conformational dynamics of Zn-substituted myoglobin observed by time-resolved hole-burning spectroscopy. *Biochemistry*, 38:1789–1801, 1999.
43. A. Ansari. Langevin modes analysis of myoglobin. *J. Chem. Phys.*, 110:1774–80, 1999.
44. P. W. Fenimore, H. Frauenfelder, B. H. McMahon, and F. G. Parak. Slaving: Solvent fluctuations dominate protein dynamics and functions. *Proc. Natl. Acad. Sci. USA*, 99:16047–51, 2002.
45. P. W. Fenimore, H. Frauenfelder, B. H. McMahon, and R. D. Young. Proteins are paradigms of stochastic complexity. *Physica A*, 351:1–13, 2005.
46. W. Doster, S. Cusack, and W. Petry. Dynamic instability of liquidlike motions in a globular protein observed by inelastic neutron scattering. *Phys. Rev. Lett.*, 65:1080–3, 1990.
47. K. Hinsen, A. J. Petrescu, S. Dellerue, M. C. Bellissent-Funel, and G. R. Keller. Liquid-like and solid-like motions in proteins. *J. Mol. Liquids*, 98-99:381–98, 2002.
48. H. Frauenfelder, P. W. Fenimore, G. Chen, and B. H. McMahon. Protein folding is slaved to solvent motions. *Proc. Natl. Acad. Sci. USA*, 103:15469–72, 2006.
49. V. Lubchenko, P. G. Wolynes, and H. Frauenfelder. Mosaic energy landscapes of liquids and the control of protein conformational dynamics by glass-forming solvents. *J. Phys. Chem. B*, 109:7488–99, 2005.
50. P. W. Fenimore, H. Frauenfelder, B. H. McMahon, and R. D. Young. Bulk-solvent and hydration-shell fluctuations, similar to α- and β-fluctuations in glasses, control protein motions and functions. *Proc. Natl. Acad. Sci. USA*, 101:14408–13, 2004.
51. R. H. Austin, K. W. Beeson, L. Eisenstein, H. Frauenfelder, and I. C. Gunsalus. Dynamics of ligand binding to myoglobin. *Biochemistry*, 14:5355–73, 1975.

52. J. A. Rupley and G. Careri. Protein hydration and function. *Adv. Protein Chem.*, 41:37–172, 1991.
53. B. Halle. Protein hydration dynamics in solution: A critical survey. *Phil. Trans. R. Soc. Lond. B*, 359:1207–24, 2004.
54. Y. Levy and J. N. Onuchic. Water mediation in protein folding and molecular recognition. *Ann. Rev. Biophys. Biomol. Struct.*, 35:398–415, 2006.
55. V. Helms. Protein dynamics tightly connected to the dynamics of surrounding and internal water molecules. *ChemPhysChem.*, 8:23–33, 2007.
56. H. Frauenfelder, G. Chen, J. Berendzen, P. W. Fenimore, H. Jansson, B. H. McMahon, I. R. Stroe, J. Swenson, and R. D. Young. A unified model of protein dynamics. *Proc. Natl. Acad. Sci. USA*, 106: 5129–5134, 2009.
57. J. Swenson, H. Jansson, and R. Bergman. Relaxation processes in supercooled confined water and implications for protein dynamics. *Phys. Rev. Lett.*, 96:247802, 1992.
58. F. Parak, E. W. Knapp, and D. Kucheida. Protein dynamics–Mössbauer spectroscopy on deoxymyoglobin crystals. *J. Mol. Bio.*, 161:177–94, 1982.
59. W. Doster, S. Cusack, and W. Petry. Dynamical transition of myoglobin revealed by inelastic neutron scattering. *Nature*, 337:754–6, 1989.
60. V. I. Goldanski and R. H. Herber, editors. *Chemical Applications of Mössbauer Spectroscopy.* Academic Press, New York, 1968.
61. K. Achterhold and F. G. Parak. Protein dynamics: Determination of anisotropic vibrations at the heme iron of myoglobin. *J. Phys. : Condens Matter*, 15:S1603–92, 2003.
62. S.-H. Chong et al. Dynamical transition in myoglobin in a crystal: Comparative studies of X-ray crystallography and Mössbauer spectroscopy. *Eur. Biophys. J.*, 30:319–29, 2001.
63. J. Gafert, H. Pschierer, and J. Friedrich. Proteins and glasses: A relaxation study in the millikelvin range. *Phys. Rev. Lett.*, 74:3704–7, 1995.
64. W. E. Moerner. *Persistent Spectral Hole-Burning: Science and Applications.* Springer, Berln, 1988.
65. P. Schellenberg and J. Friedrich. Optical spectroscopy and disorder phenomena in polymers, proteins, and glasses. In R. Richert and A. Blumen, editors, *Disorder Effects on Relaxational Processes.* Springer, Berlin, 1994.
66. A. Kurita, Y. Shibata, and T. Kushida. Two-level systems in myoglobin probed by non-Lorentzian hole broadening in a termperature-cycling experiment. *Phys. Rev. Lett.*, 74:4349–52, 1995.
67. V. V. Ponkratov, J. Friedrich, J. M. Vanderkooi, A. L. Burin, and Y. A. Berlin. Physics of proteins at low temperatures. *J. Low Temperature Phys.*, 137:289–317, 2004.
68. V. V. Ponkratov, J. Friedrich, and J. M. Vanderkooi. Hole burning experiments with proteins: Relaxations, fluctuations, and glass-like features. *J. Non-Cryst. Solids*, 352:4379–86, 2006.
69. C. Hofmann, H. Michel, T. J. Aartsma, K. D. Fritsch, and J. Friedrich. Direct observation of tiers in the energy landscape of a chromoprotein. *Proc. Natl. Acad. Sci. USA*, 100:15534–8, 2003.
70. J. Schlichter, V. V. Ponkratov, and J. Friedrich. Structural flucuations and aging process in deeply frozen proteins. *Low Temperature Physics*, 29:795–800, 2003.

71. K. Fritsch and J. Friedrich. Spectral diffusion experiments on a myoglobin-like protein: Statistical and individual features of its energy landscape. *Physica D*, 107:218–24, 1997.

16

Protein Quantum Dynamics? (R. H. Austin[1])

The puzzling and possibly profound effect we want to examine in this chapter is one model for how chemical free energy may get trapped in a protein rather than flowing ergodically into the thermodynamic limit of equal occupation of all the degrees of freedom.

Failure of equipartition into all accessible states can be achieved in at least two ways: (a) kinetically the rate into some accessible state can be so small compared to others that the relaxation rate to equipartition can be very slow. We would expect this to happen in a system with a rough free energy landscape, a protein being an example of such a system [1], which we have discussed extensively in previous chapters in this book. Such a kinetic trapping of free energy is both an incoherent process and metastable: Eventually the free energy will flow into all the possible states, maximizing the entropy of the system. The flow of free energy on the free energy surface in a linear system is a "passive" process in that the surface itself is unchanged by the presence of the occupation of energy levels that constitute the surface. (b) Coherent self-trapping in anharmonic systems [2]. In this process, excitation of a system to an excited quantum mechanical state results in the time-dependent alteration of the free energy surface itself: The free energy shape is changed by the presence of the excitation. This is inherently a nonlinear process and can result in extremely long lifetimes of the self-trapped state if the coherent localization that drives the localization does not disperse. The most famous example of this kind of coherent self-trap is the *Davydov soliton* [3].

The experimental biological physics world originally got interested in the Davydov soliton model for coherent vibrational energy flow in proteins because of the intriguing infrared temperature dependence of molecular crystals of the amino acid analog acetanilide [4, 5]. Acetanilide, a metabolic precursor to the commercial pain reliever Tylenol, has a far simpler structure than a globular protein, but since it is an amino acid analog and forms a molecular crystal there was hope that there would be "new physics" happening in such

[1] Department of Physics, Princeton University, Princeton, NJ 08544, USA.

H. Frauenfelder, *The Physics of Proteins*, Biological and Medical Physics, Biomedical Engineering, DOI 10.1007/978-1-4419-1044-8_16,

a molecular crystal that **might** map over to protein dynamics of biological relevance. Acetanilide shows a strong temperature-dependent splitting of the amide-I band, and it was conjectured by Careri et al. [4] that the new, long-wavelength shifted band that appeared below 150 K was a Davydov soliton self-trapped state. This led Alywn Scott to conjecture that indeed the interactions between the acetanilide molecules in the molecular crystal formed by acetanilide had the right parameters to form a self-trapped Davydov solition at low temperatures upon absorbing an IR photon at around 6 microns, statements we will try to explain shortly. It was certainly intriguing to hope that maybe all the complex quantum mechanical machinery of Davydov might have some relevance to biology and help explain one of the basic mysteries in biology: energy flow in a protein in the nonoptical energy scale below 1 eV.

16.1 Self-Trapped States, Classical and Quantum Mechanical

The simplest (classical) example of self-trapping of vibrational energy is two anharmonic oscillators (pendula) coupled by a linear torsional spring. The Hamiltonian for these two pendula is:

$$H(\theta_1, \theta_2) = -\mathrm{mgL}\cos(\theta_1) - \mathrm{mgL}\cos(\theta_2) - \frac{1}{2}\kappa(\theta_1 - \theta_2)^2 \qquad (16.1)$$

where κ is the torsional spring constant of the coupling spring. When $\theta_1, \theta_2 <<$ 1 and mgL $>> \kappa$ the problem reduces to a coupled harmonic oscillator with split normal modes $\theta_{\mathrm{optical}} = \theta_1 - \theta_2$ and $\theta_{\mathrm{acoustic}} = \theta_1 + \theta_2$ with frequency splitting $\Delta\omega = \frac{\kappa}{\mathrm{mgL}}$. The acoustic mode where both pendula swing with phase difference $\Delta\theta = 0$ lies lower in energy than the optical mode. There is no energy trapping here: Energy flows back and forth between the two modes, and on average each mode has the same energy. Things are different when we can no longer make the small angle approximation. When that happens the frequency ω_o of the oscillators becomes a function of the angle of the maximum excursion of the oscillator θ_o and the driving phase $\theta_1(t)$ of one the oscillators now averages to 0 torque over time if the other oscillator is initially at rest. In that case the energy in the large angle excursion oscillator is "trapped" and cannot transfer much energy to the other oscillator, in spite of the coupling between the oscillators. Viewed from a spectroscopic perspective the frequency of the large angle oscillator is shifted out of resonance (to lower frequencies) with the small angle oscillator as the energy of the oscillator is driven higher, as the two frequencies shift out of resonance, the energy is trapped in the vibrationally excited state that cannot transfer its energy to the ground state if there is no dissipation (dephasing) in the system.

The simplest quantum mechanical analogy to the dynamics of a weakly coupled set of anharmonic oscillators is a 2-level system, which, of course, is extremely anharmonic (there are only 2 levels, not an infinite ladder of them). Let the energy splitting between the ground– and excited-states be $\hbar\omega_o$, and let there be an incident driving electric field $E_o \exp(i\omega t)$ driving transitions between the ground and excited states, so that the occupation levels $p_1(t)$ and $p_2(t)$ change with time. If there is an electric dipole matrix element $p = e\,[\mathbf{r} \cdot \wedge \mathbf{n}]$ connecting the two states, the occupation of the excited $p_2(t)$ varies as [6]:

$$p_2(t) = \frac{1}{2}\,[1 - \cos(\chi t)] \tag{16.2}$$

where χ is the Rabi frequency, given by:

$$\chi = \frac{pE_o}{\hbar} \tag{16.3}$$

when the incoming photon energy is exactly resonant with the energy splitting $\hbar\omega_o$. If the photon is off resonance, let $\Delta = \hbar\omega_o - \hbar\omega$ be the energy difference between the splitting of the ground and excited states and the energy of the incident photons. When $\Delta = 0$ there is a full periodic inversion of the ground- and excited-state populations at the Rabi frequency χ; when the incident field frequency ω is detuned from splitting frequency ω_o the amplitude of $p_2(t)$ becomes smaller, which is also what happens in our simple 2-coupled anharmonic pendula when the amplitude dependence of the period detunes one pendulum from another. The off-resonance driving frequency results in a reduced transfer of energy to the other system. The classical and quantum systems have the same basic behavior, caused by the nonlinearity of the system and phase difference between the driving force and the response of the excited state.

The localization of energy on one 2-level system due to detuning from another 2-level system isn't really energy trapping because the 2-level system splitting is not a function of the driving field, but it is an example of quantum mechanical coupling between two discrete anharmonic systems. The Davydov soliton that Alwyn Scott worked so hard on [7] comes out of exciton and polaron theory, and these theories come out of semiconductor physics and the concept of delocalized identical particles: band theory. A semiconductor is a material with a filled valence band and an empty conduction band with bandgap $\hbar\omega_o$ of the order of 1 eV or less, so that typical semiconductors appear highly absorbing in the visible range due to allowed optical transitions. Unfortunately, most biological polymers are insulators: The band gap is at least 5 eV or so, and they are thus transparent in the visible range. One basic problem is then to what extent can the band theories developed for localized electronic states just below the conduction band edge be applied to systems with far bigger band gaps than semiconductors? That there are fundamental problems with applying concepts that work for small band gap semiconductors to biological polymers is unfortunately a very long and sad story in biological physics. It is sad because it would be great fun and exciting to use simple fundamental quantum mechanical concepts of delocalization and actually apply them to protein polymers and (even better) DNA polymers energy transport:

we could have very low-resistance charge transport, even superconductivity in biological polymers [8], and it goes on from there to even greater excesses of quantum mechanical splendor [9, 10]. But there are both experimental failures to observe these effects [11], and a fundamental theoretical problem: Electronic states are in fact highly localized in biopolymers and not delocalized.

There is a reason for this localization of charge where one would hope that somehow the interactions between monomers (nucleic acid bases or amino acids) would conspire to give delocalization: the Mott insulator transition. The failure of simple band theory to properly predict the insulating nature of materials that naively would be conductors is a very old problem first addressed by Mott in his Nobel Prize-winning work [12]. A Mott insulator is a metal that is an insulator: It is an insulator because the Coulomb repulsion interaction U between electrons in the material is greater than the exchange integral W, which delocalizes identical particles and hence lowers their kinetic energy. Roughly speaking, the Coulomb repulsion energy $U(a)$ for two electrons situated on a simple linear lattice of spacing a is given by:

$$U(a) \sim \frac{e^2}{a} \exp\left(-a/k_s\right) \tag{16.4}$$

where $1/k_s$ is the the Thomas-Fermi screening length for an electron gas of density n_o and Fermi energy ϵ_F:

$$\frac{1}{k_s^2} = \frac{\epsilon_F}{6\pi\rho_o e^2} \sim \frac{a_B}{\rho_o^{1/3}} \tag{16.5}$$

where a_B is the Bohr radius. Because ρ_o for biological polymers is quite small, on the order of 10^{19} cm^{-3}, the screening length is quite large, on the order of 1 nm, more than 10 times the screening length of electrons in a typical metal. It is hard to go much further here quantitatively, other than to point out that because of the low density of charge carriers in a typical biopolymer, the Coulomb energy is so large as to localize the charges, and much of the apparatus of small band gap semiconductor physics is simply not applicable. If you open a typical textbook on biopolymer infrared spectroscopy, such as Mantsch and Chapman [13], you will find no talk of semiconductors or band gaps; all the discussion is purely molecular. Most of the calculations of Davydov solition dynamics use explicitly semiconductor delocalized wave functions and terminology, creating not only a gap in understanding of the difference between the dynamics of a Mott insulator and a semiconductor, but also a complete disconnect between the community of biophysicists who read Mantsch and Chapman and those who read Kittel [14]; this is very unfortunate.

The two key ingredients in the Davydov scheme for energy trapping are anharmonicity in the mid-IR spectral region and "long" dephasing times to maintain the same phase coherence needed in the simple 2-level quantum system. Basically, the phase coherence must last as long as the energy is to be trapped , so one would guess that phase coherence times at least on the order of 10 ps are necessary. Does a globular protein have any unusual properties in these two sectors? The anharmonicity in soliton theory is typically expressed by the quantity χ:

$$\chi = \frac{dE_o}{dR} \tag{16.6}$$

where E_o is the vibrational energy of the amide I band (about 0.2 eV) and R is the length of the hydrogen bond that couples to the C=O stretch, which is the primary origin of the amide I transition. It really isn't right, but in analogy to our anharmonic coupled oscillator you could view $E_o/\hbar$ as the small angle frequency of anharmonic oscillator ω_o:

$$\omega_o = \sqrt{\frac{g \cos\theta}{L}}_{lim\theta=0} \tag{16.7}$$

and χ is:

$$\chi = \frac{dE_o}{d\theta} \sim E_o\frac{\theta}{2} \tag{16.8}$$

χ represents a detuning of the oscillator with angle and the origin of localization of the energy in the excited oscillator.

The connection between the dependence of the anharmonicity of the system on the hydrogen bond length R and the IR spectroscopy is mapped by the temperature dependence of the red shifted band, which is determined by the binding energy of the soliton in the theory expanded by Scott. The binding energy E_b of the soliton is given by:

$$E_b = \frac{\chi^2}{2K} - 2J \tag{16.9}$$

where K is the spring constant of the C=O bond (about 13 N/m) and J is the electromagnetic strength of the coupling between adjacent amide I oscillators (about 1–2 cm^{-1}) [5]. Note that J plays the role of the torsional spring in our mechanical toy, K plays the role of gravity g, and χ plays the role of the nonlinear angular dependence $\cos\theta$ of the restoring force on the pendulum bob.

16.2 What Are the Experiments?

It is fairly straightforward using IR spectroscopy to make rough estimates of the anharmonicity of the α-helix amide-I bands assuming that the temperature dependence comes from this self-trapping model, as discussed in [15]. A truly harmonic (linear response) transition has no temperature dependence since all levels are equally spaced in energy. An anharmonic transition will show a temperature dependence because with increasing temperature (assuming that the potential surface becomes softer with increasing energy) there should be a red shift. While the electronic transitions of semiconductors have enormous red shifts with increasing temperature, giving rise to the phenomena of thermal runaway, proteins are in fact rather hard harmonic oscillators within the vibrational manifold, showing a rather weak temperature dependence. Figure 16.1 shows a sketch of the temperature dependence of the amide I vibrational spectral region of the protein myoglobin from 5 K to 300 K. There is a clear anharmonicity in the amide I band (and much more in the amide II band), and a new red-shifted band appears rather similar to the acetanilide low-temperature band. If one assumes that the red-shifted band is some sort of a trapped state as Careri and his colleagues did, then the binding energy of the state is approximately 20 cm^{-1}, which translates into a χ of approximately 100 pN, of the range needed for soliton stability as discussed by Scott [16]. Note that the temperature dependence of the amide I band is such that the amide I band on the blue side of the spectrum has almost no temperature dependence. Note also that this *thermal* temperature dependence is different from the nonlinear *optical* dependence expected from driving excited states, as we discussed in Eq.(29.2).

Mb (a highly α-helical-containing protein) thus has a long wavelength shifted band that has a similar temperature dependence and binding energy (with the soliton model) as does acetanilide. If Davydov solitons truly exist in acetanilide, and if they are biologically important for self-trapping of vibrational energy in biology, then it would appear that the simple protein myoglobin would be the place to look for them. Alas, it would appear that no such long-lived states actually exist in proteins; we refer the student to the literature to judge for him- or herself [17, 18]. We can give one plot of the pump-probe dynamics in Mb, which represents the best efforts of Xie et al.

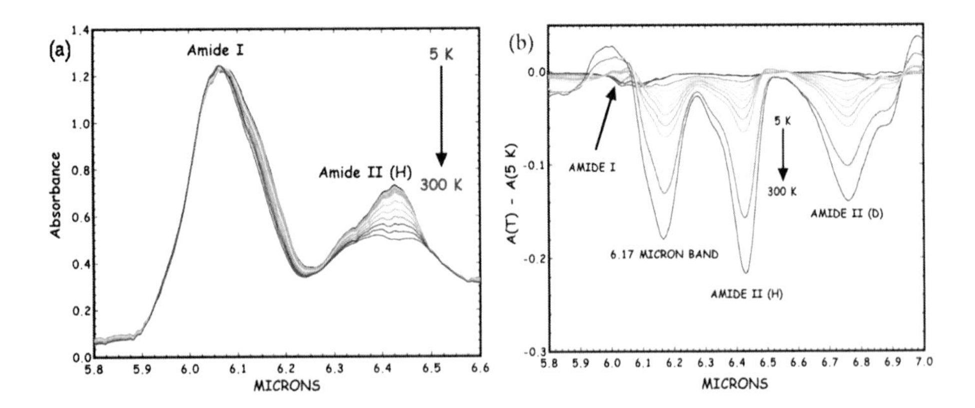

Fig. 16.1. (a) The infrared absorption spectrum of sperm whale myoglobin as a function of temperature from 5 to 300 K from 5.8 to 6.6 microns. (b) The band at 6.17 microns decreases with absorbance as the temperature increases. The difference of the amide I and amide II regions of Mb (5 K base) as a function of temperature.

[18] in Fig. 16.2. Although there is evidence for the conformational distribution of protein states, there is no spectacular long-lived (100s of picoseconds) state.

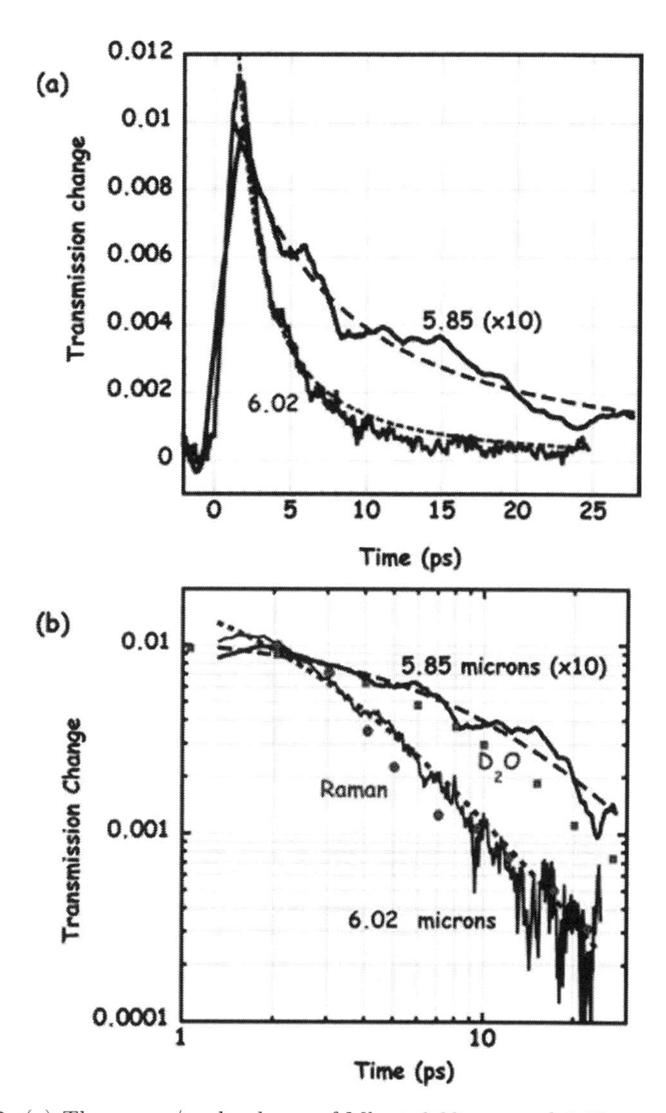

Fig. 16.2. (a) The pump/probe decay of Mb at 6.02 mm and 5.85-mm. The 5.85-mm decay amplitude has been multiplied by 10. The power law curve fits to the decay of the pump/probe signal are shown as dashed lines for the two wavelengths. The pulse width of FELIX was set to be as long as possible for these experiments, 1.5 ps or 10 cm^{-1} in linewidth. (b) Log amplitude versus log time plot of the data in (a). Taken from [18].

The far-infrared dynamics are less explored and there are some tantalizing results out in the the THz regime [19]-[21] that hint of very long lifetimes and functional coherent effects, but this is not the Davydov domain and involves long-range collective modes. An example of the long-lifetime dynamics of pump-probe experiments is given in Fig. 16.3 for the protein bacteriorhodopsin [21].

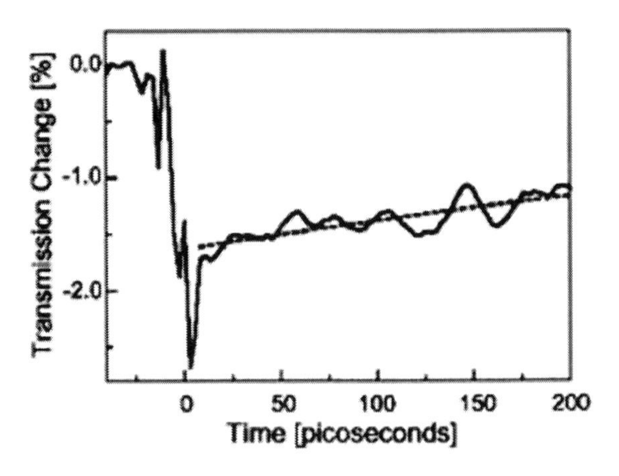

Fig. 16.3. Pump-probe response of bR at 87 microns. The data are shown as the percentage transmission change. The spot size of the pump beam was 1000 microns and the pump energy per micropulse was 1 μJ. The dotted line is a single exponential fit to the data, with a time constant of 500 $\pm$ 100 ps. Taken from [21].

The area of quantum functional effects in protein dynamics is not closed yet. It is important in science to test powerful ideas developed by the theorists. Sometimes, however improbably, the ideas are correct: Special and general relativity, the neutrino, the breaking of parity conservation in the weak interaction followed by CP violation come immediately to mind. Sometimes Mother Nature is not so cooperative; there seem to be no magnetic monopoles, no one has seen a proton decay yet, and maybe we will never find the axion. Davydov started the idea that vibrational energy could be self-trapped in proteins, and Alywn Scott devoted a good part of his life to exploring that idea. Scott catalyzed a huge theoretical effort that still continues to this day, not so many experimentalists felt inclined to test all the calculations with experiments. Perhaps someday we will see what Scott and Davydov were driving at as we gain ever deeper understanding of the possible functional and nonrivial connections between quantum mechanics and biology.

References

1. R. H. Austin, K. W. Beeson, L. Eisenstein, H. Frauenfelder, and I. C. Gunsalus. Dynamics of ligand-binding to myoglobin. *Biochem.*, 14(24):5355–73, 1975.
2. Guglielmo Lanzani, Sandro De Silvestri, and Giulio Cerullo. *Coherent Vibrational Dynamics.* CRC Press, Boca Raton, FL, 2007.
3. A. S. Davydov. Solitons and energy-transfer along protein molecules. *J. Theoret. Bio.*, 66(2):377–87, 1977.
4. G. Careri, U. Buontempo, F. Carta, E. Gratton, and A. C. Scott. Infrared-absorption in acetanilide by solitons. *Phys. Rev. Lett.*, 51(4):304–7, 1983.
5. G. Careri, U. Buontempo, F. Galluzzi, A. C. Scott, E. Gratton, and E. Shyamsunder. Spectroscopic evidence for davydov-like solitons in acetanilide. *Phys. Rev. B*, 30(8):4689–4702, 1984.
6. P. W. Milonni and J. H. Eberley. *Lasers.* John Wiley and Sons, New York, 1988.
7. A. Scott. Davydov's soliton. *Phys. Rep.*, 217:1–67, 1982.
8. T. Kawakami, Y. Kitagawa, F. Matsuoka, Y. Yamashita, H. Isobe, H. Nagao, and K. Yamaguchi. Possibilities of molecular magnetic metals and high T-c superconductors in field effect transistor configurations. *Int. J. Quant. Chem.*, 85(4–5):619–35, 2001.
9. P. C. W. Davies. Does quantum mechanics play a non-trivial role in life? *Biosys.*, 78(1–3):69–79, 2004.
10. R. Penrose. The emperors new mind–concerning computers, minds, and the laws of physics. *Behav. and Brain Sci.*, 13(4):643–54, 1990.
11. Y. Zhang, R. H. Austin, J. Kraeft, E. C. Cox, and N. P. Ong. Insulating behavior of λ-DNA on the micron scale. *Phys. Rev. Lett.*, 89(19):198102–5, 2002.
12. N. F. Mott. Metal-insulator transitions. *Contemp. Phys.*, 14(5):401–13, 1973.
13. H. H. Mantsch and D. Chapman. *Infrared Spectroscopy of Biomolecules.* Wiley-Liss, New York, 1986.
14. C. Kittel. *Introduction to Solid State Physics*, 7th Edition. Wiley and Sons, New York, 2007.
15. R. H. Austin, A. H. Xie, L. van der Meer, B. Redlich, P. A. Lindgard, H. Frauenfelder, and D. Fu. Picosecond thermometer in the amide I band of myoglobin. *Phys. Rev. Lett.*, 94(12):12810–4, 2005.
16. J. C. Eilbeck, P. S. Lomdahl, and A. C. Scott. Soliton structure in crystalline acetanilide. *Phys. Rev, B*, 30(8):4703–12, 1984.

17. W. Fann, L. Rothberg, M. Roberson, S. Benson, J. Madey, S. Etemad, and R. Austin. Dynamical test of Davydov-type solitons in acetanilide using a picosecond free-electron laser. *Phys. Rev. Lett.*, 64(5):607–10, 1990.
18. A. H. Xie, L. van der Meer, W. Hoff, and R. H. Austin. Long-lived amide I vibrational modes in myoglobin. *Phys. Rev. Lett.*, 84(23):5435–8, 2000.
19. R. H. Austin, M. K. Hong, C. Moser, and J. Plombon. Far-infrared perturbation of electron-tunneling in reaction centers. *Chem. Phys.*, 158(2–3):473–86, 1991.
20. R. H. Austin, M. W. Roberson, and P. Mansky. Far-infrared perturbation of reaction-rates in myoglobin at low-temperatures. *Phys. Rev. Lett.*, 62(16):1912–15, 1989.
21. A. Xie, L. van der Meer, and R. H. Austin. Excited-state lifetimes of far-infrared collective modes in proteins. *J. Bio. Phys.*, 28(2):147–54, 2002.

17

Creative Homework: Dynamics and Function

Biology textbooks usually do not go into details of how biomolecules work. They restrict the treatment to providing the static average structure and a phenomenological description of the function. Here we are interested in a fundamental understanding of proteins that may be valid not just for a specific case, but for essentially all proteins. We assert that the energy landscape and the fluctuation and relaxation processes resulting from the transitions among conformational substates are the concepts governing the function of proteins and nucleic acids. However, while some notable exceptions exist [1, 2], most textbooks neglect the crucial role of dynamics. Here, instead of treating some specific examples where dynamics is essential for function and the energy landscape is the concept needed for understanding the dynamics and function, we ask the reader to do the work. We introduce some proteins and their function and list some relevant references. The task of the reader is to select one or two examples and do, for example, some or all of the following tasks:

1. Find more relevant papers.
2. Understand the protein and the related biological processes.
3. Understand the experimental techniques and the data evaluation used.
4. Are the proteins involved homogeneous or inhomogeneous? If not known, which experiments would you perform to decide?
5. What is known about the energy landscape of the proteins involved in the biological process? If nothing is known, try to use the existing data to construct the landscape. If no relevant data exist, how would you get them?
6. Are thermal fluctuations and/or relaxation processes important for the biological process? Describe the dynamic processes and how they fit into the landscape.
7. Write a review.

These questions are not all inclusive. As you delve deeper into the experimental, computational, and theoretical aspects, more questions will arise.

The proteins introduced in the following sections are only a small subset of the proteins essential for life. It is not clear yet if the concepts introduced

H. Frauenfelder, *The Physics of Proteins*, Biological and Medical Physics, Biomedical Engineering, DOI 10.1007/978-1-4419-1044-8_17,

here are relevant for most or all of them. But once some proteins are quantitatively understood, extension to others may be easier, but much experimental, theoretical, and computational work remains to be done.

Another remark concerns the references. The number of papers pertinent to each of the topics selected in the following sections is extremely large. The choice is therefore largely arbitrary, and the references given here should be considered a guide to the vast literature and not as an indication that they are the most important papers.

17.1 Myoglobin as Prototype of Complexity

Small ligands such as O_2, CO, or NO bind reversibly to proteins such as myoglobin (Mb) or hemoglobin as sketched in Chapter 4. As an example consider Mb [3]. In 1972, the function of Mb was described as follows: "... a simple one-site molecule that undergoes no conformational change on ligand binding, has no Bohr effect, and of course, no cooperativity, and whose reaction with CO under all conditions can be described as a simple one-step process: Mb + CO $\Longleftrightarrow$ MbCO." Experiments performed a few years later, using a wide temperature and time range showed, however, that Mb is not simple, but that it is complex and that the earlier characterization was wrong [4]. In fact, even today, Mb is not fully understood. Complexity does not appear only in Mb, but it is a general property of biomolecules [5]. In the following sections, we briefly outline some of the aspects of this complexity and provide a few selected references. The reader should select one or two topics, try to understand them, summarize the results, and suggest new avenues and experiments. The following topics and references are incomplete; it will be a challenge to find more properties of the many-faceted Mb protein. No exhaustive review on Mb and its functions exists at present, but enough references are listed that give some insight into the complexity, dynamics, and function of this protein.

17.1.1 Kinetics and the Reaction Landscape

The low-temperature experiments provided evidence for the existence of an energy landscape and the complexity of Mb by observing the CO kinetics of CO binding [4, 6, 7]. The experiment is, in principle, simple. Mb with and without bound CO has a different optical spectrum. The bond between the CO and the heme iron in Mb is broken by a laser flash and the CO moves away from the heme iron. It then either rebinds directly (geminately) or moves farther away and rebinds later. The subsequent rebinding is followed as a function of time and temperature by recording, for instance, the intensity of a spectral line characteristic of the bound state. Crucial to understanding the result is the proper data evaluation [8]–[10] and using an extended time range with averaging over exponentially increasing time intervals. While early experiments [4] covered the time range from microseconds (μs) to about 100 s, modern techniques have moved into picoseconds (ps) [11] and femtoseconds (fs) (e.g., [12]) range. Photodissociation is also studied with a technique called TDS (temperature-derivative spectroscopy) [13]. After a light flash breaks the Fe–CO bond at a low temperature, say 3 K, the temperature is ramped up while the infrared spectrum is recorded. Information about rebinding and moving away from the heme iron after the bond breaking is obtained from the IR spectrum as a function of time (t) and temperature (T). An example is given in [14]. The experiments show that CO binding is not a one-step process as initially surmised, but involves a number of states in which the CO resides for some time [4]. Exit and entry of a ligand thus do not occur by passage over a single barrier, but by following a reaction landscape.

17.1.2 Pathways

Figure 4.11 and similar pictures in any biochemistry text show that the binding site for CO and O_2 is inside the protein, not on the outside. The question therefore becomes to find the pathways between the ligand binding site at the heme iron and the outside. The initial data [4] suggested a linear sequence B $\Longleftrightarrow$ C $\Longleftrightarrow$ D $\Longleftrightarrow$ S. B was assumed to be in the heme pocket close to the Fe, C and D somewhere in the heme pocket and S in the solvent. The motion from the pocket to the solvent raised a problem. X-ray diffraction did not show a channel through which CO or O_2 could diffuse. Entry and exit therefore must involve fluctuations ("relaxation makes access to the active center easier" [4]). Molecular dynamics calculations supported this conclusion [15]. Deeper insight into the pathways of ligands through Mb came from a number of directions. Structure determination under xenon gas pressure revealed the existence of four internal cavities, denoted Xe_1 to Xe_4 [16, 17]. Time-resolved X-ray structures show that these cavities act as docking sites for some time after photodissociation and give information about the corresponding structural changes [18]–[20]. Mutants selectively can block some pathways and thus can help to establish a pathway map (e.g., [14, 21]–[24]). Ligand binding studies as a function of Xe pressure provide additional information [25]. These experimental investigations suggest that there is not just one linear pathway, as initially assumed, but that there must be multiple paths. Computer simulations support this view [26, 27]. Despite these efforts, no unique map exists at present.

17.1.3 Dynamics

As discussed in Chapter 15, proteins are dynamic systems and constantly fluctuate. These fluctuations permit the transit of the ligand through the protein and then exit into the solvent. Two types of fluctuation processes control the dynamics, α and β (Chapter 14). The α fluctuations permit entry and exit from the solvent [6, 28]. The rate coefficient for the exit is, however, slower than the rate coefficient of the α fluctuations. The slowing is interpreted as a random walk in the energy landscape. The β fluctuations appear to control the motions from cavity to cavity [29].

17.1.4 Hydration

Proteins are surrounded by the hydration shell, one or two layers of water with properties different from bulk water [30]–[33]. Dry proteins do not work. While a great deal is already known about how the hydration shell affects the workings of proteins, much remains to be explored, especially the connection between the hydration shell and the β fluctuations. The interplay of the α and β fluctuations and the effect of hydration on this interaction are also topics of current efforts. See Chapter 24 for an introduction to the physics of water as a start.

17.2 Neuroglobin – A Heme Protein in the Brain (G. U. Nienhaus[1])

Neuroglobin (Ngb) is a small heme protein expressed in the central and peripheral nervous systems of vertebrates that plays an important role in the protection of brain neurons when oxygen supply is insufficient [34]. The primary sequences among neuroglobins of different species exhibit significantly higher agreement than Mbs and Hbs, which may indicate a specific and likely important physiological function [35]. The detailed physiological function is still under debate. Suggestions include oxygen storage and delivery, scavenging of NO and/or reactive oxygen species, oxygen sensing, and signal transduction. In recent years, the molecular structures of Ngb with CO bound to the heme iron and without an exogenous ligand have been elucidated, and interesting structural changes have been noticed after ligand binding. Moreover, equilibrium and kinetic properties of the reactions with ligands have been examined in great detail. Here we briefly outline some of the features of this most intriguing heme protein.

17.2.1 The Reaction Energy Landscape of Ligand Binding

In the absence of an exogenous ligand, the reduced heme iron of ferrous (Fe^{2+}, deoxy) Mb and Hb is bound to the proximal histidine, HisF8, and the four pyrrole nitrogens. The ligand binding site is thus vacant. In contrast, ferrous Ngb is hexacoordinate even when no exogenous ligand is bound, which can easily be inferred from the peculiar features observed in the optical spectra [34, 36]–[38]. Whenever an exogenous ligand such as O_2, CO, or NO dissociates

[1] Institute of Applied Physics and Center for Functional Nanostructures (CFN), Karlsruhe Institute of Technology (KIT), 76128 Karlsruhe, Germany

from Ngb at physiological conditions, the heme iron binds the imidazole side chain of the distal histidine, HisE7. This endogenous ligand needs to again thermally dissociate from the heme iron before an exogenous ligand can bind. The five-coordinate deoxy species thus occurs transiently under physiological conditions; only the liganded and six-coordinate deoxy species are significantly populated. Because in both these states, the heme iron is characterized by a covalent bond at the sixth coordination, their overall enthalpy is more similar compared with liganded and unliganded Mb. It has been suggested that the resulting weaker temperature dependence of the binding equilibrium may be physiologically relevant [39].

17.2.2 Molecular Structure of Ngb

The three-dimensional structure of murine [40] and human [41] ferric Ngb have been solved at resolutions of 1.50 and 1.95 Å, respectively. They display the familiar, α-helical globin fold of Mb and Hb. Although the heme iron is in the oxidized state (Fe^{3+}), these structures are assumed to be good models of the six-coordinate deoxy (Fe^{2+}) species. A huge internal volume of $\sim$290 Å^3 connects the distal and proximal sides and features a large channel to the outside, and on the distal side, there are two more cavities, one of which corresponds to the secondary ligand docking site (Xe4) of Mb [22, 42]. Note that internal volumes are detrimental to the stability of a protein, and therefore, one generally assumes that they are required for functional processes. For Ngb, nothing is known concerning the role of these cavities yet. Another noteworthy observation concerns the large structural heterogeneity (B factors) in the CD corner, and even more so in the EF corner. The murine Ngb structure in the CO-ligated state revealed a most unexpected and novel control mechanism for ligand binding [43]: To bind CO at the heme iron, the HisE7 side chain does not move away but essentially retains its conformation. Instead, the heme prosthetic group slides deeper into the heme crevice, which appears to be preshaped to accommodate this shift. Concomitantly, the internal volume changes shape and its total volume increases slightly. A significant decrease of the structural disorder in the EF loop communicates the state of heme ligation to the protein surface, which may form the basis of a potential sensor function [44].

17.2.3 Kinetic Experiments and Ligand Binding

The kinetics of ligand binding of NgbCO has been studied in detail by flash photolysis. The kinetic analysis is more complex, as there are three principal species that need to be considered, i.e., the bound state and the five-coordinate and six-coordinate deoxy species. At ambient temperature, two distinct kinetic phases are observed, a pronounced step on the millisecond time scale that is assigned to binding of exogenous CO as well as the endogenous imidazole ligand to the pentacoodinate ferrous protein. Free ligands bind from the solvent in a bimolecular reaction, as confirmed by the ligand concentration dependence of the kinetics [45]). The binding coefficients for the diatomic gases NO, CO, and O_2 are rather high ($4 - 15 \times 10^7 M^{-1} s^{-1}$) [36, 38, 46], reflecting a highly reactive heme iron. The HisE7 imidazole group competes with exogenous CO ligands for the sixth coordination, giving rise to a fraction of hexacoordinate deoxy species. This fraction can be distinguished from the pentacoordinate species by its blue-shifted optical absorption, which persists out to the second time scale [46]. The weak covalent bond between the heme iron and the imidazole ultimately breaks, so that the more strongly binding exogenous CO ligand can bind in the second kinetic step, and the preflash equilibrium is finally restored. There is yet another kinetic phase at fast times that is clearly visible at 2 °C but escapes the observation time window of nanosecond flash photolysis at room temperature. In this geminate rebinding process, CO ligands ($\sim$ 65% at 20 °C) do not escape from the protein after dissociation from the heme iron but rather rebind directly [46]. For O_2 the geminate yield is even higher (> 90% at 20 °C). Ligand escape is much more efficient in Mb, mainly because of the much higher rebinding enthalpy barrier at the heme iron [22, 47].

17.2.4 Conformational Energy Landscape

Nonexponential rebinding in NgbCO at low temperatures ($T < 160$ K) can be modeled with a spectrum of enthalpy barriers at the heme iron [46], which provides direct evidence of conformational substates in this protein in the same way as for MbCO [4].

More specific information on the so-called taxonomic substates [48] can be obtained by vibrational spectroscopy. Especially the infrared stretching bands of heme-bound CO have been recognized as excellent reporters of active site fine structure and heterogeneity due to their exquisite sensitivity to local electric fields created by amino acid residues surrounding the active site. In MbCO, HisE7 is the main determinant of active site heterogeneity sensed by the CO stretching vibration [49, 50] and the resulting major conformations, called A substates, have been characterized in great detail [51]. The Fourier transform infrared spectrum of murine NgbCO displays two different A substates at neutral pH. The frequency of the A_1 band at 1938 cm^{-1} suggests a strong interaction of the positive partial charge on the Nε2 hydrogen of HisE7 with the CO. The A_2 band at the comparatively high frequency of 1980 cm^{-1} may represent a tautomeric form in which the Nδ1 nitrogen of the neutral HisE7 is protonated so that the nonbonded electron pair of the Nε2 nitrogen places a negative partial charge in close vicinity to the heme-bound ligand. Actually, it is exclusively this (usually negligibly populated) Nδ1 tautomer [52] that can form the hexacoordinate ferrous species after dissociation of the exogenous ligand [45]. Two new A substate bands, denoted A_3 (1923 cm^{-1}) and A_o (1968 cm^{-1}), appear at low pH at the expense of A_1 and A_2 [45]. The A_o band is at the typical frequency of a heme protein with the protonated HisE7 that has swung out of the distal pocket [53, 54]. The rather low frequency of the A_3 conformation implies a large positive partial charge near the bound CO ligand, which may indicate that a fraction of protonated HisE7 side chains stays within the distal heme pocket.

Further information on the rebinding behavior within the individual taxonomic substates of NgbCO has been obtained by temperature-derivative spectroscopy, including evidence of ligand migration to alternative docking sites at cryogenic temperatures [55].

17.2.5 What Is Ngb Good for in the Body?

Initially, Ngb was suggested to be an O_2 storage and transport protein, performing a function in neural tissue similar to that of Mb in muscle [34]. Three observations are in conflict with this suggestion, though: (i) The high autoxidation rate of ferrous Ngb argues against a function as an O_2 storage protein. (ii) The more recently reported P_{50} values in the range of 7.5–20 torr [56] also shed doubt on this conjecture. To supply O_2 to the tissues, the protein should have a P_{50} below normal tissue pressures. (iii) Moreover, the average Ngb concentration in the brain is in the micromolar range [34] and thus is way too small to contribute to the oxygen supply. Mb concentrations are much higher in heart and red muscle tissue, $\sim 200\ \mu$M [57] and 100–400 μM [58], respectively.

Ngb could play an important role as a scavenger of NO and/or ROS as it is upregulated during hypoxic episodes, in which these compounds accumulate in tissues. Ferrous NgbNO can associate with peroxynitrite, which is produced under hypoxic conditions by the reaction of NO with O_2 [59]. The association rate of HbNO and peroxynitrite is almost two orders of magnitude smaller, which may indicate that NgbNO may function in vivo as an efficient scavenger of this harmful compound [60].

The intermediate O_2 affinity, which is in the range of the partial pressures measured in the brain, suggests yet another function for Ngb, namely that of an oxygen sensor that reports hypoxic conditions to signal transduction chains, which trigger processes that protect the cell against the detrimental effects of low oxygen pressure. A number of reports [61]–[63] pointed to a possible role of Ngb in cell signaling. The structural changes after binding an exogenous ligand mentioned earlier suggest that it could indeed function as a sensor molecule of diatomic ligands.

Notwithstanding the as-yet-unclear functional role of Ngb, the protein significantly contributes to a protection of cells under oxygen deprivation via a specific mechanism that is not shared by Mb [64]. There is more work ahead of us to elucidate the physiological role of this intriguing protein.

17.3 Protein Folding and Unfolding—An Introduction (R. D. Young[2])

The first definitive studies of protein folding and unfolding were performed by Christian Anfinsen, who looked at reversible folding and unfolding of ribonuclease using chemical denaturants. Anfinsen clearly showed that all the information for a protein to fold to biologically active conformations is contained in the primary sequence of amino acides. This fundamental fact informs all protein studies. Anfinsen's Nobel lecture is a wonderful historical introduction for students new to the field [65].

Cyrus Levinthal noted another seminal fact regarding the time scale for the linear amino acide sequence to fold to a functional conformation after construction in the ribosome [66]. Levinthal realized that an amino acid sequence with N residues and only two possible environments for each residue has 2^N states. If only one of these 2^N states is properly folded and the search time is 1 ns per state, then searching all states to find the one properly folded functional state would take on average $2^N \times 10^{-9}$ s. For even a small protein of 100 residues, this time is of the order of 10^{13} years, a time of cosmological scale. Biologically relevant amino acid sequences fold on physiological time scales, say minutes or less. One or, most likely, both of the conditions underlying the absurd result of the Levinthal observation must be irrelevant to biologically functional amino acid sequences. As recognized by Onuchic, Luthey-Schulten, and Wolynes [67], molecular evolution selects only amino acid sequences that fold on a biologically meaningful time scale. One can expect that biologically active proteins have a large number of functional conformations, that is, conformational substates, rather than a single functional conformation. Furthermore, one can anticipate that there is a "bias," probably energetic, that mitigates the effect of the large entropy of the unfolded state and reduces the time for folding. These issues are explored in this section.

The concepts of conformational energy landscape and conformational substate are central to the protein dynamics of folded, functioning proteins (Chapter 15). These concepts a fortiori are valid for protein folding and unfolding and lead to the model pictured in Fig. 17.1 [68]. Figure 17.1(a) is a cartoon of a 1D cross section through the high-dimensional EL of a protein. Figure 17.1(b) is a 2D cross section. Each valley represents a conformational substate of the unfolded basin of the EL (U), the transition state ensemble (TSE), or the folded basin (N or F). The discussion in Chapter 15 is confined to the folded basin of the EL. Protein dynamics in the EL extended to the unfolded basin are complex. An unfolded protein starts out in U and makes a random walk in the U basin until it reaches the TSE. Once in the TSE, the protein either returns to U or falls into the folding funnel that leads to N. The concept of a folding funnel responds to the issue of bias in the EL to speed folding [67, 69]. Figure 17.1(a) is misleading because it suggests that there is only one pathway

---

[2] Center for Biological Physics, Arizona State University, Tempe, AZ 85287, USA.

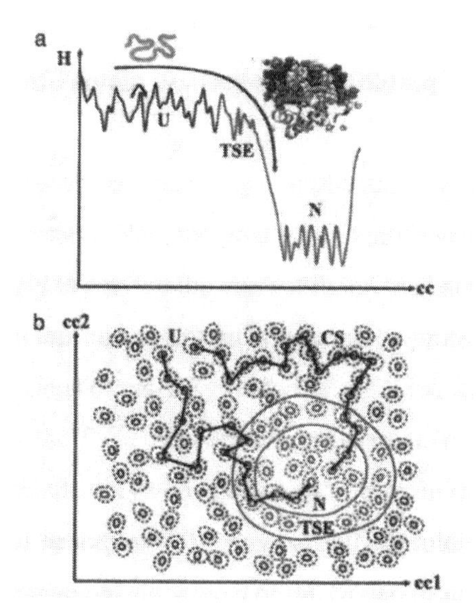

Fig. 17.1. The conformational energy landscape of a protein extended to include both the folded and unfolded basins. See text for discussion. Figure adapted from [68].

for folding. The 2D cross section in Fig. 17.1(b) shows that there are many pathways and that the density of substates can differ in different parts of the EL. A dense region in U or the TSE can act as an intermediate state. The extended energy landscape provides a unifying concept for understanding protein folding and unfolding studies, as well as the dynamics of folded proteins. The EL in Fig. 17.1 is a potential energy landscape. It is frequently convenient to consider a free energy landscape with a reduced set of conformational coordinates. Since misfolded proteins have been implicated in several diseases, the EL takes on even clinical significance [70].

17.3.1 Two-State Model of Protein Folding and Unfolding

The introductory comments suggest the many complications in understanding protein folding and unfolding processes. Here the goal is to identify general principles of protein folding and unfolding likely to survive the current theoretical and experimental onslaught. To accomplish this end, attention is focused on proteins that fold and unfold without significant populations of partially folded or unfolded structures, that is, intermediate states in U or the TSE. An experimental sample of proteins in equilibrium can then be assumed to include only two ensembles of protein structures, the folded (F or N) structures and unfolded structures (U). Any partially unfolded structures, that is, those structures not having experimental signatures of the folded or unfolded structures, have such short lifetimes that they are not populated to an appreciable extent. Such proteins can be described by two-state models. In addition, only monomeric proteins are considered for simplicity. The quantitative description of two-state models for monomeric proteins leads to clear analysis of protein folding and unfolding.

The probabilities $p_i(t)$, $i = F, U$, of the monomeric protein molecule being in the folded or unfolded state at time t are normalized, $p_F(t) + p_U(t) = 1$. These probabilities are related dynamically by a rate equation at temperature T,

$$\frac{dp_U}{dt} = -\lambda_F(T)\, p_U(t) + \lambda_U(T)\, p_F(t)\ . \tag{17.1}$$

The rate coefficients $\lambda_i(T)$ in Eq. (17.1) are not microscopic rate coefficients for transitions between conformational substates in the free energy landscape of the protein, as indicated by the smaller arrow in Fig. 17.1. The $\lambda_i(T)$ are, in fact, complex quantities representing the flow of protein molecules between large basins in the free energy landscape representing the folded and unfolded states, as indicated by the larger arrow in Fig. 17.1. At infinite time, the probabilities approach time-independent equilibrium values, $p_i(\infty) \equiv p_i^{eq}(T)$, which are related by the expression for the equilibrium coefficient $K_{\mathrm{eq}}(T)$,

$$K_{\mathrm{eq}}(T) = \frac{p_U^{\mathrm{eq}}(T)}{p_F^{\mathrm{eq}}(T)} = \frac{\lambda_U(T)}{\lambda_F(T)} \equiv \exp\left(-\frac{\Delta G_{UF}(T)}{\mathrm{k_B}T}\right)\ . \tag{17.2}$$

Here the free energy difference $\Delta G_{UF}(T) = \Delta H_{UF}(T) - T\Delta S_{UF}(T)$, where H is the average enthalpy and S the average entropy of the protein molecules in the EL, and $\Delta G_{UF} = G_U - G_F$, for example. The folding transition temperature T_f (also called the midpoint temperature) is defined by the condition that $K_{\mathrm{eq}}(T_f) = 1$, resulting in several conditions: $p_F^{\mathrm{eq}}(T_f) = p_U^{\mathrm{eq}}(T_f) = 1/2$ so that the populations of folded and unfolded structures are equal; equality of rate coefficients, $\lambda_F(T_f) = \lambda_U(T_f)$; and relations on free energies, $\Delta G_{UF}(T_f) = 0$ and $\Delta H_{UF}(T_f) = T_f \Delta S_{UF}(T_f)$. Although the definition of T_f is arbitrary, it focuses on the region where the folded and unfolded proteins coexist in about equal populations with about equal folding and unfolding times in the EL. Indeed, T_f can also be called the unfolding transition temperature.

The possibility of a temperature dependence in the free energy, enthalpy, and entropy is included in the previous expressions. The necessity for this temperature dependence follows from the temperature dependence of the heat capacities of the folded and unfolded protein [71]. The heat capacities are usually taken at constant pressure and will be in this chapter. The heat capacities for the folded and unfolded states are taken to be linear in temperataure, $c_i = c_{0,i} + c_{1,i}T$, where $i = F,\ U$. Higher-order temperature dependence for the unfolded states may be needed over a large temperature range. The difference in heat capacity between the folded and unfolded states is $\Delta c_{UF}(T) = \Delta c_0 + \Delta c_1 T$. It is convenient to express this difference in terms of a reference temperature, which is frequently taken to be the folding transition temperature T_f,

$$\Delta c_{UF}(T) = \Delta c_{UF}(T_f) + \Delta c_1(T - T_f) \tag{17.3}$$

where $\Delta c_{UF}(T_f) = \Delta c_0 + \Delta c_1 T_f$. The enthalpy and entropy are related to the heat capacity by the thermodynamic relations, $\Delta c_{UF}(T) = (\partial \Delta H_{UF}/\partial T)_p = T(\partial \Delta S_{UF}/\partial T)_p$. Integration between T and T_f gives enthalpy and entropy differences,

$$\Delta H_{UF}(T) = T_f \Delta S_{UF}(T_f) + \Delta c_{UF}(T_f)(T - T_f) + \frac{\Delta c_1}{2}(T - T_f)^2 \tag{17.4}$$

and

$$\Delta S_{UF}(T) = \Delta S_{UF}(T_f) + \Delta c_{UF}(T_f) \ln \frac{T}{T_f}$$

$$+\Delta c_1 \left[(T - T_f) - T_f \ \ln \frac{T}{T_f}\right]. \tag{17.5}$$

The Gibbs free energy is obtained from the two expressions as

$$\Delta G_{UF} = -\Delta S_{UF}(T_f)(T - T_f) + \Delta c_{UF}(T_f)\left[(T - T_f) - T \ln \frac{T}{T_f}\right]$$

$$-\frac{\Delta c_1}{2}\left[(T^2 - T_f^2) - 2T \ T_f \ \ln \frac{T}{T_f}\right]. \tag{17.6}$$

These last three expressions, and all equations, should be derived for homework. If there is no heat capacity difference between the folded and unfolded states, the enthalpy and entropy differences are constant and the Gibbs free energy difference is linear in temperature, $\Delta G_{UF} = -\Delta S_{UF}(T_f)(T - T_f)$, being zero at the folding transition temperature.

The unfolded equilibrium probability $p_U^{\rm eq}(T)$ exhibits a characteristic sigmoidal temperature dependence. The sharpness of the sigmoidal shape can be quantified by the slope of the probability $p_U^{\rm eq}(T_f)$ at the folding transition temperature providing insight into a general principle of protein folding. Differentiation of the unfolded equilibrium probability $p_U^{\rm eq}(T)$ with respect to temperature and use of the thermodynamic relation, $\partial \Delta G_{UF}/\partial T = -\Delta S_{UF}$, establishes that

$$\Delta S_{UF}(T_f) = 4\,\mathrm{k_B}\ T_f \left.\frac{\partial p_U^{eq}}{\partial T}\right|_{T_f}. \tag{17.7}$$

See Eq. (17.11) for the general expression. A representative value for the derivative is $\partial p_U^{\rm eq}/\partial T|_{T_f} \approx 0.08$ giving $\Delta S_{UF}/\mathrm{k_B} \approx 100$ at 320 K, an enormous entropy difference between the folded and unfolded states at the folding transition temperature, here selected to be a representative 320 K. The ratio of the number of accessible states in the unfolded and folded states is huge, $\Omega_U/\Omega_F \approx 2.6 \times 10^{43}$, using the Boltzmann relation for entropy, $S = k_B \ln \Omega$. This is another example underlying the Levinthal observation for folding times. In the next section, representative examples of experimental data add specificity and additional insight.

17.3.2 Thermodynamic Equilibrium Studies of Folded and Unfolded Proteins

The conditions for applicability of the two-state model are more easily met for equilibrium studies than for kinetic studies because of the much longer experimental time scale for equilibrium. To keep the discussion of concepts close to reality, representative data are adapted for a small protein, the DNA-binding domain of bacterial integrase Tn916 consisting of 74 residues with mainly β-sheet structure [72]. A representative data set from equilibrium experiments on integrase Tn916 is shown in Fig.17.2. Figure 17.2(a) presents representative data coming from fluorescence measurements. The data show the sigmoidal trace of the fluorescence wavelength $\lambda_{\mathrm{max}}(T)$ of maximum emission between a baseline corresponding to the folded state (F) and one corresponding to the unfolded state (U). Figure 17.2(b) shows representative data adapted from differential scanning calorimetry (DSC) experiments showing a highly peaked transition in the heat capacity of the sample between baselines corresponding to the folded (F) and unfolded (U) states. The solid dark line shows that characteristic sigmoidal trace between baselines also seen in Fig.17.2(a).

The two experimental data sets describe different, but related, information. Sigmoidal data such as the fluorescence data in Fig. 17.2(a) and the DSC signal indicated by the solid line in Fig. 17.2(b) are described by a simple linear relation for the equilibrium probability of the unfolded state,

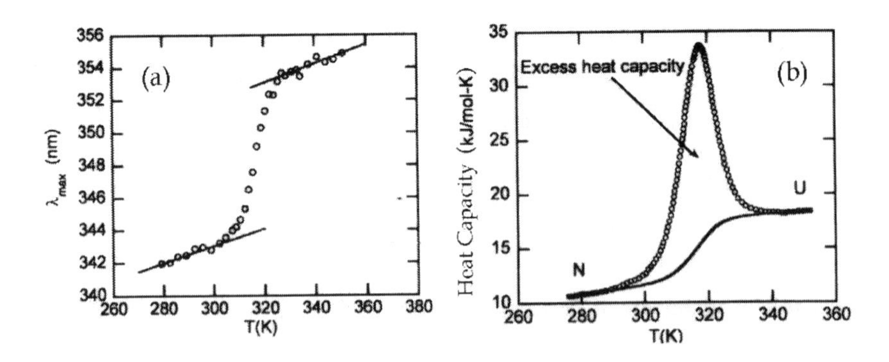

Fig. 17.2. Thermal folding and unfolding of integrase Tn916 followed by (a) fluorescence spectroscopy and (b) DSC. See text for discussion. Data adapted from Milev et al. [72].

$$p_U^{\mathrm{eq}}(T) = \frac{O(T) - O_N(T)}{O_U(T) - O_N(T)} , \tag{17.8}$$

where O represents the experimental observable, for example, the wavelength of maximum emission in the fluorescence experiment or the heat capacity in the DSC experiment. The fluorescence baselines $O_N(T)$ and $O_U(T)$ follow linear relations in temperature shown as the solid lines in Fig. 17.2(a). The resulting sigmoidal function for $p_U^{\mathrm{eq}}(T)$ is in Fig. 17.3. In Fig. 17.3(b), $p_U^{\mathrm{eq}}(T)$ is linear in the region of about ± 3 K around the folding transition tempearture T_f. The linearity allows a good estimate of the folding transion temperature $T_f = 316.4\,\mathrm{K}$ and the derivative $\partial p_U^{\mathrm{eq}}/\partial T|_{T_f} = 0.075\ \mathrm{K}^{-1}$ yielding the $\Delta S_{UF}(T_f) = 95\ \mathrm{k_B} = 0.79\,\mathrm{kJ/(mol{\cdot}K)}$. The heat capacity data are analyzed next.

Here the representative results for the heat capacity parameters, $\Delta c_{UF}(T_f) = 5.48\ \mathrm{kJ/(mol{\cdot}\,K)}$ and $\Delta c_1 = -0.0186\,\mathrm{kJ/(mol{\cdot}\,K^2)}$, are used to analyze the data in Fig. 17.2. The solid line in Fig. 17.3(a) comes from the unfolded equilibrium probability $p_U^{\mathrm{eq}}(T) = [1 + \exp(\Delta G_{UF}/\mathrm{k_B}T)]^{-1}$ with parameters T_f, $\Delta S_{UF}(T_f)$, $\Delta c_{UF}(T_f)$, Δc_1. But, with the resolution of Fig. 17.3(a), there is no noticeable difference in the sigmoidal shape when $p_U^{\mathrm{eq}}(T)$ is plotted with only the two variables T_f, $\Delta S_{UF}(T_f)$. This suggests another general feature of proteins that fold and unfold according to a two-state model. The sigmoidal shape of the experimental data mainly arises from two physical effects: (i) the position of the sigmoidal shape characterized by the folding transition temperature T_f is determined by the equality of the Gibbs energies for the folded and unfolded states, $G_F(T_f) = G_U(T_f)$; (ii) the sharpness of the sigmoidal shape is mainly determined by the entropy difference $\Delta S_{UF}(T_f)$ between the folded

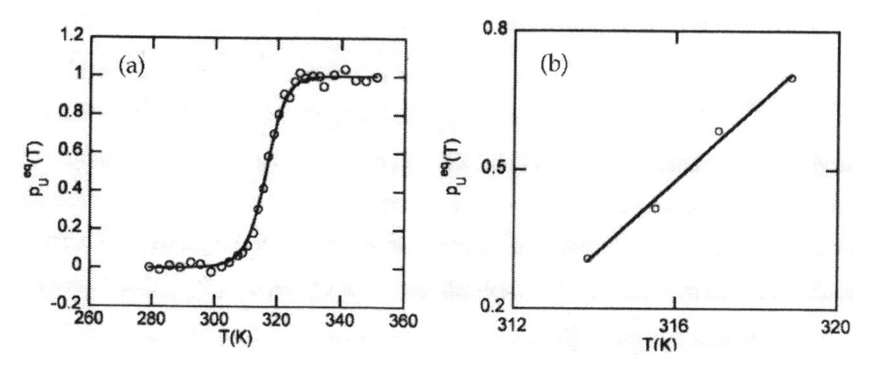

Fig. 17.3. (a) Sigmoidal data and fit for the equilibrium probability $p_U^{\mathrm{eq}}(T)$ for the integrase Tn916 fluorescence data in Fig. 17.2(a). (b) Expanding the $p_U^{\mathrm{eq}}(T)$ plot in (a) from 313 to 319 K, the unfolding transition is approximately linear with T.

and unfolded states at the folding transition temperature. The effects of heat capacity, in particular, $\Delta c_{UF}(T_f)$ and Δc_1, are of secondary importance.

Returning to the heat capacity data, in Fig. 17.2(b), two features stand out: (i) the heat capacity of the unfolded protein is larger than that of the folded protein in the temperature range shown; (ii) there is a large contribution to the heat capacity, sometimes called the excess heat capacity, above the solid line that describes the transition from the folded to unfolded protein following the sigmoidal shape of $p_U^{\mathrm{eq}}(T)$. The first feature is understood by an extension of the Dulong-Petit law for metals to say that the additional degrees of freedom in the unfolded protein result in a larger heat capacity. This is closely related to the equipartition principle for energy where each degree of freedom contributes $\mathrm{k_B}T/2$ to the internal energy of the system. The excess heat capacity is explained by the two-state model. To see this, write the average enthalpy for the ensemble of folded and unfolded proteins as

$$\langle H \rangle = H_F(T) + p_U^{\mathrm{eq}}(T) \Delta H_{UF}(T). \tag{17.9}$$

The temperature derivative gives the total heat capacity as measured in a DSC experiment,

$$c(T) = c_F(T) + p_U^{\mathrm{eq}}(T) \Delta c_{UF}(T) + \Delta H_{UF}(T) \frac{\partial p_U^{\mathrm{eq}}}{\partial T} \tag{17.10}$$

where

$$\frac{\partial p_U^{\mathrm{eq}}}{\partial T} = p_U^{\mathrm{eq}}\, p_F^{\mathrm{eq}} \frac{\Delta H_{UF}(T)}{\mathrm{k_B} T^2} \ . \tag{17.11}$$

The first two terms correspond to the solid line in Fig. 17.2(b), where the two baselines, $c_F(T)$ and $c_U(T)$, are described by linear relations for temperatures below about 285 K for the folded state and above about 345 K for the unfolded state. The last term corresponds to the contribution to the heat capacity caused by additional degrees of freedom resulting from the unfolding process as temperature is increased. The peak in Fig. 17.2(b), that is, the heat capacity above the solid line showing the sigmoidal behavior between the two baselines $c_F(T)$ and $c_U(T)$, is the excess heat capacity and is shown in Fig. 17.4(a) and expressed theoretically as follows.

$$\begin{aligned} \text{excess heat capacity} &= \Delta H_{UF}(T)\, \frac{\partial p_U^{\mathrm{eq}}}{\partial T} \\ &= \mathrm{k_B}\, p_U^{\mathrm{eq}}\, p_F^{\mathrm{eq}} \left(\frac{\Delta H_{UF}(T)}{\mathrm{k_B} T} \right)^2 . \end{aligned} \tag{17.12}$$

The good agreement of the data for excess heat capacity with the preceding theoretical expression is strong evidence that the protein folds and unfolds

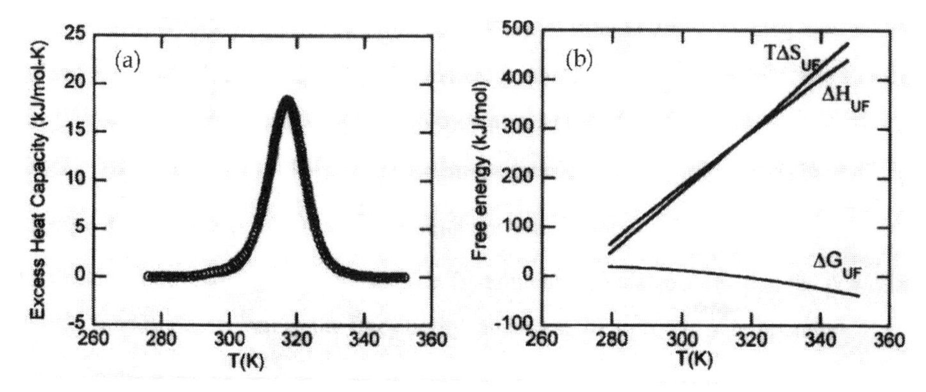

Fig. 17.4. (a) Excess heat capacity of integrase Tn916. (b) Gibbs free energy, enthalpy, and entropic free energy difference between the folded and unfolded states of integrase Tn916. See text for discussion.

according to a two-state model. A simplification in the excess heat capacity again demonstrates the importance of the entropy difference for protein folding and unfolding, according to the two-state model. Neglecting the terms dependent on heat capacity so that the enthalpy difference is temperature independent and given by $\Delta H_{UF}(T) \approx T_f \Delta S_{UF}(T_f)$ and the Gibbs free energy is given by $\Delta G_{UF}(T) \approx -(T - T_f)\Delta S_{UF}(T_f)$ results in the solid line in Fig. 17.4(a) which is very close to the exact theoretical expression including the heat capacity in Eq. (17.12).

Figure 17.4(b) shows the Gibbs energy difference $\Delta G_{UF}(T) = \Delta H_{UF}(T) - T\Delta S_{UF}(T)$, enthalpy difference $\Delta H_{UF}(T)$, and entropic free energy difference $T\Delta S_{UF}(T)$ for the model two-state folder integrase Tn916. The slope of the entropic free energy difference $T\Delta S_{UF}(T)$ is larger than that of the enthalpy difference $\Delta H_{UF}(T)$. At 340 K, for example, the entropic free energy difference is larger than the enthalpy difference resulting in a negative Gibbs free energy difference, ΔG_{UF} (340 K) < 0, and a larger population of unfolded proteins. But the larger slope of the entropic free energy difference results in the magnitude of the Gibbs free energy difference decreasing as temperature decreases until at the folding transition temperature the Gibbs free energy difference is zero, ΔG_{UF} ($\sim$316.3 K) $= 0$, and the folded and unfolded protein populations are equal. As temperataure decreases below the folding transition temperature, the enthalpy difference is now larger than the entropic free energy difference so that the Gibbs free energy difference is positive, resulting in a larger population of folded protein. The difference in slope between the enthalpy difference and entropic free energy difference can be understood by basic thermodynamics. The slope of the enthalpy difference is equal to the heat capacity difference, $\partial \Delta H_{UF}/\partial T = \Delta c_{UF}(T)$. The slope of the entropic free energy difference is given by $\partial(T\Delta S_{UF})/\partial T = \Delta c_{UF}(T) + \Delta S_{UF}(T)$. The larger slope of the entropic free energy difference is clearly a result of the entropy difference $\Delta S_{UF}(T)$ between the unfolded and folded states that experiment and theoretical arguments show is positive. The slope of the Gibbs free energy is also determined by the entropy difference since $\partial \Delta G_{UF}/\partial T = -\Delta S_{UF}(T)$. This analysis demonstrates the central role played by the entropy difference $\Delta S_{UF}(T)$ in determining the temperature dependence of the thermodynamic stability of proteins. The experimental data also show that the potential energy landscape, as well as the free energy landscape, of a protein are temperature dependent.

17.3.3 Kinetic Studies of Protein Folding and Unfolding

Much of the current interest in protein folding and unfolding arises from the Levinthal observation that protein folding would occur on cosmological time scales if folding was determined solely by the most straightforward consideration of the entropy difference between the folded and unfolded states. Since proteins fold on much shorter time scales, the free energy landscape of proteins must have properties, most likely resulting from molecular evolution, that enable protein folding on physiological time scales. The earlier discussion already places one condition on the rate coefficients for folding and unfolding, that is, the ratio of the rate coefficients is determined by the Gibbs free energy difference,

$$\frac{\lambda_U}{\lambda_F} = \frac{\exp(G_F/\mathrm{k_B}T)}{\exp(G_U/\mathrm{k_B}T)} , \tag{17.13}$$

which has been rewritten in a suggestive form. The unfolding rate coefficient refers to transitions from the folded basin in the free energy landscape to the unfolded basin, and the folding rate coefficient refers to transitions from the unfolded basin in the free energy landscape to the folded basin. The detailed balance condition, $\lambda_F \; p_U^{\mathrm{eq}} = \lambda_U \; p_F^{\mathrm{eq}}$, means that, at equilibrium, the flow of proteins from the folded to the unfolded basin of the free energy landscape equals the flow of proteins from the unfolded to folded basins. To make further progress toward general principles underlying the kinetics of protein folding and unfolding requires more detailed consideration of the kinetic phenomena. Temperature-jump experiments provide much insight into the kinetics of protein folding and unfolding.

Modern temperature-jump experiments employ fast heating of the protein sample by short (nanosecond) laser pulses. After the heating laser pulse is over, the protein sample has populations of folded and unfolded proteins characteristic of the lower temperature before laser heating so the populations of folded and unfolded proteins relax toward the new equilibrium probabilities characteristic of the high temperature. This relaxation is measured by monitoring some property of the protein. Proteins that fold and unfold according to a two-state model follow Eq. (17.1), and the probabilities and experimentally measured protein properties show time courses that are described by a single exponential. The observed rate coefficient λ_{obs} is related to the folding and unfolding rate coefficients by eliminating $p_F(t)$ from Eq. (17.1) to get

$$\frac{dp_U}{dt} = -\lambda_{\text{obs}} \left[p_U(T) - p_U^{\text{eq}}(T)\right] , \tag{17.14}$$

which has a general solution $p_U(t) = a\exp(-\lambda_{\text{obs}}t) + b$, where $\lambda_{\text{obs}} = \lambda_U + \lambda_F$. The constants a and b are determined by the equilibrium probabilities characteristic of the sample temperatures before and after the laser pulse. The important point here is the kinetics experiments measure only the sum of the folding and unfolding rate coefficients through the observed rate coefficient λ_{obs}. The folding and unfolding rate coefficients can be found, however, from the observed rate coefficient and the equilibrium probabilities using $\lambda_U(T) = p_U^{\text{eq}}(T)\lambda_{\text{obs}}(T)$ and $\lambda_F(T) = \left[1 - p_U^{\text{eq}}(T)\right]\lambda_{\text{obs}}(T)$ obtained by using λ_{obs} with the basic equilibrium condition, Eq. (17.2). The observed rate coefficient λ_{obs} approaches the unfolding rate coefficient at temperatures significantly above the folding transition temperature and the folding rate coefficient at temperatures significantly below the folding transition temperature. Representative data are adapted for a small protein, Pin WW domain consisting of 24 residues with an antiparallel β-sheet structure, that is important as a mediator in protein–protein recognition by binding Pro-rich ligands [73]. Figure 17.5(a) shows an Arrhenius plot of the rate coefficients obtained by kinetic fluorescence experiments, and Fig. 17.5(b) shows the equilibrium fraction of unfolded protein measured by circular dichroism at 226 nm. The shape of the Arrhenius plot of the observed rate coefficient λ_{obs} in Fig. 17.5(a) is cupped upward and has been dubbed a chevron plot.

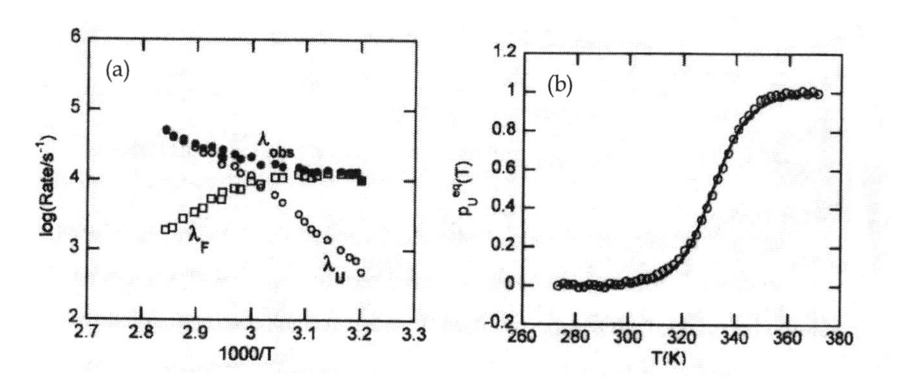

Fig. 17.5. (a) Observed rate coefficient and folding and unfolding rate coefficients for Pin WW domain from kinetic fluorescence experiments. (b) Sigmoidal behavior of the probability of the unfolded proteins from circular dichroism experiments. See text for discussion. Data adapted from Jager et al. [73].

The chevron shape arises because the observed rate coefficient is always the sum of the folding and unfolding rate coefficients. The chevron plot in Fig. 17.5(a) has several characteristic features of two-state folders:

1. The unfolding and folding rate coefficients are equal at the folding temperature with the unfolding rate coefficient being the larger above T_f and the folding rate coefficient being the larger below T_f.

2. The unfolding rate coefficient λ_U follows approximately an Arrhenius law, $\log \lambda_U \approx \log \lambda_{U,0} - \log H_U^\dagger/\mathrm{k_B T}$, suggesting an enthalpy barrier $H_U^\dagger$ that controls unfolding. However, the preexponential $\lambda_{U,0}$ and barrier $H_U^\dagger$ are too large for a simple kinetic process such as gas-phase kinetics. For example, Fig. 17.5(a) is best fit by $\log \lambda_{U,0} \approx 20$ and $H_U^\dagger \approx 105$ kJ/mol suggesting that the unfolding kinetics reflect a cooperative dynamics on the free energy landscape. Similar phenomena were seen for the dynamics of folded proteins and glasses in previous chapters.

3. The folding rate coefficient λ_F has no resemblance to an Arrhenius law. The shape of $\log \lambda_F$ in the Arrhenius plot is positive at higher temperatures and approximately constant at lower temperatures. The behavior points to entropic control of folding on the free energy landscape. The positive slope of a portion of the Arrhenius plot of the folding rate coefficient suggests an energetic "bias." Said another way, downhill folding, using a gravitational analogy, speeds the rate of folding.

As pointed out during the discussion of Levinthal's observation, the folding and unfolding rate coefficients are reduced from the rate coefficients for elementary processes, but not to cosmological scales. It is useful to express the folding and unfolding rate coefficients in terms of two slowing coefficients $n_F(T)$ and $n_U(T)$ as $\lambda_i = k/n_i$, $i = F, U$, where k is a rate coefficient for transitions between conformational substates in the EL. It is crucial to note that the folding parameters λ_F and $n_F(T)$ are properties of the free energy landscape corresponding to the *unfolded* basin while the unfolding parameters λ_U and $n_U(T)$ correspond to the *folded* basin. From Eq. (17.13), $n_F/n_U = \exp[-G_U/\mathrm{k_B}T]/\exp[-G_F/\mathrm{k_B}T]$, suggesting that the slowing coefficients satisfy the proportionalities, $n_F \propto \exp[-G_U/\mathrm{k_B}T]$ and $n_U \propto \exp[-G_F/\mathrm{k_B}T]$. The proportionality coefficients must come from a theoretical model. In the next section, a simple model of protein folding and unfolding kinetics developed by Zwanzig confirms the dependence of the slowing coefficients on the free energy through the appropriate Boltzmann factor, $\exp[-G/\mathrm{k_B}T]$, and provides a specific expression for the proportionality factor.

17.3.4 Zwanzig Model of Protein Folding and Unfolding

The Zwanzig model [74] is a simple physical model of the free energy landscape and dynamics in the free energy landscape that makes no contact with the specifics of the primary amino acid sequence of any real protein. It uses N discrete parameters that may be, for example, the environment of a particular amino acid. There is only one correct value for the folded configurations and ν_i incorrect values. It is easy to generalize the theoretical model to allow for a larger number ν_c of correct values. The number of parameters that have incorrect values is denoted by σ, which is a measure of correctness with $\sigma = 0$ denoting a correctly folded protein. The property of folding correctness was introduced by Bryngelson and Wolynes [75] and is a basic property of any protein configuration. It is selected as a crucial conformational coordinate for expressing the free energy landscape. The energy of the protein is taken to be a function of correctness $E_\sigma = \sigma U - \varepsilon\delta_{\sigma 0}$, $\sigma = 0, 1, \ldots, N$, where U and ε are positive. The energy spacing between neighboring values of σ is a constant U except for the larger energy gap $U + \varepsilon$ between $\sigma = 0$ and $\sigma = 1$. The degeneracy g_σ of the state specified by σ is the number of ways to choose σ incorrect values,

$$g_\sigma = \nu_c^{N-\sigma}\ \nu_i^\sigma \binom{N}{\sigma}. \tag{17.15}$$

The expectation is that $\nu_c < \nu_i$. The original Zwanzig model had $\nu_c = 1$, which is adopted here for simplicity. The Zwanzig model has all the qualitative features important to two-state folding and unfolding. Indeed, the model with four parameters (k, ν_i, U, and ε if N is taken as the number of amino acids) fits the data in Fig. 17.5 well. The focus here is on the insight to the slowing coefficients. The elementary rate, labeled k, is the rate of changing an incorrect to a correct parameter in the Zwanzig model. The slowing coefficients are given by

$$n_F = \left[\frac{\exp(U/\mathrm{k_B}T)}{N\nu_i}\right] \exp\left(-\frac{G_U}{\mathrm{k_B}T}\right)$$

and

$$n_U = \left[\frac{\exp(U/\mathrm{k_B}T)}{N\nu_i}\right] \exp\left(-\frac{G_F}{\mathrm{k_B}T}\right). \tag{17.16}$$

With free energies,

$$G_F = -\varepsilon \tag{17.17}$$

and

$$G_U = -\mathrm{k_B}T \ln Q_0, \tag{17.18}$$

we obtain

$$Q_0 = [1 + \nu_i \exp(-U/\mathrm{k_B}T)]^N - 1. \tag{17.19}$$

The results of the Zwanzig model confirm the concepts extracted from the experimental data for the kinetics of folding and unfolding. Unfolding is controlled by an energy (that is, enthalpy) barrier, $\varepsilon + U$. Folding is more subtle, with entropy playing an important role, as expected. The energy bias U is also important because it reduces Q_0 from the purely entropic factor $(1+\nu_i)^N$, that comes from the total number of states in the limit of zero bias energy, $U = 0$. There also can be a temperature dependence in the elementary rate k. The theoretical Zwanzig model confirms the expectations coming from experimental studies of the folding and unfolding kinetics of proteins that follow the two-state model. In particular, the slowing coefficient is proportional to the Boltzmann factor, $\exp[-G/k_BT]$, where G is the appropriate Gibbs free energy. The proportionality factor, $[\exp(U/k_BT)/N\nu_i]$, is model dependent and not easily determined from experiment alone.

17.3.5 Role of Solvent Dynamics in Protein Folding and Unfolding – A Research Perspective

Large-scale conformational fluctuations of folded proteins are known to be strongly coupled to solvent fluctuations. The effect of this strong coupling is a slowing of protein processes relative to the rate coefficient for the large–scale α fluctuations of the solvent, a phenomenon referred to as slaving since solvent fluctuations drive protein fluctuations [76]. If λ_{prot} is the rate coefficient for a protein conformational process and k_{solv} the rate coefficient for α fluctuations of the solvent, then the slaving condition between these two rate coefficients can be expressed in terms of a slowing coefficient n_{prot} by the relation $\lambda_{\text{prot}} = k_{\text{solv}}/n_{\text{prot}}$. Experiments that observe fluctuations between the A substates of MbCO involving partial unfolding of the protein show that the slowing coefficient can be as large as 10^5. The slowing coefficient n_{prot} can arise from two distinct processes: (i) slowing due to protein transitions in the EL described, for example, by the Zwanzig model in the previous section. This slowing coefficient is designated n_{EL}; (ii) slowing because the effects of solvent α fluctuations are screened by the amino acid environment comprising the protein [77]. This slowing coefficient is designated n_{scr}, and the total slowing coefficient is $n_{\text{prot}} = n_{\text{scr}}\, n_{EL}$. If the motions of proteins, both folded and partially unfolded, are slaved to solvent fluctuations, then the motions involved in protein folding and unfolding should also be slaved to solvent fluctuations. The elementary rate coefficient, introduced earlier to describe slowing of protein folding and unfolding, is then the α relaxation rate coefficient of the solvent giving $\lambda_i = k_{\text{solv}}/n_i$, $i = U, F$ where the slowing coefficients include both effects just described, $n_i = (n_{\text{scr}} n_{EL})_i$. Although there have been several studies of the effects of solvent viscosity on protein folding (see [68] for a list), the measurements involve varying the chemical composition of the solvent to vary the viscosity, a procedure that can introduce additional chemical effects to the folding process [68]. There have been no systematic studies of protein folding involving direct measurement of the α relaxation processes of the solvent and protein folding and unfolding over a wide temperature range. The kinetic and equilibrium data for the small protein Pin WW domain were taken in water with 10 mM sodium phosphate, pH 7.0. See Section 17.3.3 for a brief description of the Pin WW domain and Fig. 17.5 for the data. But the dynamic properties of the solvent are not characterized to provide a full picture. To give a feel for the issues, relaxation data for pure water

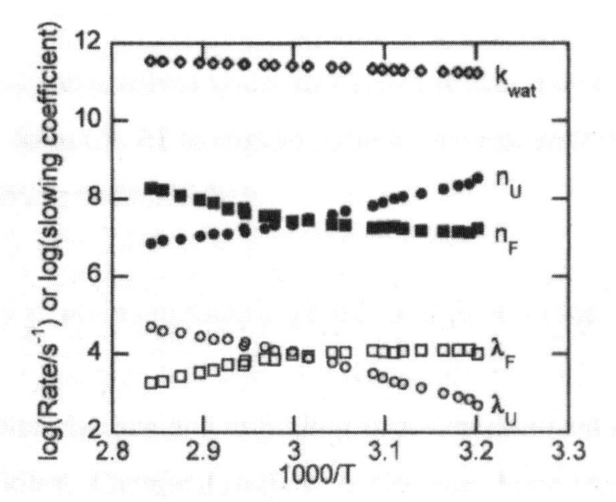

Fig. 17.6. Folding and unfolding coefficients, water relaxation rate coefficient, and protein slowing coefficients for folding and unfolding of Pin WW domain (Sec. 17.3.3 and Fig. 17.5(a)). See text for discussion. Data adapted from Jaeger et al. [73].

[78] is substituted for the water-buffer solution. Figure 17.6 shows the slowing coefficients, $n_i = k_{\text{water}}/\lambda_i,\ i = U, F$ for Pin WW domain.

The slowing coefficients in Fig. 17.6 exhibit several features:

1. The slowing coefficients for folding and unfolding of Pin WW domain range between about 10^7 and 10^9 and are significantly larger than those for folded proteins (maximum $\sim 10^5$). The slowing coefficients cross and are equal at the folding transition temperature.

2. The slowing coefficient for unfolding n_U is approximately linear in the Arrhenius plot but with a positive slope. If the effect of screening of solvent fluctuations is temperature independent, then the positive slope corresponds to an average stabilization energy (enthalpy) of the folded basin in the free energy landscape of 88 kJ/mol ($\varepsilon + U$ in the Zwanzig model after the temperature dependence of k_{water} is accounted for).

3. The slowing coefficient for folding n_F has less variation than the slowing coefficient for unfolding with some curvature, suggesting a combination of entropy and enthalpy effects for the folding process. The enthalpy effects can come from roughening of the free energy landscape for the unfolded states either from an unfavorable energy bias for bad amino acid contacts (U in the Zwanzig model) or a distribution of enthalpy barriers between conformational substates in the unfolded basin of the free energy landscape [67, 69, 75].

The analysis in this section involves issues that are of research level. Read ers can follow the suggested outline from this chapter to engage in their own research on these issues in the kinetics of protein folding and unfolding.

17.3.6 Chemical and Protein Engineering Studies of Protein Folding and Unfolding

The discussion of protein folding and unfolding emphasized physical techniques, especially thermal studies. Chemical studies are also very important, involving, for example, denaturant chemicals and pH. Protein engineering involving laboratory mutation of proteins has also contributed to the understanding of protein folding and unfolding. The goal of these studies is frequently to understand intermediates in the transition state ensemble or even the unfolded basin. Students can begin their studies of these very important resesarch activities in the many papers and textbooks cited [69, 72, 73, 79]–[81]. Molecular dynamics simulation also has been extensively applied to protein folding and unfolding employing the free energy landscape paradigm [82]. A new development in protein folding and unfolding studies involves the existence of other proteins that assist in folding. These helper proteins have been dubbed *chaperone* proteins. The textbook by Whitford is a good start for further reading regarding *chaperone* proteins [81].

17.4 Farewell

The rest of this book concerns itself with the tools of the biological physicist and their applications to studying the physics of proteins. To a considerable extent, the study of biomolecules is similar to that of atoms, molecules, solids, and even nuclei. Understanding these simple systems help in investigations of the more complex biomolecules. Many features are alike in all these quantum-mechanical systems; ideas and concepts from molecular and solid-state physics can thus be taken over directly. Biomolecules, however, show some beautiful characteristics that cannot be found in simpler systems. It is even possible that some laws of physics and chemistry can be studied in biomolecules more easily and directly than in other places.

So far, the discussion has often used the temperature dependence of processes to get insight. The pressure dependence of Mb phenomena is actually just as important [83], but many problems remain unsolved. In textbooks, the role of Mb is simply O_2 storage. It turns out, however, that Mb is far more potent and is involved in O_2 transport and in NO enzymatics [84, 85]. Experiments at very low temperatures give information about the low tiers of the energy landscape [86, 87], but it is not yet clear if these substates play a role in the protein function at ambient temperature. After photodissociation, the ligand does not leave the heme pocket immediately, but performs some gymnastics [88]. Single-molecule experiments have verified the sample-average data presented in this book and have yielded new insight in protein dynamics [89, 90]. These examples show that the field is rich and that many unexpected results can still make our understanding even deeper. We still have a great deal to learn about how proteins function at a deep level.

References

1. M. Daune. *Molecular Biophysics: Structures in Motion.* Oxford University Press, Oxford, 1999.
2. P. Nelson. *Biological Physics.* W. H. Freeman, New York, 2004.
3. E. Antonini and M. Brunori. *Hemoglobin and Myoglobin in Their Reactions with Ligands.* North-Holland Pub. Co., Amsterdam, 1971.
4. R. H. Austin, K. W. Beeson, L. Eisenstein, H. Frauenfelder, and I. C. Gunsalus. Dynamics of ligand binding to myoglobin. *Biochem.*, 14:5355–73, 1975.
5. H. Frauenfelder, J. Deisenhofer, and P. Wolynes, editors. *Simplicity and Complexity in Proteins and Nucleic Acids.* Dahlem University Press, Berlin, 1999.
6. T. Kleinert, W. Doster, H. Leyser, W. Petry, V. Schwarz, and M. Settles. Solvent composition and viscosity effects on the kinetics of CO binding to horse myoglobin. *Biochem.*, 37:717–33, 1998.
7. C. Tetreau and D. Lavalette. Dominant features of protein reaction dynamics: Conformational relaxation and ligand migration. *Biochemica and Biophysica Acta*, 1724:411–24, 2005.
8. P. J. Steinbach, K. Chu, H. Frauenfelder, J. B. Johnson, D. C. Lamb, G. U. Nienhaus, T. B. Sauke, and R. D. Young. Determination of rate distributions from kinetic experiments. *Biophys. J.*, 61:235–45, 1992.
9. E. R. Henry and J. Hofrichter. Single value decomposition: Application to analysis of experimental data. *Methods in Enzymology*, 210:129–92, 1992.
10. A. T. N. Kumar, L. Zhu, J. F. Christian, A. A. Demidov, and P. M. Champion. On the rate distribution analysis of kinetic data using the maximum entropy method: Applications to myoglobin relaxation on the nanosecond and femtosecond timescales. *J. Phys. Chem. B*, 105:7847–56, 2001.
11. R. H. Austin and L. J. Rothberg. Picosecond infrared spectroscopy of hemoglobin and myoglobin. *Methods in Enzymology*, 232:176–204, 1994.
12. F. Rosca, A. T. N. Kumar, X. Ye, T. Sjodin, A. A. Demidov, and P. Champion. Investigations of coherent vibrational oscillations in myoglobin. *J. Phys. Chem. A*, 104:4280–90, 2000.
13. J. Berendzen and D. Braunstein. Temperature-derivative specroscopy: A tool for protein dynamics. *Proc. Natl. Acad. Sci. USA*, 87:1–5, 1990.
14. D. C. Lamb, K. Nienhaus, A. Arcovito, F. Draghi, A. E. Miele, M. Brunori, and G. U. Nienhaus. Structural dynamics of myoglobin-ligand migration among

protein cavities studied by Fourier transform infrared/temperature derivative spectroscopy. *J. Bio. Chem.*, 277:11636–44, 2002.

15. D. A. Case and M. Karplus. Dynamics of ligand binding to heme proteins. *J. Mol. Bio.*, 132:343–68, 1979.
16. R. F. Tilton, I. D. Kuntz, and G. A. Petsko. Cavities in proteins: Structure of a metmyoglobin-xenon complex solved to 1.9 Å. *Biochem.*, 23:2849–57, 1984.
17. M. Brunori and Q. H. Gibson. Cavities and packing defects in the structural dynamics of myogoblin. *EMBO Reports*, 2:647–79, 2001.
18. S. Srajer et al. Protein conformational relaxation and ligand migration in myoglobin: A nanosecond to millisecond molecular movie from time-resolved Laue X-ray diffraction. *Biochem.*, 40:13802–15, 2001.
19. M. Schmidt, K. Nienhaus, R. Pahl, A Krasselt, S. Anderson, F. Parak, G. U. Nienhaus, and V. Srajer. Ligand migration pathway and protein dynamics in myoglobin: A time-resolved crystallographic study on L29WMbCO. *Proc. Natl. Acad. Sci. USA*, 102:11704–9, 2005.
20. D. Bourgeois, B. Vallone, A. Arcovito, G. Sciara, F. Schotte, P. A. Anfinrud, and M. Brunori. Extended subnanosecond structural dynamics of myoglobin revealed by Laue crystallography. *Proc. Natl. Acad. Sci. USA*, 103:4924–9, 2006.
21. E. E. Scott, Q. H. Gibson, and J. S. Olson. Mapping the pathways for O_2 entry and exit from myoglobin. *J. Bio. Chem.*, 276:5177–88, 2001.
22. K. Nienhaus, P. Deng, J. M. Kriegl, and G. U. Nienhaus. Structural dynamics of myoglobin: Effect of internal cavities on ligand migration and binding. *Biochem.*, 42:9647, 2003.
23. W. Cao, X. Ye, T. Sjodin, J. F. Christian, A. A. Demidov, S. Berezhna, W. Wang, D. Barrick, J. T. Sage, and P. M. Champion. Investigations of photolysis and rebinding kinetics in myoglobin using proximal ligand replacements. *Biochem.*, 43:11109–17, 2004.
24. D. Dantsker, U. Samuni, A. J. Friedman, M. Yang, A. Ray, and J. M. Friedman. Geminate rebinding in trehalose-glass embedded myoglobins reveals residue-specific control of intramolecular trajectories. *J. Mol. Bio.*, 315:239–51, 2002.
25. C. Tetreau, Y. Blouquit, E. Novikov, E. Quiniou, and D. Lavalette. Competition with xenon elicits ligand migration and escape pathways in myoglobin. *Biophys. J.*, 86:435–47, 2004.
26. D. Vitkup, G. A. Petsko, and M. Karplus. A comparison between molecular dynamics and X-ray results for dissociated CO in myoglobin. *Nature Structural Biology*, 4:202–8, 1997.
27. J. Cohen, A. Arkhipov, R. Braun, and K. Schulten. Imaging the migration pathways for O_2, CO, NO, and Xe inside myoglobin. *Biophys. J.*, 91:1844–57, 2006.
28. P. Fenimore, H. Frauenfelder, B. H. McMahon, and F. G. Parak. Slaving: Solvent fluctuations dominate protein dynamics and functions. *Proc. Natl. Acad. Sci. USA*, 99:16047–51, 2002.
29. P. W. Fenimore, H. Frauenfelder, B. H. McMahon, and R. D. Young. Bulk–solvent and hydration-shell fluctuations, similar to α and β-fluctuations in glasses, control protein motions and functions. *Proc. Natl. Acad. Sci. USA*, 101:14408–13, 2004.
30. J. A. Rupley and G. Careri. Protein hydration and function. *Adv. Protein Chem.*, 41:37–172, 1991.
31. C. Mattos. Protein-water interactions in a dynamic world. *Trends in Biochemical Sciences*, 27:203–8, 2002.

32. B. Halle. Protein hydration dynamics in solution: A critical survey. *Phil. Trans. R. Soc. Lond. B*, 359:1207–24, 2004.
33. S. K. Pal and A. H. Zewail. Dynamics of water in biological recognition. *Chem. Rev.*, 104:2099–2123, 2004.
34. T. Burmester, B. Weich, S. Reinhardt, and T. Hankeln. A vertebrate globin expressed in the brain. *Nature*, 407:520–3, 2000.
35. S. Wystub, B. Ebner, C. Fuchs, B. Weich, T. Burmester, and T. Hankeln. Interspecies comparison of neuroglobin, cytoglobin and myoglobin: Sequence evolution and candidate regulatory elements. *Cytogenet. Genome Res.*, 105:65–78, 2004.
36. S. Dewilde, L. Kiger, T. Burmester, T. Hankeln, V. Baudin-Creuza, T. Aerts, M. C. Marden, R. Caubergs, and L. Moens. Biochemical characterization and ligand binding properties of neuroglobin, a novel member of the globin family. *J. Bio. Chem.*, 276:38949–55, 2001.
37. J. T. Trent, A. N. Hvitved, and M. S. Hargrove. A model for ligand binding to hexacoordinate hemoglobins. *Biochem.*, 40:6155–63, 2001.
38. J. T. Trent, R. A. Watts, and M. S. Hargrove. Human neuroglobin, a hexacoordinate hemoglobin that reversibly binds oxygen. *J. Bio. Chem.*, 276:30106–10, 2001.
39. J. Uzan, S. Dewilde, T. Burmester, T. Hankeln, L. Moens, D. Hamdane, M. C. Marden, and L. Kiger. Neuroglobin and other hexacoordinated hemoglobins show a weak temperature dependence of oxygen binding. *Biophys. J.*, 87:1196–1204, 2004.
40. B. Vallone, K. Nienhaus, M. Brunori, and G. U. Nienhaus. The structure of murine neuroglobin: Novel pathways for ligand migration and binding. *Proteins*, 56:85–92, 2004.
41. A. Pesce, S. Dewilde, M. Nardini, L. Moens, P. Ascenzi, T. Hankeln, T. Burmester, and M. Bolognesi. Human brain neuroglobin structure reveals a distinct mode of controlling oxygen affinity. *Structure (Camb)*, 11:1087–95, 2003.
42. M. Brunori, B. Vallone, F. Cutruzzola, C. Travaglini-Allocatelli, J. Berendzen, K. Chu, R. M. Sweet, and I. Schlichting. The role of cavities in protein dynamics: crystal structure of a photolytic intermediate of a mutant myoglobin. *Proc. Natl. Acad. Sci. USA*, 97:2058–63, 2000.
43. B. Vallone, K. Nienhaus, A. Matthews, M. Brunori, and G. U. Niehaus. The structure of carbonmonoxy neuroglobin reveals a heme-sliding mechanism for control of ligand affinity. *Proc. Natl. Acad. Sci USA*, 101:17351–6, 2004.
44. M. Brunori, A. Giuffre, K. Nienhaus, G. U. Nienhaus, F. M. Scandurra, and B. Vallone. Neuroglobin, nitric oxide, and oxygen: Functional pathways and conformational changes. *Proc. Natl. Acad. Sci. USA*, 102:8483–8, 2005.
45. K. Nienhaus, J. M. Kriegl, and G. U. Nienhaus. Structural dynamics in the active site of murine neuroglobin and its effects on ligand binding. *J. Bio. Chem.*, 279:22944–52, 2004.
46. J. M. Kriegl, A. J. Bhattacharyya, K. Nienhaus, P. Deng, O. Minkow, and G. U. Nienhaus. Ligand binding and protein dynamics in neuroglobin. *Proc. Natl. Acad. Sci. USA*, 99:7992–7, 2002.
47. J. S. Olson and G. N. Phillips Jr. Kinetic pathways and barriers for ligand binding to myoglobin. *J. Bio. Chem.*, 271:17593–96, 1996.
48. H. Frauenfelder, S. G. Sligar, and P. G. Wolynes. The energy landscapes and motions of proteins. *Sci.*, 254:1598–1603, 1991.

49. T. Li, M. L. Quillin, G. N. Phillips Jr., and J. S. Olson. Structural determinants of the stretching frequency of CO bound to myoglobin. *Biochem.*, 33:1433–46, 1994.
50. K. M. Vogel, P. M. Kozlowski, M. Z. Zgierski, and T. G. Spiro. Determinants of the feXO (X = C, N, O) vibrational frequencies in heme adducts from experiment and density function theory. *J. Amer. Chem. Soc.*, 121:9915–21, 1999.
51. A. Ansari, J. Berendzen, D. Braunstein, B. R. Cowen, H. Frauenfelder, M. K. Hong, I. E. Iben, J. B. Johnson, P. Ormos, T. B. Sauke et al. Rebinding and relaxation in the myoglobin pocket. *Biophys. Chem.*, 26:337–55, 1987.
52. S. Bhattacharya, S. F. Sukits, K. L. MacLaughlin, and J. T. Lecomte. The tautomeric state of histidines in myoglobin. *Biophys. J.*, 73:3230–40, 1997.
53. J. D. Müller, B. H. McMahon, E. Y. Chien, S. G. Sligar, and G. U. Nienhaus. Connection between the taxonomic substates and protonation of histidines 64 and 97 in carbonmonoxy myglobin. *Biophys. J.*, 77:1036–51, 1999.
54. F. Yang and G. N. Phillips Jr. Crystal structures of CO-, deoxy- and met-myoglobins at various pH values. *J. Mol. Bio.*, 256:762–74, 1996.
55. J. M. Kriegl, K. Nienhaus, P. Deng, J. Fuchs, and G. U. Nienhaus. Ligand dynamics in a protein internal cavity. *Proc. Natl. Acad. Sci. USA*, 100:7069–74, 2003.
56. A. Fago, C. Hundahl, S. Dewilde, K. Gilany, L. Moens, and R. E. Weber. Allosteric regulation and temperature dependence of oxygen binding in human neuroglobin and cytoglobin: Molecular mechanisms and physiological significance. *J. Bio. Chem.*, 279:44417–26, 2004.
57. K. E. Conley, G. A. Ordway, and R. S. Richardson. Deciphering the mysteries of myoglobin in striated muscle. *Acta Physi. Scand.*, 168:623–34, 2000.
58. B. A. Wittenberg and J. B. Wittenberg. Transport of oxygen in muscle. *Ann. Rev. Phys.*, 51:857–78, 1989.
59. P. Lipton. Ischemic cell death in brain neurons. *Phys. Rev.*, 79:1431–1568, 1999.
60. S. Herold, A. Fago, R. E. Weber, S. Dewilde, and L. Moens. Reactivity studies of the Fe(III) and Fe(II)NO forms of human neuroglobin reveal a potential role against oxidative stress. *J. Bio. Chem.*, 279:22841–47, 2004.
61. K. Wakasugi, T. Nakano, and I. Morishima. Oxidized human neuroglobin acts as a heterotrimeric Gα protein guanine nucleotide dissociation inhibitor. *J. Bio. Chem.*, 278:36505–12, 2003.
62. K. Wakasugi, C. Kitatsuji, and I. Morishima. Possible neuroprotective mechanism of human neuroglobin. *Ann. NY Acad. Sci.*, 1053:220–30, 2005.
63. A. Fago, A. J. Mathews, L. Moens, S. Dewilde, and T. Brittain. The reaction of neuroglobin with potential redox protein partners cytochrome b5 and cytochronome *c*. *FEBS Lett*, 580:4884–8, 2006.
64. A. A. Khan, Y. Wang, Y. Sun, X. O. Mao, L. Xie, E. Miles, J. Graboski, S. Chen, L. M. Ellerby, K. Jin, and D. A. Greenberg. Neuroglobin-overexpressing transgenic mice are resistant to cerebral and myocardial ischemia. *Proc. Natl. Acad. Sci. USA*, 103:17944–8, 2006.
65. C. B. Anfinsen. Studies on the principles that govern the folding of protein chains, Nobel Lecture, Dec. 11, 1972, www.nobelprize.org.
66. C. Levinthal. In P. Debrunner, J. C. M. Tsibris, and E. Munck, editors, *Mossbauer Spectroscopy in Biological Systems, Proceedings of a Meeting Held at Allerton House, Monticello, IL.* University of Illinois Press, Urbana, IL, 1969. p. 22.
67. J. N. Onuchic, Z. Luthey-Schulten, and P. G. Wolynes. Theory of protein folding: The energy landscape perspective. *Ann. Rev. Phys. Chem.*, 48:545–600, 1997.

68. H. Frauenfelder, P. W. Fenimore, G. Chen, and B. H. McMahon. Protein folding is slaved to solvent motions. *Proc. Natl. Acad. Sci. USA*, 103:15469–72, 2006.
69. M. Olivberg and P. G. Wolynes. The experimental survey of protein-folding energy landscapes. *Q. Rev. Bioph.*, 38:245–88, 2005.
70. C. M.Dodson. Protein folding and misfolding. *Nature*, 426:884–90, 2003.
71. P. L. Privalov and G. I. Makhatdze. Heat capacity of proteins II. *J. Mol. Bio.*, 213:385–91, 1990.
72. S. Milev, A. A. Gorfe, A. Karshikoff, R. T. Chubb, H. R. Bosshard, and I. Jelesarov. Energetics of sequence-specific protein-DNA association: Conformational stability of the DNA binding domain of integrase Tn916 and its cognate DNA duplex. *Biochem.*, 42:3492–3502, 2003.
73. M. Jager, H. Nguyen, J. C. Crane, J. W. Kelly, and M. Gruebele. The folding mechanism of a β-sheet: The WW domain. *J. Mol. Bio.*, 311:373–93, 2001.
74. R. Zwanzig. Simple model of protein folding kinetics. *Proc. Natl. Acad. Sci. USA*, 92:9801–4, 1995.
75. J. D. Byrngelson and P. G. Wolynes. Spin glasses and the statistical mechanics of folding. *Proc. Natl. Acad. Sci. USA*, 84:7524–8, 1987.
76. P. W. Fenimore, H. Frauenfelder, B. H. McMahon, and F. G. Parak. Slaving: Solvent fluctuations dominate protein dynamics and folding. *Proc. Natl. Acad. Sci. USA*, 99:16047–51, 2002.
77. V. Lubchenko, P. G. Wolynes, and H. Frauenfelder. Mosaic energy landscapes of liquids and the control of protein conformational dynamics by glass-forming solvents. *J. Phys. Chem. B*, 109:7488–99, 2005.
78. C. Ronne, L. Thrane, P.-O. Astrand, A. Wallqvist, K. V. Mikkelson, and S. R. Keiding. Investigation of the temperature dependence of dielectric relaxation in liquid water by THz reflection spectroscopy and molecular dynamics. *J. Chem. Phys.*, 107:5319–31, 1997.
79. A. Fersht. *Structure and Mechanism in Protein Science: A Guide to Enzyme Catalysis and Protein Foldings.* W. H. Freeman, New York, 1998.
80. K. Sneppen and G. Zocchi. *Physics in Molecular Biology.* Cambridge University Press, Cambridge, 2005, chapter 5.
81. D. Whitford. *Proteins Structure and Function.* John Wiley, West Sussex, 2005, chapter 11.
82. A. R. Dinner, A. Sali, L. J. Smith, C. M. Dobson, and M. Karplus. Understanding protein folding via free energy surfaces from theory and experiment. *Trends Bioch. Sci.*, 25:331–9, 2000.
83. H. Frauenfelder et al. Proteins and pressure. *J. Phys. Chem.*, 94:1024–37, 1990.
84. H. Frauenfelder, B. H. McMahon, R. H. Austin, K. Chu, and J. T. Groves. The role of structure, energy landscape, dynamics, and allostery in the enzymatic function of myoglobin. *Proc. Natl. Acad. Sci. USA*, 98:2370–4, 2001.
85. J. B. Wittenberg and B. A.Wittenberg. Myoglobin function reassessed. *J. Exp. Bio.*, 206:2011–20, 2003.
86. V. V. Ponkratov, J. Friedrich, and J. M. Vanderkooi. Hole burning experiments with proteins: Relaxations, fluctuations, and glass-like features. *J. Non-Cryst. Solids*, 352:4379–86, 2006.
87. V. V. Ponkratov, J. Friedrich, J. M. Vanderkooi, A. L. Burin, and Y. A. Berlin. Physics of proteins at low temperatures. *J. Low Temp. Phys.*, 137:289–317, 2004.

88. M. Lim, T. A. Jackson, and O. A. Anfinrud. Femtosecond near-IR absorbance study of photoexcited myoglobin: Dynamics of electronic and thermal relaxation. *J. Phys. Chem.*, 100:12043–51, 1996.
89. X. S. Xie. Single-molecule approach to dispersed kinetics and dynamic disorder: Probing conformational fluctuations and enzymatic dynamics. *J. Chem. Phys.*, 117:11024–32, 2002.
90. W. Min, G. Luo, B. J. Cherayil, S. C. Kou, and X. S. Xie. Observation of a Power-law memory kernel for fluctuations within a single protein molecule. *Phys. Rev. Lett.*, 94:198302-6, 2005.

Part V

APPENDICES: TOOLS AND CONCEPTS FOR THE PHYSICS OF PROTEINS

The appendices describe some information that can come in handy when studying the biological physics of proteins but that are not necessary for the main flow of the text. The chapters do not follow any particular order. As an introduction, we make a few remarks about units and typical energies.

In different subfields of physics and chemistry, different units for energy are used, and this can lead to confusion. We therefore summarize some of the units and conversion factors here. We first note that we can state the energy either per atom or molecule or per mole. The corresponding Boltzmann factors and the relations are given in Table A.1.

We sometimes switch without warning from "per atom" or "per molecule" to "per mole." No harm should be done! In the chemical literature, energies are frequently given in kJ/mol. In atomic physics the electron Volt, eV, is a handy unit; every physicist remembers the binding energy of the first Bohr orbit of the hydrogen atom, –13.6 eV. Spectroscopists either measure energy in terms of frequency, using the Hertz, Hz, or kHz, MHz, GHz as the basic unit, or in terms of wavenumbers, cm^{-1}, i.e. the inverse of the wavelength. These units are related to the energy proper by the following expressions

$$E = h\nu = \hbar\omega = \frac{\hbar c}{\lambda} \tag{A.1}$$

where h is Planck's constant, $\hbar = h/2\pi$, ν is the frequency in Hz (= 1 cycle/sec), ω is angular frequency in radians/sec, c is the speed of light, and λ is the wavelength, $\lambda = \lambda/2\pi$, $\hbar c = 197\,\mathrm{eV\,nm}$. Energy can also be expressed in terms of absolute temperature T in Kelvin, K.

$$E = k_B T \qquad E' = RT \;. \tag{A.2}$$

Conversion expressions are summarized in Table A.2. Note that the conversion switches without warning from "per mole" to "per molecule." Moreover,

Table A.1.

Per Molecule	Per Mole
$\exp(E/k_BT)$	$\exp(E'/RT)$
$E = N\,E$	
$R = N\,k_B$	
$N = 6.022 \times 10^{23}\mathrm{mol}^{-1}$	
$k_B = 1.381 \times 10^{-23}\,\mathrm{J\,K}^{-1}$	
$R = 8.314\,\mathrm{J\,mol}^{-1}K^{-1}$	

Table A.2. Conversions of energy units.

1 kJ/mol = 0.239 kcal/mol = 10.36 meV = 2.50 GHz = 83.5 cm^{-1} = 120 K.

the equations are dimensionally incorrect. The usefulness of the table is nevertheless clear – it permits changes from one value to another.

In Fig. A.1, we compare energy, frequency, and wavelength scales; we label the various spectroscopic regions; and we indicate typical atomic or molecular transitions.

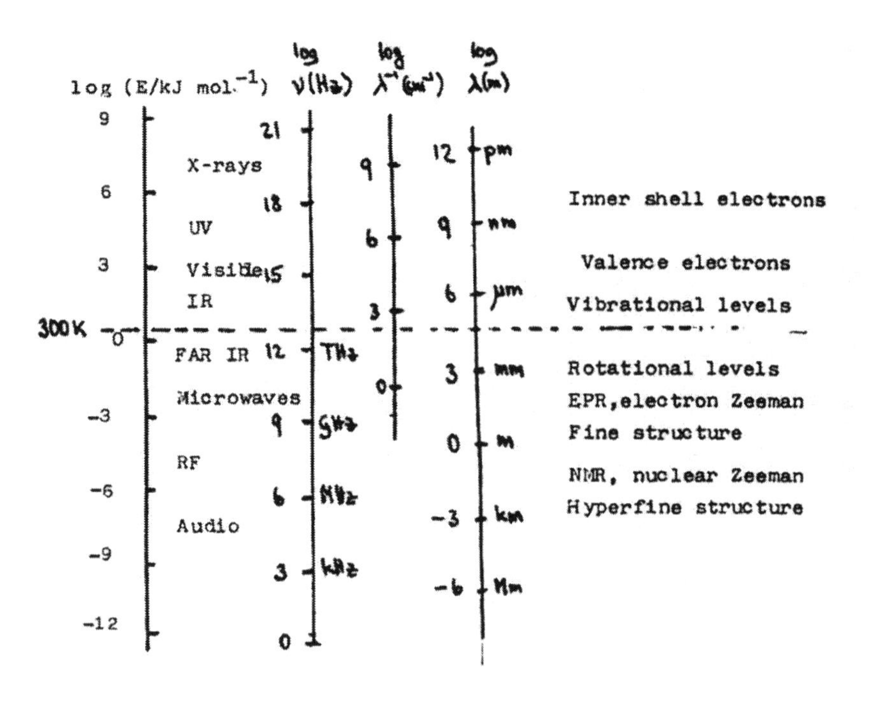

Fig. A.1. Energies, frequencies, and wavelengths.

As is evident from the diagram, the various types of excitation are widely separated in frequency. In particular, there are many bands in which the molecule does not absorb any energy at all. If we tune in to a specific absorption band of a biomolecule, chances are that the absorption is due to a localized group only, a so-called chromophore. A typical example is the heme group in myoglobin, which accounts for all the visible absorption and therefore the color of Mb.

The fact that the electronic, vibrational, and rotational energies of a molecule are so far apart is due to the large ratio of the nuclear to the electronic mass

$$M/m \approx 10^4 \ldots 10^5. \tag{A.3}$$

The zero-point motion of the nuclei is relatively small; nuclei have fairly well defined equilibrium positions about which they oscillate slowly compared to the electronic motion. The order of magnitude of the electronic, vibrational, and rotational energies can easily be estimated. The uncertainty relation yields for an electron of mass m confined to a linear dimension $a \sim 0.1\,\text{nm}$,

$$E_e \sim \hbar^2/2ma^2 \sim \text{few eV}. \tag{A.4}$$

The ratio of vibrational to electronic energies is

$$\frac{E_{\text{vib}}}{E_e} \sim \left(\frac{m}{M}\right)^{1/2}, \quad \text{i.e., } E_{\text{vib}} \sim 0.01, \ldots, 0.1\,\text{eV}. \tag{A.5}$$

Rotational energy levels are the result of quantized rotation of the entire molecule about its center of mass; the energies are given by

$$E_{\text{rot}} \sim \frac{\hbar^2 \ell(\ell+1)}{2j} \sim 10^{-4} \ldots 10^{-5}\text{eV}. \tag{A.6}$$

Here $j \sim Ma^2$ is the moment of inertia of the molecule. Equations (A.4) and (A.6) give

$$\frac{E_{\text{rot}}}{E_e} \sim m/M. \tag{A.7}$$

The energy of an excited state is thus given by

$$E = E_e + E_{\text{vib}} + E_{\text{rot}}. \tag{A.8}$$

If the E_{rot} levels are closely spaced, band spectra result.

18

Chemical Forces

"Strong" and "weak" forces are crucial for biomolecular structure and function. The "strong" ones hold the backbone permanently together; the "weak" ones allow the protein to form its three-dimensional structure and to fluctuate around the main structure. Much biological action is intimately related to movements. A knowledge of the various forces is therefore necessary for an understanding of structure and function.

All chemical forces are of electromagnetic origin. In principle, one can write the Schrödinger equation and solve the problem. In practice, however, this approach is often impossible, and knowledge of the characteristic properties of the various forces is useful. (Of course, nature does not separate the forces as we do; the entire interaction is unified. The separation that we perform simply is an indication of our inability to treat the problem as a whole.)

18.1 Survey

Some characteristics of the various forces are given in Table 18.1.

Table 18.1. Chemical Forces.

Force	Strength, $kJmol^{-1}$	Remarks
Covalent	200–1000	"Strong"
Electrostatic	10–30	
Hydrogen bond	5–20	
Van der Waals	1–10	Dispersion force
Hydrophobic	0–20	Not a true force

H. Frauenfelder, *The Physics of Proteins*, Biological and Medical Physics, Biomedical Engineering, DOI 10.1007/978-1-4419-1044-8_18,

A more detailed overview is given in Fig. 18.1.

TYPE OF INTERACTION		INTERACTION ENERGY
COVALENT	H_2, H_2O	COMPLICATED, SHORT RANGE
CHARGE–CHARGE	Q_1 D Q_2	$U = +Q_1 Q_2/(4\pi\varepsilon_0 D)$
CHARGE–NEUTRAL	Q D α	$U = -Q^2\alpha/(8\pi\varepsilon_0 D^4)$
DIPOLE–DIPOLE	p_1 θ_1 D ϕ θ_2 p_2 FIXED DIPOLES (LOW T)	$U = \frac{-p_1 p_2}{4\pi\varepsilon_0 D^3}(2\cos\theta_1 \cos\theta_2 - \sin\theta_1 \sin\theta_2 \cos\phi)$
	p_1 D p_2 FREELY ROTATING DIPOLES (HIGH T)	$U = \frac{-p_1^2 p_2^2}{6\pi\varepsilon_0 kT D^6}$
CHARGE–DIPOLE	p θ D Q FIXED	$U = -Qp\cos\theta/(4\pi\varepsilon_0 D^2)$
	p D Q ROTATING	$U = -Q^2p^2/(12\pi\varepsilon_0 kT D^4)$
DIPOLE–NEUTRAL	p θ D α FIXED	$U = -p^2\alpha(1+3\cos^2\theta)/(8\pi\varepsilon_0 D^6)$
	p D α ROTATING	$U = -p^2\alpha/(4\pi\varepsilon_0 D^6)$
HYDROGEN BONDING	···H–O(H)···H–O(H)···H–O···	$U \approx -$ CONSTANT$/D^2$ SHORT RANGE
NEUTRAL–NEUTRAL (VAN DER WAALS OR DISPERSION)	α D α	$U = -(3/4)h\nu\alpha^2/[(4\pi\varepsilon_0)^2 D^6]$

Fig. 18.1. Common type of interactions between atoms and molecules. U = interaction energy (in J), Q = electric charge (C), p = electric dipole moment (C m), k = Boltzmann constant (1.38×10^{-23} J K^{-1}), T = absolute temperature (K), D = distance between interacting atoms or molecules (m), θ = angle, α = electric polarizability (F m), h = Planck constant (6.626×10^{-34} J s), ν = electron orbiting frequency (s^{-1}), ϵ_o = permittivity of free space (8.854×10^{-12} F m^{-1}). The force is obtained by differentiating the energy U with respect to distance (from J. I. Israelachvili, *Contemp. Phys.* 15:159-77, 1974). Units: 1 kJ/mol = 239 cal/mol = 1.036 $\times10^{-2}$ eV = 120 K.

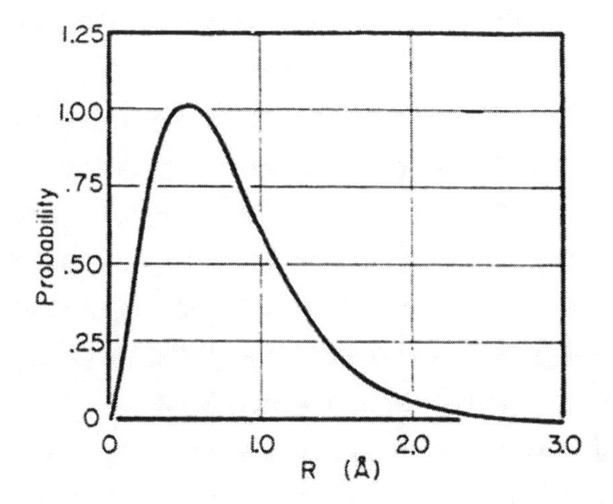

Fig. 18.2. Probability of finding the electron of a hydrogen atom at distance R.

18.2 Atomic Radii

Before turning to the individual forces, we look at atomic radii, because they determine packing and possible motions in biomolecules. In Fig. 18.2, we plot the probability distribution for a hydrogen atom. No clear edge is observable; the distance of approach of two atoms depends on the strength of the mutual interactions. The atomic "radius" therefore is not a universal number, but it varies greatly.

A number of typical radii are given in Table 18.2. Note that 100 pm = 1 Å. "Noble" means the structure of the corresponding noble gas, "univ." the univalent ion.

In Table 18.2, approximate "models" of molecules can be built. Consider CO. In a first approximation, the bond between C and O can be described

Table 18.2. Atomic Radii, in pm.

Element	Van der Waals	Ionic				Covalent		
		+		−				
		Noble	Univ.	Noble	Univ.	Single	Double	Triple
1 H	117	0	—	208	208	30	—	—
6 C	170	15	29	260	414	77	67	63
7 N	157	11	25	174	247	74	62	55
8 O	150	9	11	140	176	74	62	55

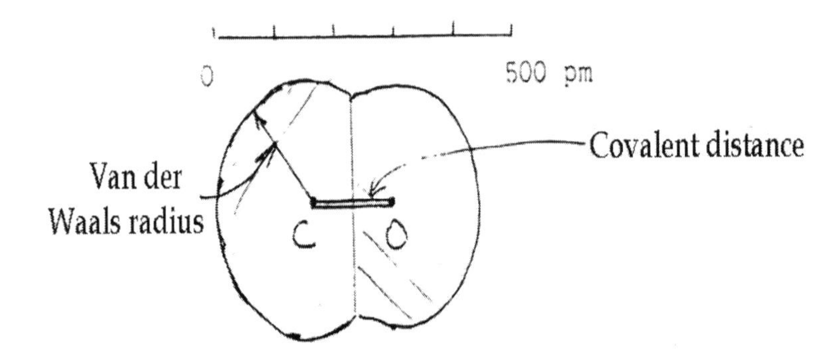

Fig. 18.3. Simple construction of C=O.

as a double bond, C=O. The distance between the centers of C and O then would be 129 pm, the radius around C 170 pm, around O 150 pm. The actual CO distance is somewhat smaller, 113 pm [1] so that the molecules can be pictured as shown in Fig. 18.3.

The covalent bond leads to an interpenetration of the shells of the two atoms.

18.3 Electric Dipole Moment and Polarizability

Two more properties are important when looking at binding forces: the electric dipole moment and the polarizability. Two charges q and $-q$, separated by a distance r, form a dipole with dipole moment $\mathbf{p} = q\,\mathbf{r}$.

Here $\mathbf{r}$ is the vector from the negative to the positive charge. Dipole moments are measured in debye (D):

$$1\mathrm{D} = 3.338 \times 10^{-30}\mathrm{C\,m}.$$

1 D was originally defined as 10^{-18}esu cm. With $q = -e$ and $r = 100\,\mathrm{pm}$, $p = 4.8\,\mathrm{D}$. (Note that D is also used as an abbreviation for dalton, the mass unit). Nuclei and atoms do not have permanent dipole moments, but many molecules do. Molecules with permanent electric dipole moments are called polar [2]; some examples are given in Table 18.3.

Nonpolar molecules acquire an induced dipole moment in an external electric field $\mathbf{E}$:

$$\mathbf{P}_{\mathrm{ind}} = \alpha\,\mathbf{E}. \tag{18.1}$$

Here α is called polarizability and values are also given in Table 18.3. The polarizability gives information about the distortion produced by $\mathbf{E}$.

Table 18.3. Dipole Moment and Polarizability.

Molecule	p/D	$\alpha/10^{-24}\text{cm}^3$
H_2	0	0.79
N_2	0	1.76
CO	0	2.65
O_2	0	—
CO	0.10	1.95
H_20	1.85	1.48

Proteins can have large electric dipole moments [3, 4]. The geometry and dipole moment of a single peptide unit are shown in Fig. 18.4 [5, 6].

In an α-helix the peptide dipole moments are aligned nearly parallel to the helix axis. The field of the dipole can be approximated by the field of a continuous line dipole with a dipole density of 2–3 D/Å [6]. The field of such a dipole is equal to the field of a positive charge at the amino end and a negative charge at the carboxyl end, each of magnitude of $e/2$. Unlike single charges in proteins, which are either solvated or compensated, the fictive charges exert their full effect.

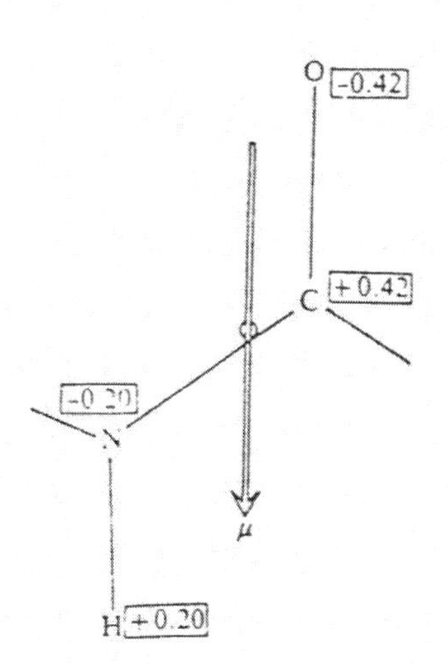

Fig. 18.4. Geometry and dipole moment of the peptide unit. Numbers in boxes give the approximate fractional atomic charges (in units of the elementary charge, e) with the charge of carbon changed into +0.42 in order to preserve charge neutrality (Table 18.5). The dipole moment μ has a value of 0.72 e Å= 3.46 D = 1.155 $\times 10^{-29}$C m.

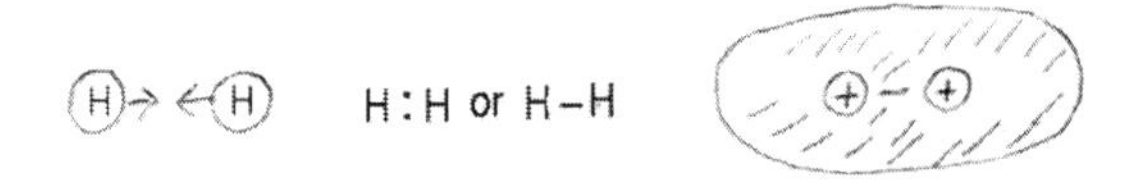

Fig. 18.5. The molecule H_2.

18.4 The Covalent Bond

Since the covalent bond [1, 7]–[9] is treated essentially everywhere, only a few remarks will be made here. The covalent bond is a typical quantum effect and results from the sharing of electrons by two atoms. Consider two neutral hydrogen atoms. Classically, they exert no force and do not bind. Quantum mechanics shows, however, that there is a state in which the two electrons form a bond, as indicated in Fig. 18.5.

Semiclassicially, we explain the bond by saying that the electron pair has a large probability of being between the two nuclei. There therefore exists an attraction + − + that binds the two atoms. Symbolically, the two paired electrons are written as two dots or one line.

The chemical or covalent bond is *directional* and *saturated.* A given atom can only establish a certain number of bonds; the number is called the *valence.* Since a state can consist of a superposition, a particular bond can assume a character intermediate between "pure" states. In Table 18.4 bond energies and lengths are summarized.

If the covalent bond occurs between two nonidentical neutral atoms, it results in a charge shift and each atom of the newly formed molecule has a "partial charge."

In some proteins, the disulfide bond (sulfur bridge), formed between the sulfurs of two cystines (–S–S–), adds to the stability. The bond energy is 220 kJ/mol. The covalent bonding situation of paramount interest in proteins is the peptide bond that forms the backbone.

18.5 Electrostatic Forces

Covalent forces are short range. If no overlap between orbitals exist, the electrostatic forces become important. The force laws are given in Fig. 18.1 so we can restrict the discussion here to pointing out some essential features.

(a) *The ionic bond (salt bond).* Ionic bonds occur between charged atoms or molecules. In a vacuum they can be very strong. The high dielectric coefficient

Table 18.4. Bond Energies and Lengths.

Bond	Energy (kJ/mol)	Length (pm)
H–H	450	74
C–C	350	154
C=C	680	134
C≡C	960	121
C–H	410	110
C–N	340	147
C=N	660	129
C≡N	1000	116
C–O	380	142
C=O	730	121
C≡O	1100	113
O–H	460	96
N–H	390	100

of water (about 80) reduces the bond strength considerably so that in water the ionic bond becomes about as strong as a hydrogen bond. Within proteins, the value of the macroscopic dielectric coefficient, ϵ, of amide polymers, is about 4, and salt bonds can become strong.

Charges in proteins are caused by the covalent bonds that lead to an asymmetric bond electron distribution. Some typical values are given in Table 18.5. Such charges can indeed stabilize proteins [10, 11].

(b) *Other electrostatic forces.* Expression for the interaction between charged and uncharged polar and nonpolar molecules are also given in Fig. 18.1. As an example in Table 18.6 we give the interaction energy between two dipoles, each with a moment 1 D at a distance of 0.5 nm.

18.6 Hydrogen bond

In a hydrogen bond [12]–[15], a hydrogen atom acts as a link between two other atoms, for instance, in HF_2^-: $(F\text{–}H\text{–}F)^-$. To explain some of its features, consider an H bond between two unequal atoms. The bond will be strong to one of the two, the one that is more electronegative. (*Electronegativity* is vaguely defined by Pauling as the power of an atom in a molecule to attract electrons to itself.) The bond is then shown as

$$\text{Donor} - \text{H} \quad \cdots \quad \text{Acceptors.}$$

Table 18.5. Partial Atomic Charges.

Position	Atom	Charge/e
Backbone	N	–0.36
Backbone	C_α	+0.06
Backbone	C	+0.45
Backbone	O	–0.38
Side chain	C_β	–0.12
Side chain	C_γ	+0.46
Side chain	N_δ	–0.45

Table 18.6. Dipole-Dipole Interaction.

Orientation	E/kJmol^{-1}
$\longrightarrow\longrightarrow$	–5.6
$\longrightarrow\longleftarrow$	+5.6
$\uparrow\uparrow$	+2.8
$\uparrow\downarrow$	–2.8

We can describe the bond as a combination of covalent and ionic bonds:

$$\begin{array}{cc} \delta^- \ \delta^+ & \epsilon^- \\ \text{D:H} \ \cdots & \text{A} \\ \text{covalent} & \text{ionic} \end{array}$$

The covalent bond produces a charge shift and the hydrogen atom obtains a net charge δ^+. Then there can be an ionic bond to the acceptor atom with charge ϵ^-. The computation of the properties of the H bond is in general difficult because there are five contributions of about equal magnitude, electrostatic energy, exchange repulsion, polarization energy, covalent contribution, and van der Waals (dispersion) energy.

The hydrogen bond possesses a number of features that are important for biological systems:

1. Only the most electronegative atoms (F,O,N) form hydrogen bonds. The strength increases with electronegativity. Relative electronegativities are

$$\text{F} \quad 4.0 \qquad \text{O} \quad 3.5 \qquad \text{N} \quad 3.0$$

2. A hydrogen bond is formed only between two atoms.

Table 18.7. Hydrogen Bonds.

Systems	Substance	Distance Acceptor-Donor in pm	Bond Energy in $kJmol^{-1}$
O–H···O	H_2	276	19
O–H···O	$(HCOOH)_2$	267	30
N–H···N	NH_3	340	6
N–H···O	amide-carbonyl	290	

3. Hydrogen bonds are shorter than the sum of the relevant van der Waals radii. Some values of the distance between donor and acceptor are given in Table 18.7. Bond energies in general decrease rapidly with increasing bond length.

4. Hydrogen bonds are linear. The positive hydrogen ion is located between two negative charges. The optimum interaction occurs if the charges are collinear. Nonlinear bonds are weaker. Some hydrogen bonds of biological importance are shown in Fig. 18.6.

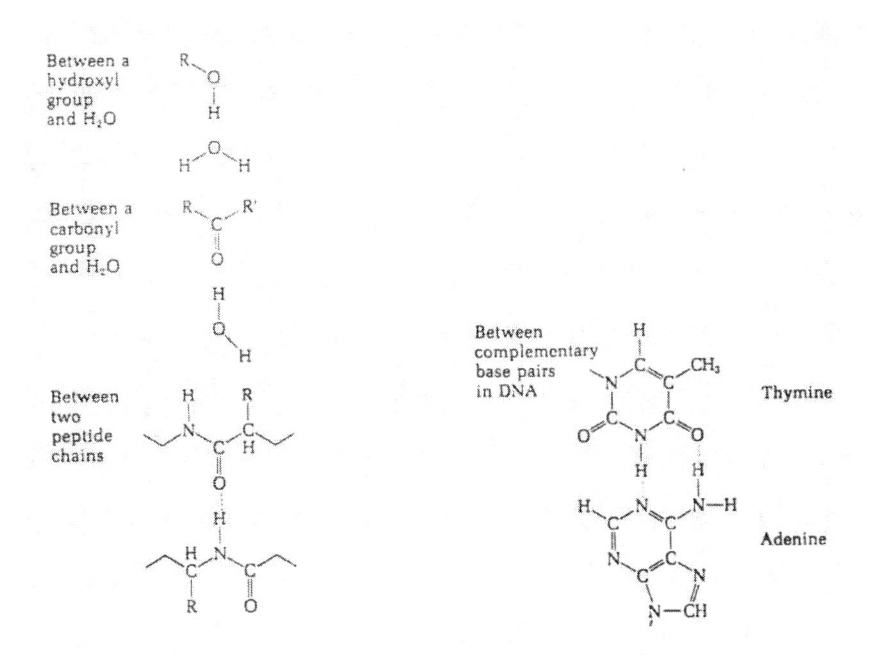

Fig. 18.6. Some hydrogen bonds of biological importance.

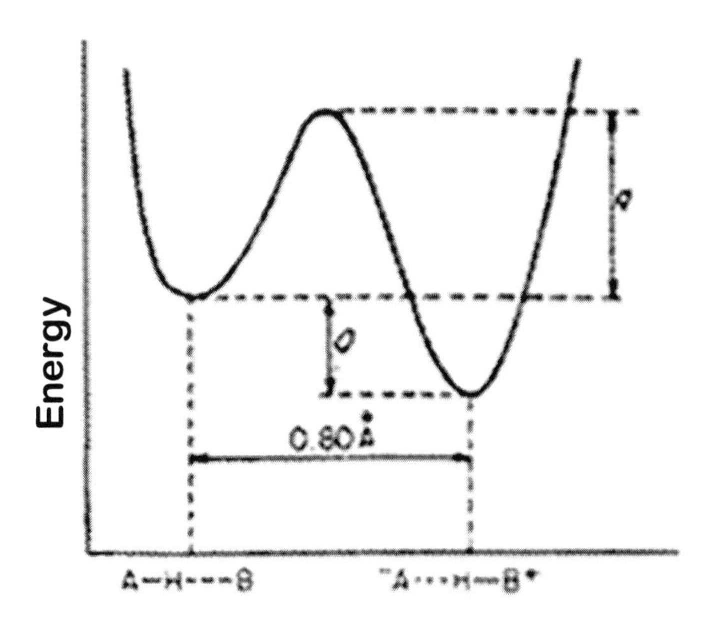

Fig. 18.7. A potential energy diagram for proton transfer within an H-bonded complex. The proton transfer complex or ionized state, is chosen, arbitrarily, to have a minimum lower than that of the un-ionized state.

In a hydrogen-bonded complex, proton transfer can occur and the transfer may be crucial for many biological functions. The potential energy diagram corresponding to the proton transfer is shown in Fig. 18.7. Transfer will occur by quantum-mechanical tunneling.

18.7 Van der Waals (Dispersion) Force

Van der Waals forces [16, 17], like gravitational forces, are always present. They arise from mutual polarization. Each atom can be considered as a fluctuating dipole that induces a polarization in the other atom. The net result is generally an attractive force. For crude estimates, the interaction energy can be written as

$$E_{VdW} = -\alpha^2 I/r^6 \qquad (18.2)$$

where α is the polarizability and I the average ionization energy. Between small molecules, E_{VdW} lies between 1 and 10 kJmol^{-1}. It is possible that fluctuations in biomolecules can lead to much larger forces. The force is actually much more complicated than stated here, and for details we refer to [16, 17].

18.8 Electron Shell Repulsion and Van der Waals Potential

At small distances, atoms and molecules repel each other. The repulsion can be described by a steeply rising potential at about the van der Waals radius. For calculations, it is convenient to combine the repulsion, the van der Waals attraction, and the electrostatic interactions into one potential. In the simplest case, the van der Waals potential is isotropic and written as

$$E = \frac{A}{R^{12}} - \frac{B}{R^6} + \frac{q_1 q_2}{R}. \tag{18.3}$$

Here R is the distance between the partners, q_1 and q_2 their effective charges, and A and B two empirical parameters. The first term describes the repulsion, the second the van der Waals attraction, and the third the electrostatic interaction.

18.9 The Hydrophobic (Apolar) Interaction

There is no real hydrophobic force [18]–[20]. The amino acids assume a configuration with the lowest free energy. Since

$$G = H - TS,$$

a minimum in the Gibbs energy G can be obtained by making H large and negative or by making S large and positive. In the first case, a real force can bind the groups, leading to a stable configuration. In the second case, the protein configuration allows the solvent to assume a state of low order of high positive entropy. Such a reaction is called entropy driven, and one speaks about a hydrophobic interaction.

To see how such an interaction can occur, we picture a hydrophobic side group in water (Fig. 18.8). The water forms a cagelike clathrate structure around it. The water is in a state of high order. Upon forming a globular protein, water can return to the normal state of disorder. The resulting increase in entropy leads to a state of lower free energy and stabilizes the protein. Further details on this can be found Chapter 23, where we go into more detail on water.

It is difficult to calculate the "hydrophobic interaction," and thus measurements are needed. Tanford has observed the Gibbs energy change of amino acids upon transfer from water to ethanol. Ethanol is taken to represent the interior of protein molecules. The free energy change is shown in Table 18.8 [21]. The first column represents the measured change; the second is obtained

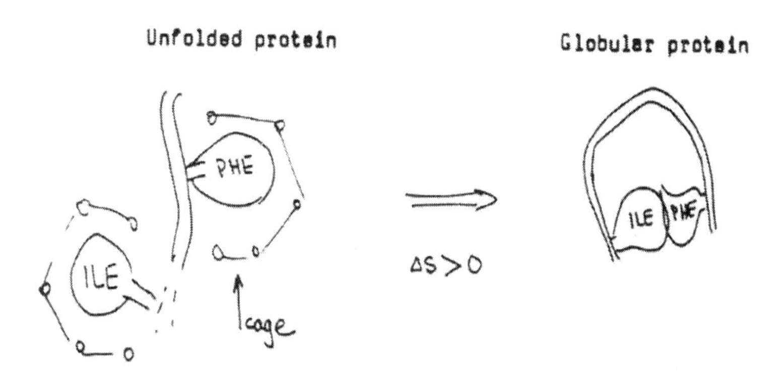

Fig. 18.8. Folding of a protein can lead to increased entropy and consequently lower Gibbs energy.

after subtraction of the contribution of groups that are not in the side chain, by using glycine. Further details on the protein-folding problem can be found in Chapter 17.

Table 18.8. Gibbs energy change in kilocalories per mole for transfer from water to 100% ethanol at 25°C [21].

	$\Delta G_{\text{transfer}}$	
	Whole Molecule	Side Chain Only
Glycine	+4.63	0
Alanine	+3.90	–0.73
Valine	+2.94	–1.69
Leucine	+2.21	–2.42
Isoleucine	+1.69	–2.97
Phenylalanine	+1.98	–2.65
Proline	+2.06	–2.60
Methionine	+3.33	–1.30
Tyrosine	+1.76	–2.87
Threonine	+4.19	–0.44
Serine	+4.59	–0.04
Asparagine	+4.64	+0.01
Glutamine	+4.73	+0.10

References

1. L. Pauling. *The Nature of the Chemical Bond.* Cornell Univ. Press, Ithaca, 1960. The standard work. Somewhat difficult to read for physicists because many concepts have to be translated into standard quantum mechanical language. Still, the first book to look at when problems arise.
2. P. Debye. *Polar Molecules.* Dover, New York, 1945. Reprint of the 1929 work. The classic book, still useful.
3. R. Pethig. *Dielectric and Electronic Properties of Biological Materials.* Wiley, New York, 1979.
4. E. H. Grant, R. J. Sheppard, and G. P. South. *Dielectric Behavior of Biological Molecules in Solution.* Clarendon Press, Oxford, 1978.
5. A. Wada. The α-helix as an electric macro dipole. *Adv. Biophys.*, 9:1–63, 1976.
6. W. G. J. Hol, P. T. van Duijnen, and H. J. C. Berendsen. The alpha-helix dipole and the properties of proteins. *Nature*, 273(5662):443–6, 1978.
7. Information and detailed calculations on the chemical bond can be found in all texts on quantum chemistry and most texts on quantum mechanics. A particularly clear introduction is in W. Heitler, *Elementary Wave Mechanics*, 2nd edition. Clarendon Press, Oxford, 1956.
8. C. A. Coulson. *Valence*, 2nd edition. Oxford Univ. Press, New York, 1961.
9. T. Shida. *The Chemical Bond: A Fundamental Quantum-Mechanical Picture.* Springer, Berlin, 2004.
10. C. Tanford and J. G. Kirkwood. Theory of protein titration curves. I.: General equations for impenetrable spheres. *J. Amer. Chem. Soc.*, 79:5333–9, 1957.
11. S. J. Shire, G. I. H. Hanania, and F. R. N. Gurd. Electrostatic effects in myoglobin: Hydrogen ion equilibriums in sperm whale ferrimyoglobin. *Biochem.*, 13:2967–74, 1974.
12. G. A. Jeffrey. *An Introduction to Hydrogen Bonding.* Oxford Univ. Press, New York, 1997.
13. G. C. Pimentel and A. L. McCellan. *Hydrogen Bond.* W. A. Freeman, San Francisco, 1960.
14. C. Tanford. Protein denaturation. *Adv. Protein Chem.*, 23:121–282, 1968 and *Adv. Protein Chem.*, 24:1–95, 1970.
15. P. Schuster, G. Zundel, and C. Sandorfy, editors. *The Hydrogen Bond.* North Holland, Amsterdam, 1976, 3 vol.

16. V. A. Parsegian. *Van der Waals Forces: A Handbook for Biologists, Chemists, Engineers and Physicists.* Cambridge Univ. Press, New York, 2005.
17. D. Langbein. *Theory of Van der Waals Attraction*, Springer Tracts in Modern Physics, vol. 72, Springer, Berlin, 1974.
18. A. Ben-Naim. *Hydrophobic Interactions.* Plenum Press, New York, 1980.
19. C. Tanford. *The Hydrophobic Effect.* Wiley, New York, 1980.
20. P. R. ten Wolde. Hydrophobic interactions: An overview. *J. Phys. Condens. Matter*, 14:9445–60, 2002.
21. C. Tanford. Contribution of hydrophobic interactions to the stability of the globular conformation of proteins. *J. Amer. Chem. Soc.*, 84:4240–7, 1962.

19

Acids and Bases for Physicists

Charges are essential for the stability and function of many proteins. To understand the effect of the environment, particularly its pH on stability and function, a few concepts are needed.

19.1 Acids and Bases

An *acid* is a proton donor, a *base* a proton acceptor; they always occur in conjugated pairs [1] :

$$\text{acid} \leftrightarrow \text{base} + \mathrm{H}^+. \tag{19.1}$$

Examples are

$$\begin{array}{lll} \mathrm{NH}_4^+ & \leftrightarrow & \mathrm{NH}_3 & + \mathrm{H}^+ \\ \mathrm{H}_2\mathrm{SO}_4 & \leftrightarrow & \mathrm{HSO}_4^- & + \mathrm{H}^+ \\ \mathrm{H}_3\mathrm{O}^+ & \leftrightarrow & \mathrm{H}_2\mathrm{O} & + \mathrm{H}^+ \\ \mathrm{H}_2\mathrm{O} & \leftrightarrow & \mathrm{OH}^- & + \mathrm{H}^+ \end{array}$$

Some compounds can act as either base or acid.

Physicists often believe that the H^+ appears as a proton. In reality, however, protons have the tendency to sneak into the electron shell of neighboring molecules. In an aqueous solution, protons appear in the form of hydronium ions, H_3O^+. The dissociation of an acid, Eq. (19.1), thus occurs as shown in Fig. 19.1.

The acid transfers its proton directly onto a neighboring water molecule. The correct way to describe the acid-base equilibrium hence is

$$\mathrm{AH} + \mathrm{H}_2\mathrm{O} \leftrightarrow \mathrm{A}^- + \mathrm{H}_3\mathrm{O}^+. \tag{19.2}$$

For "simplicity" (and to confuse physicists), the reaction is usually written in the form of Eq. (19.1).

H. Frauenfelder, *The Physics of Proteins*, Biological and Medical Physics, Biomedical Engineering, DOI 10.1007/978-1-4419-1044-8_19,

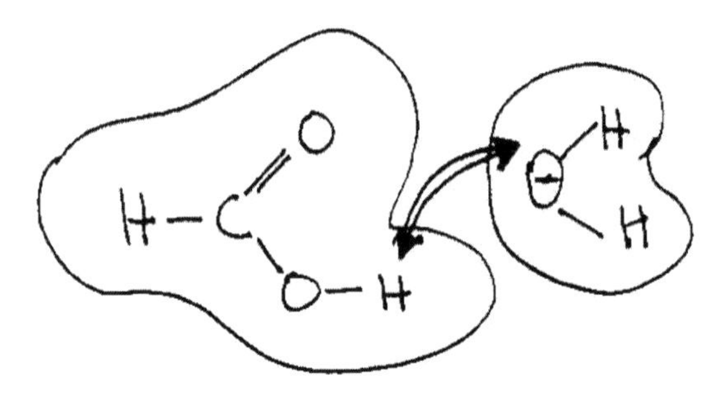

Fig. 19.1. Acid-base equilibrium.

19.2 The pH Scale

In water, Eq. (19.2) becomes

$$H_2O + H_2O \leftrightarrow HO^- + H_3O^+ \tag{19.3}$$

or

$$H_2O \leftrightarrow OH^- + H^+.$$

The equilibrium coefficient of this reaction is given by

$$K = [H^+][OH^-]/[H_2O]. \tag{19.4}$$

Usually, the concentration $[H_2O]$ of water is very large, about 55 M, and the concentrations of the H^+ and OH^- ions very small, $[H_2O]$ is essentially constant during the reaction (19.4). A new coefficient K_{eq} is consequently introduced through the definition

$$[H^+][OH^-] = K_{\mathrm{eq}}. \tag{19.5}$$

The coefficient K_{eq} is called the ion product of water. Its value at 25°C is $1.0 \times 10^{-14}\mathrm{M}^2$. In an acid solution, the concentration $[H^+]$ is relatively high; in a basic solution, $[OH^-]$ is relatively high.

Equation (19.5) is the basis for the pH scale. Since $\mathrm{K_{eq}}$ can change over many orders of magnitude, it is more convenient to use logarithms than powers. The term pH is defined as [2]

$$\mathrm{pH} = -\log[H^+].$$

In pure water, Eq. (19.4) tells us that $[H^-] = [OH^+] = 10^{-7}\,\mathrm{M}$ and hence pH (pure water) = 7. The pH of some solutions is given in Table 19.1.

Table 19.1.

Blood plasma	pH 7.4
Intracellular fluids	
Muscle	pH 6.1
Liver	pH 6.9
Interstitial fluid	pH 7.4

19.3 The Henderson-Hasselbalch Equation

Consider a conjugate acid-base pair, $AH \leftrightarrow A^- + H^+$. The equilibrium coefficient K for this reaction is called the dissociation coefficient:

$$K = \frac{[A][H^+]}{[AH]}. \tag{19.6}$$

Again, it is customary to use its negative logarithm,

$$pK = -\log K. \tag{19.7}$$

Equation (19.7) can then be written as

$$pH = pK - \log\{[AH]/[A^-]\}. \tag{19.8}$$

In this form, it is called the Henderson-Hasselbalch equation. At a fixed pH, it gives the ratio of protonated (AH) to deprotonated (A^-) form of the acid AH.

19.4 Amino Acids

The charges on amino acids are crucial for the function of proteins. Amino acids with neutral side chains can assume three charge states; A^+, A^0, and A^-. For the reactions $A^+ \leftrightarrow A^0 + H^+$ and $A^0 \leftrightarrow A^- + H^+$, two equilibrium coefficients K_1 and K_2 can be defined in analogy to Eq. (19.4). At low pH (pH $< pK_1$) the amino acid will be predominantly positive; at pH $> pK_2$, it will be predominantly negative. At pH $= pK_1$, the probabilities of finding A^+ or A^0 are equal. For amino acids with charged side chains, pK_R similarly indicates the pH at which charged and neutral sidechains have equal probabilities. Some values are given in Table 19.2.; pI is the pH of an aqueous solution such that the net charge on the molecule is zero [3].

Table 19.2. Values of pKs for some amino acids.

Amino Acid	pK_1 COOH	pK_2 NH_3	pK_R Residue	pI
ASP	2.10	9.82	3.89	2.98
GLU	2.1	9.47	4.07	3.08
GLY	2.35	9.78	—	6.06
HIS	1.77	9.18	6.10	7.64
LYS	2.18	8.95	10.53	9.47
TYR	2.20	9.11	10.07	5.63

References

1. G. Adam, P. Läuger, and G. Stark. *Physikalische Chemie und Biophysik*, 5. Aufl. Springer, Berlin, 2009.
2. The correct definition of pH involves the concept of activity (I. M. Klotz and R. M. Rosenberg, *Chemical Thermodynamics,* W. A. Benjamin, Menlo Park, 1972). For details, see, for instance, R. G. Bates, *Determination of pH: Theory and Practice,* Wiley, New York, 1973.
3. C. Tanford. *Physical Chemistry of Macromolecules*. Wiley, New York, 1961.

20

Thermodynamics for Physicists

Physicists know thermodynamics, but many use it only rarely in their research work. Knowledge therefore is not supplemented by intuition [1]. In biomolecular physics, thermodynamics is necessary. Equilibrium problems and dynamic questions call for thermodynamic concepts. In particular, entropy and volume changes during biomolecular reactions may be among the most important clues to the mechanism of a reaction. In the present section, an outline is given of the aspects of thermodynamics that we will need most. The discussion will be brief; further details and generalizations can be found in many texts [2]–[4].

20.1 Variables

Consider a homogeneous equilibrium system with one mechanical and one thermal degree of freedom. What is necessary to describe such a system completely? Only experiment can tell. Two hundred years of experiments have shown that four variables are necessary and sufficient: volume (V), entropy (S), pressure (P), and temperature (T). Volume and entropy are extensive; pressure and temperature are intensive quantities [5]: Of the four variables,

	Mechanical	Thermal
Extensive	V	S
Intensive	P	T

entropy is usually least familiar to physicists [6]. Many conceptual problems disappear if the Boltzmann relation is remembered

$$S/k_B = \ln W. \tag{20.1}$$

H. Frauenfelder, *The Physics of Proteins*, Biological and Medical Physics, Biomedical Engineering, DOI 10.1007/978-1-4419-1044-8_20,

Here, $k_B = 1.38 \times 10^{-23}$J/K $= 199$ cal mol^{-1}K^{-1} $= 8.62 \times 10^{-5}$eV K^{-1}, is the Boltzmann constant and W is the number of states. Quantum mechanically, W is well defined and can in principle be calculated for any system.

Two of the variables V, S, P, and T can be taken as independent, the other two are then *conjugate*. The choice depends on the experimental situation. At this point, an analogy with quantum mechanics may be helpful. A quantum mechanical system can be described in terms of a complete set of commuting operators. Only experiment can tell what these operators are and when a set is complete. The particular choice of operators depends on the experimental situation.

20.2 The Internal Energy

The "standard choice" is to take the extensive quantities V and S as independent. The first law of thermodynamics then states that the internal energy E of the system can change either by adding heat (dQ) to the system or letting the system do work (dW):

$$dE = dQ - dW. \tag{20.2}$$

The second law of thermodynamics relates the heat change to the change dS in entropy by

$$dQ = T\,dS, \tag{20.3}$$

and the work done by the system is given by

$$dW = P\,dV. \tag{20.4}$$

For irreversible processes, the equality sign in Eq. (20.3) is replaced by $\leq$. Equations (20.2) to (20.4) together give

$$dE = T\,dS - P\,dV. \tag{20.5}$$

The conjugate variables T and P can now be expressed in term of the independent variables V and S as

$$T = \left(\frac{\partial E}{\partial S}\right)_V, \quad P = -\left(\frac{\partial E}{\partial V}\right)_S. \tag{20.6}$$

From Eq. (20.6), the Maxwell relation

$$\left(\frac{\partial P}{\partial S}\right)_V = -\left(\frac{\partial T}{\partial V}\right)_S \tag{20.7}$$

follows. Because Eqs. (20.6) are formally similar to the relation between force and potential, $E(V, S)$ is called a *thermodynamic potential.*

20.3 Thermodynamic Potentials

The standard choice of V and S as the independent variables is not particularly useful for experiments. Experimentally, it is much easier to vary temperature and/or pressure rather than entropy and volume. To describe these situations, additional thermodynamic potentials are introduced. The situation is again similar to quantum mechanics, where different experimental conditions require different sets of commuting operators. Transformation theory allows us to go from one complete set to another. In thermodynamics, the change from one set of independent variables to another is effected by *Legendre transformations.*

To see the properties of one particular Legendre transformation, we introduce the *enthalpy* H through the definition

$$H = E + PV. \tag{20.8}$$

Its differential is

$$dH = dE + P\,dV + V\,dP$$

or, with Eq. (20.5),

$$dH = T\,dS + V\,dP. \tag{20.9}$$

The Legendre transformation in Eq. (20.8) thus results in a differential that depends on changes in S and P; the enthalpy H can be considered a function of the independent variables S and P; the conjugate variables are T and V, and they are given by equations that are analogous to Eqs. (20.6). A Maxwell relation similar to Eq. (20.7) can also be written immediately. The enthalpy thus is the proper thermodynamic potential if the experiment permits variation of S and P.

The *Helmholtz energy* F and the *Gibbs free energy* G are introduced through the Legendre transformations

$$F = E - TS \tag{20.10}$$

and

$$G = E + PV - TS. \tag{20.11}$$

The differentials

$$dF = dE - TdS - SdT = -PdV - SdT$$

$$dG = dE + PdV + VdP - TdS - SdT = VdP - SdT$$

lead immediately to expressions for the conjugate variables and to Maxwell relations. The properties of all four thermodynamic potentials are collected in Table 20.1.

Table 20.1. Thermodynamic Potentials [7].

Types of Potential Energy		Independent Variables	Conjugate Variables	Maxwell Relations
Internal energy	E	V,S	$T=\left(\frac{\partial E}{\partial S}\right)_V$, $P=-\left(\frac{\partial E}{\partial V}\right)_S$	$\left(\frac{\partial T}{\partial V}\right)_S=-\left(\frac{\partial P}{\partial S}\right)_V$
Enthalpy	$H=E+PV$	P,S	$T=\left(\frac{\partial H}{\partial S}\right)_P$, $V=\left(\frac{\partial H}{\partial P}\right)_S$	$\left(\frac{\partial T}{\partial P}\right)_S=\left(\frac{\partial V}{\partial S}\right)_P$
Helmholtz energy	$F=E-TS$	V,T	$S=-\left(\frac{\partial F}{\partial T}\right)_V$, $P=-\left(\frac{\partial F}{\partial V}\right)_T$	$\left(\frac{\partial S}{\partial V}\right)_T=\left(\frac{\partial P}{\partial T}\right)_V$
Gibbs energy	$G=E+PV-TS$	P,T	$S=-\left(\frac{\partial G}{\partial T}\right)_P$, $V=\left(\frac{\partial G}{\partial P}\right)_T$	$\left(\frac{\partial S}{\partial P}\right)_T=-\left(\frac{\partial V}{\partial T}\right)_P$

20.4 Equilibrium Conditions

For a particular experimental situation, the equilibrium condition can be established by considering the proper thermodynamic potential. Assume first that the extensive variables V and S are taken as the independent variables; the energy E then is the proper potential. Now, if all extensive variables are held constant, so that $dV = dS = 0$, the system can spontaneously move toward equilibrium. The system will move to a state where E has reached a minimum so that we can write

$$\text{at equilibrium}: \ E = \text{minimum}, \quad V = \text{const}, \quad S = \text{const}. \tag{20.12}$$

Similar equilibrium conditions hold for the other choices of independent variables. For the physicist it is useful to note that it is not always the internal energy that attains a minimum at equilibrium; depending on the experimental conditions, it can be any of the four thermodynamic potentials. The conditions are summarized in Table 20.2.

As an example, consider a system kept at constant temperature and pressure that is allowed to come to equilibrium. Then the proper thermodynamic potential to be used for its description is the Gibbs energy, G, and the equilibrium condition is G = minimum.

Table 20.2. Equilibria.

System Conditions	Equilibrium Condition
$dV = dS = 0$	E minimum
$dP = dS = 0$	H minimum
$dV = dT = 0$	F minimum
$dP = dT = 0$	G minimum

20.5 Thermodynamic Quantities and Molecular Structure

To find the thermodynamic quantities for an actual molecular system we assume that the energy levels of the system are known, either from experiment or from solving the Schrödinger equation (Fig. 20.1).

All thermodynamic quantities can then be computed from the *partition function* Z, defined by

$$Z = \sum_i g_i \, e^{-E_i/k_B T}. \tag{20.13}$$

The sum extends over all energy levels, and g_i is the weight (degeneracy) of level i. The internal energy E at temperature T is given by

$$E = \frac{1}{Z} \sum_i E_i \, g_i \, e^{-E_i/k_B T} \tag{20.14}$$

Fig. 20.1. Schematic diagram of the energy levels of a biomolecule.

or, with Eq. (20.13),

$$E = k_B T^2 \frac{d(\ln Z)}{dT}. \tag{20.15}$$

The entropy in terms of the partition function is

$$S = k_B \ln Z + \frac{E}{T} \tag{20.16}$$

or

$$S = k_B \frac{d}{dT}(T \ln Z). \tag{20.17}$$

To verify Eq. (20.16), we calculate dS by varying T and keeping the other parameters fixed:

$$dS = k_B \frac{d(\ln Z)}{dT} dT - \frac{E}{T^2}\, dT + \frac{1}{T} dE.$$

With Eq. (20.15), the first two terms cancel and $dS = dE/T$, as required for a process with $dW = 0$.

The Helmholtz energy $F = E - TS$, with Eq. (20.16), takes on a simple form

$$F = -k_B T \ln Z. \tag{20.18}$$

Equations (20.15), (20.17), and (20.18) show that the thermodynamic quantities can be obtained in a straightforward way once the partition function is known. Partition functions for some simple situations are given in Table 20.3.

20.6 Fluctuation

So far, we have treated thermodynamics as if each variable had a unique and sharp value. This approach is justified if we deal with systems consisting of many particles, for instance, a large solid or a gas in a large container. Biomolecules, however, as systems of molecular dimensions and fluctuations [2]–[4] must be considered. We start by looking at a protein embedded in a cell, as sketched in Fig. 20.2.

For simplicity, we consider only one observable, the energy E. In the isolated system, the total energy of the system is classically a constant and does not fluctuate. In the realistic case of Fig. 20.2(b), however, the protein interacts with the cell and its energy can fluctuate. Before treating these fluctuations, we return to the isolated system of Fig. 20.2(a) and note that even its energy is subject to quantum uncertainties. Quantum uncertainties can be neglected if

$$k_B T \gg \hbar/\tau, \tag{20.19}$$

Table 20.3. Molecular Partition Functions.

Motion	Degrees of Freedom	Partition Function[a]	Order of Magnitude
Translational	3	$\frac{(2\pi mkT)^{3/2}}{h^3}V$	10^{24}–10^{25} V
Rotational (linear molecules)	2	$\frac{8\pi^2 IkT}{\sigma h^2}$	10–10^2
Rotational (nonlinear molecules)	3	$\frac{8\pi^2(8\pi^3 ABC)^{1/2}(kT)^{3/2}}{\sigma h^3}$	10^2–10^3
Vibrational (per normal mode)	1	$\frac{1}{1-\exp(-h\nu\, kT)}$	1–10

[a] The term σ is a symmetry number equal to the number of indistinguishable positions into which a molecule can be turned by rigid rotations; A, B, and C are moments of inertia.

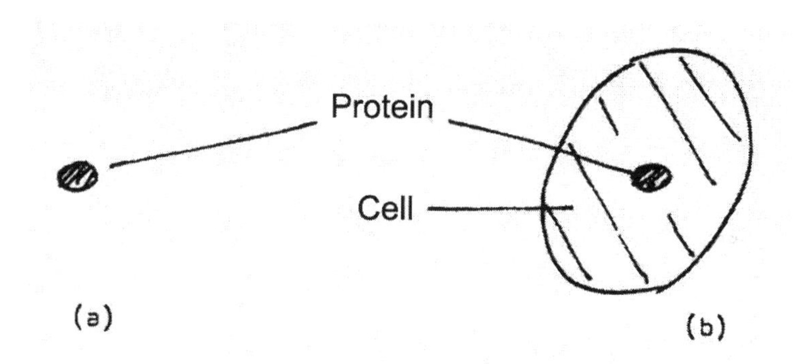

Fig. 20.2. (a) Isolated protein, (b) protein embedded in a cell (heat bath).

where τ is a characteristic time with which the energy E changes because of interactions with the surroundings [2]. In the following we assume that condition (20.19) is satisfied. We thus deal entirely with thermodynamical fluctuations.

The energy of the individual biomolecule sketched in Fig. 20.2 will not be constant but will fluctuate about an average value $\overline{E_i}$ because energy will be continuously exchanged with the entire cell. To get an idea of how the fluctuation ΔE depends on size, we assume that the cell consists of N approximately equal biomolecules. The total energy of the cell then is approximately given by $\overline{E} \approx N\overline{E_i}$. We further assume that the fluctuations in the individual biomolecules are statistically independent; the fluctuation of the cell then is given by $\overline{(\Delta E)^2} = \overline{\sum_i (\Delta E_i)^2} \cong \sum_i \overline{(\Delta E_i)^2} \approx N\overline{(\Delta E_i)^2}$. The relative fluctuation is therefore inversely proportional to $\sqrt{N}$:

$$\frac{\overline{((\Delta E)^2)^{1/2}}}{\overline{E}} \propto N^{-1/2}. \qquad (20.20)$$

For a very large system, $N \to \infty$, the fluctuation disappears and the energy can be taken as sharp. For a small system, however, fluctuations can become important. The energy spreads for a small system and a large system are sketched in Fig. 20.3. Without loss of generality, we have chosen the average energy, $\overline{E}$, to be zero.

To obtain a numerical prediction for the energy spread, we must introduce an implicit expression for the distribution function $w(E)$, where $w(E)dE$ denotes the probability of finding the subsystem (the biomolecule shown in Fig. 20.2(b)) with energy between E and $E + dE$. We assume that $w(E)dE$ is proportional to the number of states, $W(E)dE$, with energies between E and $E + dE$. Such an assumption is reasonable; the more states available, the more a particular energy will be occupied. The Boltzmann principle, Eq. (20.1), then gives $W(E) = \exp[S(E)/k_B]$ and the distribution function thus becomes

$$w(E) = \text{const} \cdot \exp[S(E)/k_B]. \qquad (20.21)$$

The constant is determined by normalization, and the entropy $S(E)$ is considered to be a function of the energy E. The normalization is

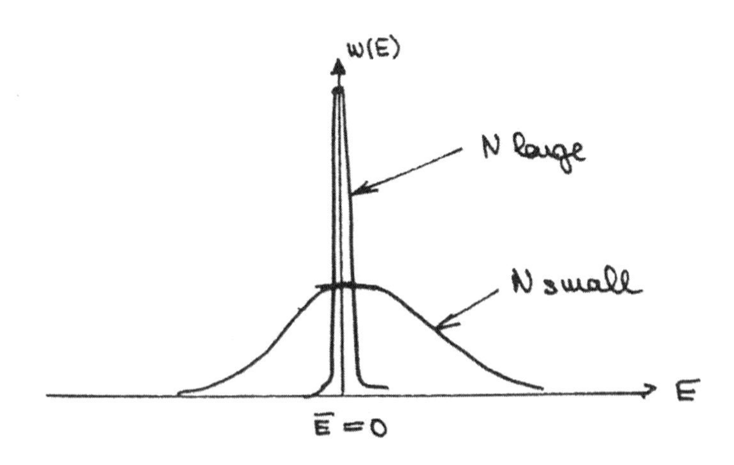

Fig. 20.3. The energy distribution for a system consisting of many components is sharp (N large); for a small system, the energy can fluctuate widely. $w(E)dE$ gives the probability of finding a system with energy between E and $E + dE$.

$$\int_{-\infty}^{+\infty} w(E)dE = 1. \tag{20.22}$$

The entropy $S(E)$ takes on its maximum value at the average energy $\overline{E} = 0$ so that $\partial S/\partial E = 0$ and $\partial^2 S/\partial E^2 < 0$ for $E = 0$. Since fluctuations are very small, $S(E)$ can be expanded,

$$S(E) = S(0) - \frac{1}{2}\beta\, k_B\, E^2$$

with

$$\beta \equiv -\frac{1}{k_B}\frac{\partial^2 S}{\partial E^2}. \tag{20.23}$$

With the normalization Eq. (20.22), the distribution function Eq. (20.21) becomes

$$w(E) = (\beta/2\pi)^{1/2}\, e^{-\beta E^2/2}. \tag{20.24}$$

The energy distribution thus is Gaussian; it assumes a maximum at the average energy and decreases rapidly and symmetrically on either side of the maximum.

The width of the energy distribution shown in Fig. 20.3 is characterized by the second moment (mean square fluctuation),

$$\overline{(\Delta E)^2} = \int w(E)E^2 dE = (\beta/2\pi)^{1/2}\int_{-\infty}^{+\infty} dE\, E^2\, \exp[-\beta E^2/2] = \beta^{-1}. \tag{20.25}$$

To express the fact that we measure the fluctuation, we have written $\overline{(\Delta E)^2}$ rather than $\overline{E^2}$. Because we have taken $\overline{E} = 0$, the two are identical. To further evaluate $\overline{\Delta E^2}$, we note that we have implicitly assumed constant volume; we are considering the energy fluctuations with all other parameters kept constant. We thus write, with Eqs. (20.6) and (20.23),

$$\beta = -\frac{1}{k_B}\frac{\partial}{\partial E}\left(\frac{\partial S}{\partial E}\right)_V = -\frac{1}{k_B}\frac{\partial}{\partial E}\left(\frac{1}{T}\right) = \frac{1}{k_B}\frac{1}{T^2}\left(\frac{\partial T}{\partial E}\right)_V.$$

Here,

$$\left(\frac{\partial E}{\partial T}\right)_V = C_V \tag{20.26}$$

is the heat capacity at constant value. C_V is an extensive property, and it is convenient to introduce the *specific heat*, the heat capacity per unit mass (c_V). Equation (20.25) then becomes

$$\overline{(\Delta E)^2} = k_B T^2 m\, c_V , \tag{20.27}$$

where m is the mass of the biomolecule under consideration.

The approach that has led to Eq. (20.27) can be used for quantities other than E. We can, for instance, consider fluctuations in the volume V of a biomolecule at constant temperature; the result is

$$\overline{(\Delta E)^2} = k_B T V \beta_T , \tag{20.28}$$

where

$$\beta_T = \frac{1}{V}\left(\frac{\partial V}{\partial P}\right)_T \tag{20.29}$$

is the isothermal compressibility. The entropy fluctuation at constant pressure becomes

$$\frac{\overline{(\Delta S)^2}}{k_B^2} = \frac{1}{k_B} m\, c_p . \tag{20.30}$$

We can apply the expressions for the fluctuations to a typical protein [8, 9]. Assuming a molecular weight of 25 kdalton, a mass $m = 4 \times 10^{-20}$g, a volume $V = 3 \times 10^{-20}\text{cm}^3$, $c_V = 1.3\,\text{Jg}^{-1}\text{K}^{-1}$, $\beta_T \equiv 2 \times 10^{-6}\text{atm}^{-1}$ (0.2 Pa), and $k_B = 1.38 \times 10^{-23}\text{JK}^{-1}$, 1 eV$= 1.6 \times 10^{-19}$J, we find, with Eqs. (20.27), (20.28) and (20.30),

$$\overline{[(\Delta E)^2]^{1/2}} \cong 1.6\,\text{eV} \cong 150\,\text{kJ/mol}$$

$$\frac{\overline{[(\Delta V)^2]^{1/2}}}{V} \cong 2 \times 10^{-3} \tag{20.31}$$

$$\overline{[(\Delta S)^2]^{1/2}}/k_B \cong 60 .$$

These fluctuations are large and again show that a biomolecule is not a static system. In particular, the fluctuations in energy and entropy are comparable to the energies and entropies that have been measured in protein unfolding. The actual fluctuations may be larger because we have neglected the hydration shell around the protein.

One additional remark is in order here. A solid of the same size as the protein would have very similar material coefficients and consequently similar fluctuations. The fluctuations would be less important because the probability that the entire fluctuations would be concentrated in one atom is small. Fluctuations in a small number of atoms would have a small effect because each atom is solidly anchored to about six others. In biomolecules the fluctuations can lead to large motions because the bonds across them are weak.

One last remark concerns the rate of fluctuations. While thermodynamics gives expressions for the magnitude of equilibrium fluctuations, it does not provide information about the rate of fluctuations.

20.7 Fluctuation-Dissipation Theorems

So far we have discussed the fluctuations of a system in equilibrium with a heat bath. These considerations apply to a protein in equilibrium with its surrounding. We thus expect that a protein in its resting state will fluctuate; it will move and breathe. The corresponding *equilibrium fluctuations* (EF) can be studied with many different tools [2, 4, 10]. Of more importance than these EF are, however, protein reactions. A protein in action will move from a nonequilibrium to an equilibrium state. We will call the corresponding motions *FIMs* for *"functionally important motions."* These FIMs are nonequilibrium phenomena. They can be caused by the absorption of a photon in photosynthesis, by the binding of a dioxygen molecule in myoglobin or hemoglobin, or by the arrival of an electron in an electron transport system. We can ask: Are the EF and the FIMs related?

The connection between the equilibrium and a nonequilibrium property was first noticed by Einstein [11], who derived the relation

$$D = k_B T / m\gamma \tag{20.32}$$

between the diffusion coefficient D of a particle with mass m and a friction coefficient $m\gamma$. Even in a homogeneous medium, a particle will undergo Brownian motion. The mean-square distance $\langle x^2 \rangle$ traversed during a time t is given by

$$\langle x^2 \rangle = 2\,D\,t. \tag{20.33}$$

D characterizes a fluctuation, the friction $m\gamma$ is a dissipation.

A major step in understanding the connection between fluctuations and dissipation was made by Nyquist [12]. Johnson had found experimentally that a resistor produces spontaneous voltage fluctuations (Johnson noise) with $\langle V^2 \rangle \alpha T$. Nyquist showed that the mean-square voltage, $\langle V^2 \rangle$, across a resistor of resistance R is equilibrium at temperature T is given by

$$\langle V^2 \rangle = 4\,k_B T\,R\,\Delta f. \tag{20.34}$$

Here Δf is the frequency bandwidth with which $\langle V^2 \rangle$ is measured. The Nyquist theorem, in its simplest form given by Eq (20.34), thus connects a fluctuation $\langle V^2 \rangle$ with a dissipative element R. The connection was further explored by Onsager and Machlup [13] and the fluctuation-dissipation theorem was formulated by Callen and Welton [14].

The connection between a nonequilibrium process and equilibrium fluctuations can be seen in the simple case of a two-well model (Fig. 20.4). In equilibrium, the ratio of populations in the two states A and B is given by the Boltzmann factor, $N_B/N_A = \exp(-\Delta/k_B T)$. The population will, however, not be constant but will fluctuate around their mean values as sketched in Fig. 20.5(a). If we establish a nonequilibrium situation as in Fig. 20.4(b), by putting all systems in state B, the system will move toward equilibrium as indicated in Fig. 20.5(b), with rate coefficient $k = k_{AB} + k_{BA}$. The rate coefficients for fluctuations and for dissipation thus are closely connected.

The connection between fluctuations and dissipation exists also in more complex situations far from equilibrium [10, 15, 16].

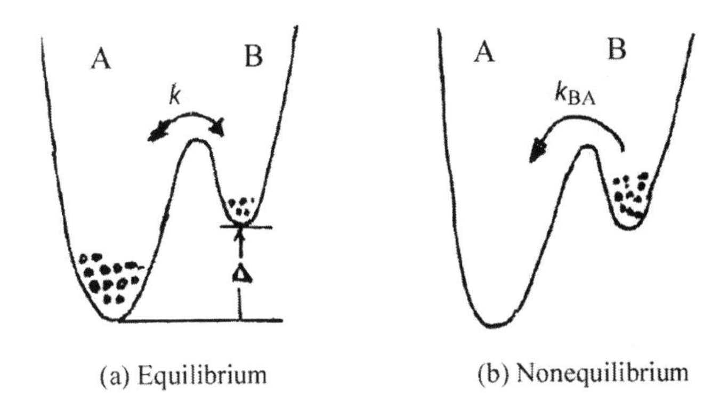

Fig. 20.4. Fluctuation and dissipation in a two-well model.

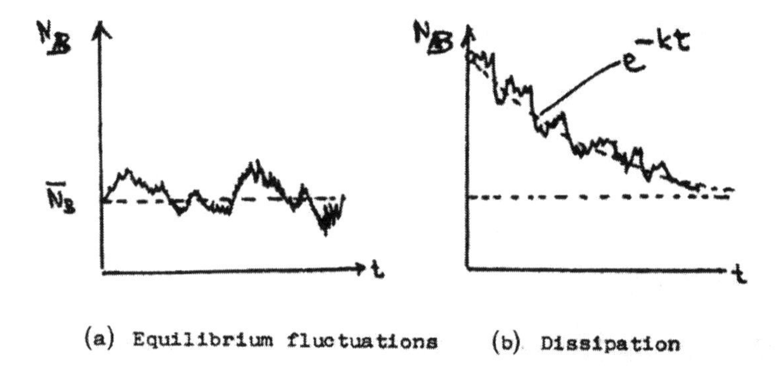

Fig. 20.5. Fluctuations and dissipation.

Of importance to biomolecular physics is the connection between equilibrium and nonequilibrium situation (EF and FIMs) and the fact that fluctuation spectroscopy [17] may be used to extract information on rate coefficients.

20.8 Imperfections and Substates

As an application of thermodynamic concepts, we consider a solid. The equilibrium of a solid at fixed T and P is determined by the minimum value of the Gibbs free energy G. This condition necessarily leads to the existence of a certain amount of disorder in the solid at all nonzero temperatures. The same conditions exist also for biomolecules.

The simplest example of lattice disorder is a vacancy-interstitial pair in which an atom originally in a regular lattice site moves to a position that is not part of the ideal lattice (Fig. 20.6). To discuss the probability of having a certain fraction of vacancy-interstitial pairs, we consider the change in enthalpy and entropy. The creation of each pair requires an enthalpy H_p. The entropy consists of two parts, *thermal* and *configurational entropy*:

$$S = S_{\text{th}} + S_{\text{con}}. \tag{20.35}$$

The thermal entropy is given by the logarithm of the total number of ways in which the lattice phonons can be distributed among all possible modes. The configuration entropy is determined by the logarithm of the total number of ways in which the interstitial can be distributed among all possible sites. If the solid consists of N atoms, and if there are n vacancy-interstitial pairs, S_{con} is given by

$$S_{\text{con}} = k_B \ln \left[\frac{(N+n)!}{N!\, n!} \right]. \tag{20.36}$$

The vacancy-interstitial pairs also change the spectrum of lattice energies so that the thermal entropy is slightly affected. The Gibbs energy thus

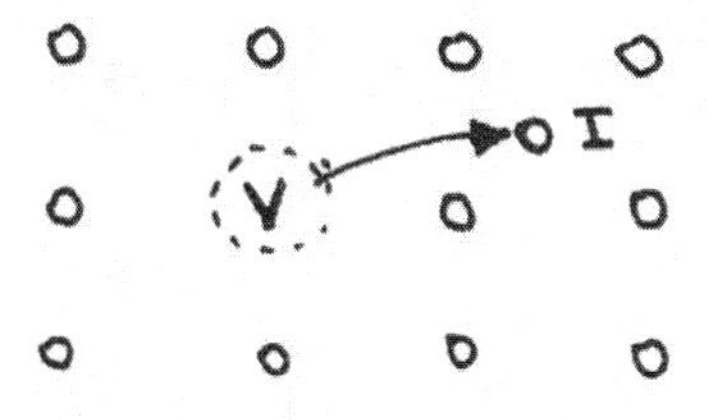

Fig. 20.6. An atom moves from its regular lattice position to an interstitial (I) and creates a vacancy (V).

becomes, with n vacancy-interstitial pairs present,

$$G(n) = G(0) + n\,H_p - n\,T\,\Delta S_{th} - k_B T \ln\left[\frac{(N+n)!}{N!\;n!}\right]. \tag{20.37}$$

The behavior of the enthalpy, entropy, and Gibbs energy as a function of (n/N) is shown in Fig. 20.7. The figure shows that at any finite temperature, the state of lowest Gibbs energy, in other words the equilibrium state, does not represent the ideal solid, but one with a finite number of dislocations. The fraction of dislocations can be calculated from the condition

$$\left(\frac{\partial G}{\partial n}\right)_T = 0. \tag{20.38}$$

With Sterling's approximation, $\ln x! \cong x\ \ln x$, and with $n \ll N$, Eqs. (20.37) and (20.38) give

$$\frac{n}{N} = \exp\left[\Delta S_{th}/k_B\right] \exp\left[-H_p/k_B\,T\right]. \tag{20.39}$$

Since a vacancy-interstitial pair affects the phonon spectrum in such a way that it increases the number of possible modes, the contribution ΔS_{th} is positive, the first factor is larger than 1, and it increases the fraction n/N at equilibrium.

For a solid, H_p is of the order of 1 eV or 100 kJ/mol. $k_B T$ at 300 K is 2.5 kJ/mol. For a solid at 300 K, n/N is of the order of 10^{-18}. For a biomolecule, H_p is much smaller and can be as small as 1 kJ/mol. n/N consequently is close to 1.

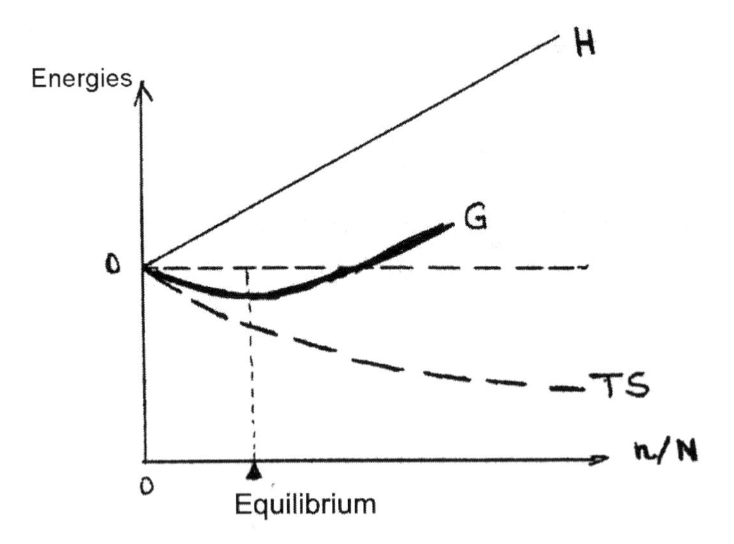

Fig. 20.7. Enthalpy, entropy, and Gibbs energy as a function of n/N, the fraction of vacancy-interstitial pairs.

References

1. H. F. once sat in Pauli's office when Otto Stern walked in and said, "Pauli, one can really see that you are a student of Sommerfeld. You don't understand thermodynamics either."
2. L. D. Landau and E. M. Lifshitz. *Statistical Physics*, 3rd edition. Pergamon Press, 1980. 2 vols.
3. P. M. Morse. *Thermal Physics*, 2nd edition. W. A. Benjamin, New York, 1969.
4. H. B. Callen. *Thermodynamics*, 2nd edition. Wiley, New York, 1985.
5. Extensive quantities are proportional to the number, n, of moles of the substance present, intensive ones independent of n. In equilibrium, intensive quantities have the same value throughout the system.
6. For an extensive discussion of entropy, see A. Wehrl. General properties of entropy, *Rev. Mod. Phys.*, 50:221-60, 1978.
7. A procedure to derive additional relations between thermodynamic quantities is described in [3], pp. 96–101.
8. A. Cooper. Thermodynamic fluctuations in protein molecules. *Proc. Natl. Acad. Sci. USA*, 73:2740–1, 1976.
9. A. Cooper. Protein fluctuations and the thermodynamic uncertainty principle. *Prog. Biophys. Mol. Bio.*, 44:181–214, 1984.
10. R. Kubo. Fluctuation-dissipation theorem. *Rep. Prog. Phys.*, 29:255–84, 1966.
11. A. Einstein. Über die von der molekularkinetischen Theorie der Wäme geforderte Bewegung von in ruhenden Flüssigkeiten suspendierten Teilchen. *Ann. Phys. Leipzig*, 17:549–60, 1905.
12. H. Nyquist. Thermal agitation of electric charge in conductors. *Phys. Rev.*, 32:110–13, 1928.
13. L. Onsager and S. Machlup. Fluctuations and irreversible processes. *Phys. Rev.*, 91:1505–12, 1953.
14. H. B. Callen and T. A. Welton. Irreversibility and generalized noise. *Phys. Rev.*, 83:34–40, 1951.
15. M. Lax. Fluctuations from the nonequilibrium steady state. *Rev. Mod. Phys.*, 32:25–64, 1960.
16. M. Suzuki. Scaling theory of non-equilibrium systems near the instability point. II. *Prog. Theor. Phys.*, 56:477–93, 1976.
17. M. B. Weissman. Fluctuation spectroscopy. *Ann. Rev. Phys. Chem.*, 32:205–32, 1981.

21

Quantum Chemistry for Physicists

The laws of quantum mechanics completely determine the wave function of any given molecule, and, in principle, we can calculate the eigenfunctions and energy eigenvalues by solving Schrödinger's equation. In practice, however, only the simplest systems, such as the hydrogen atom, have an explicit, exact solution, and for the more interesting complex systems we have to resort to approximations. It is an art to develop appropriate approximations. Depending on the purpose, many different schemes have been used, ranging from ab-initio calculations to semiempirical methods. Here we start from atomic orbitals and then show how they can be combined with molecular orbitals. Although we keep the discussion elementary, we will be able to understand the stability of chemical bonds, the type of wave functions involved, and the approximate energies of the various electronic states. Starting from our simple results we can get more realistic approximations by adding correction terms.

Among the many texts on quantum chemistry and molecular physics, we mention three [1]–[3].

21.1 Atomic Orbitals

The Schrödinger equation for a hydrogenlike atom of charge Ze

$$\frac{\hbar^2}{2m_e}\nabla^2\psi + V(r)\psi = E\psi\ , \quad V = \frac{Ze^2}{4\pi\epsilon_o r} \tag{21.1}$$

has solutions of the form

$$\psi_{n\ell m} = R_{n\ell}(r)\, Y_{\ell m}(\theta, \phi)\ , \tag{21.2}$$

where the spherical harmonics $Y_{\ell m}$ are eigenfunctions of the orbital angular momentum operators

H. Frauenfelder, *The Physics of Proteins*, Biological and Medical Physics, Biomedical Engineering, DOI 10.1007/978-1-4419-1044-8_21,

$$L_{op}^2 Y_{\ell m} = \hbar^2 \ell(\ell+1)\, Y_{\ell m}, \quad L_{z,op}\, Y_{\ell m} = m\hbar\, Y_{\ell m}. \tag{21.3}$$

Note that m_e is the mass of electron; and ℓ, m, and n are quantum numbers of energy levels. The energy eigenvalues are

$$E_n = \frac{E_1}{n^2},$$

$$E_1 = -\frac{1}{2}\left(\frac{e^2 Z}{4\pi\epsilon_o \hbar c}\right)^2 m_e c^2 \approx -13.6\, Z^2 \mathrm{eV}, \tag{21.4}$$

and the radial eigenfunctions are conveniently expressed in terms of the dimensionless parameter $\rho = Zr/a_o$, where $a_o = \hbar^2/m_e e^2 = 0.53$ Å is the Bohr radius:

$$R_{n\ell} \propto e^{-\rho/n}\left(\frac{2\rho}{n}\right)^{\ell} L_{n\ell}\left(\frac{2\rho}{n}\right). \tag{21.5}$$

Here $L_{n\ell}$ is Laguerre polynomial and a_o is the first Bohr orbit. Note that R and E_n depend on the nuclear charge Z, which is 1 for hydrogen.

We can build up many-electron atoms by placing electrons in the various orbits characterized by the quantum numbers (n, ℓ, m) starting from the lowest energies E_n. In doing so, we have to take into account two things:

1. Electrons have an intrinsic spin $s = 1/2$. They are thus fermions and obey Pauli's exclusion principle. The simplest formulation of this principle is to say that no two electrons can have the same quantum number, where the spin quantum number $m_s = \pm 1/2$ is included. A more general statement is: The total wave function of the system must be antisymmetric with respect to exchange of any two electrons. In our nonrelativistic independent-particle approximation the total wave function ψ is the product (or rather, a linear superposition of products) of the wave functions $\psi(\mathbf{r_i})\sigma(s_o)$ of each individual electron, where $\psi(\mathbf{r_i})$ denotes the space function of $\sigma(s_i)$ the spin function of electron i.

2. The independent-particle model is a crude approximation and needs a number of corrections. Most importantly, we have ignored the electron-electron repulsion,

$$V_{ij} = \frac{e^2}{4\pi\epsilon_o}\,\frac{1}{|\mathbf{r}_i - \mathbf{r}_j|}. \tag{21.6}$$

Combined with the exclusion principle, this term gives rise to *Hund's rule*, according to which states with larger spin multiplicity have lower energies, other things remaining unchanged.

Other smaller corrections arise from spin-orbit, spin-spin, and magnetic orbit-orbit interactions.

We base our discussion of molecules on atomic orbitals, concentrating, in particular, on the angular dependence of the various orbitals. Instead of the

Table 21.1. Spherical harmonics $Y_{\ell m}$, p-orbitals for $\ell = 1$, and d-orbitals for $\ell = 2$.

$$Y_{00} = \sqrt{1/4\pi}$$

$$Y_{10} = \sqrt{3/4\pi}\cos\theta = \sqrt{(3/4\pi)}\, z/r$$

$$Y_{1\pm1} = \mp\sqrt{(3/8\pi)}\sin\theta\, e^{\pm i\phi} = \mp\sqrt{(3/8\pi)}\,(x \pm iy)/r$$

$$Y_{20} = \sqrt{(5/16\pi)}(3\,\cos^2\theta - 1) = \sqrt{(5/16\pi)}\,(3z^2 - r^2)/r^2$$

$$Y_{2\pm1} = \mp\sqrt{(15/8\pi)}\sin\theta\cos\theta\, e^{\pm i\phi} = \mp\sqrt{(15/8\pi)}\, z(x \pm iy)/r^2$$

$$Y_{2\pm2} = \sqrt{(15/32\pi)}\sin^2\theta\, e^{\pm 2i\phi} = \sqrt{(15/32\pi)}\,(x \pm iy)^2/r^2$$

$\ell = 1$

$$p_x = x/r = \sin\theta\cos\phi$$

$$p_y = y/r = \sin\theta\sin\phi$$

$$p_z = z/r = \cos\theta$$

$\ell = 2$

$$d_{x^2-y^2} = \tfrac{\sqrt{3}}{2}\left(x^2 - y^2\right)/r^2 = \tfrac{\sqrt{3}}{2}\sin^2\theta(\cos^2\phi - \sin^2\phi)$$

$$d_{z^2} = (3z^2 - r^2)/2r^2 = \tfrac{1}{3}(3\cos^2\theta - 1)$$

$$d_{xy} = \sqrt{3}xy/r^2 = \sqrt{3}\sin^2\theta\cos\phi\sin\phi$$

$$d_{yz} = \sqrt{3}yz/r^2 = \sqrt{3}\sin\theta\cos\theta\sin\phi$$

$$d_{zx} = \sqrt{3}xz/r^2 = \sqrt{3}\sin\theta\cos\theta\cos\phi$$

spherical harmonics $Y_{\ell m}$, which are appropriate solutions for the spherically symmetric Hamiltonian (Eq. (21.1)), we use the real functions p_x, p_y, p_z for the $\ell = 1$ wave function and the corresponding d-orbitals for $\ell = 2$, listed in Table 21.1 and shown in Figs. 21.1 and 21.2.

21.2 Bonds—General Aspects

Bonds [4] are essential for the formation of biomolecules and for their function. We have discussed the essential features of chemical bonds in Chapter 18 and return here to some additional aspects. A chemical interaction takes place if two atoms come close together. A few typical possibilities are sketched in Fig. 21.3. If only one electron is present, a one-electron bond results. Two electrons lead to a strong bond. Two atoms with two orbitals and three elec-

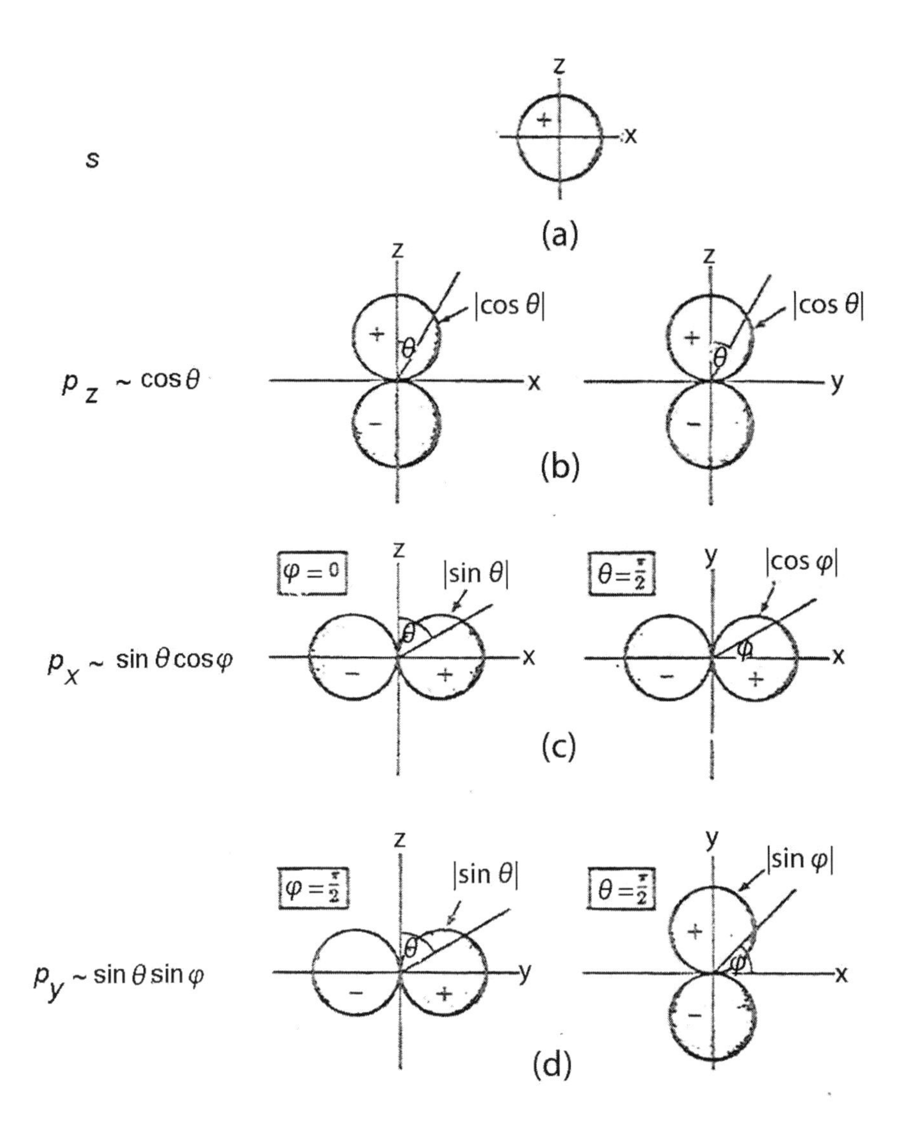

Fig. 21.1. Hydrogen atom orbitals, $\ell = 0$ and $\ell = 1$ (after Karplus and Porter [1]).

trons can still bind each other. Four electrons in two orbitals will always lead to repulsion.

Figure 21.3 shows that the orbitals of the individual atoms are essential in the bond formation. We have shown the orbitals of the individual atoms in Figs. 21.1 and 21.2. These orbitals are, however, not the only ones that can be formed. By mixing orbitals together, new orbitals are formed that are just as

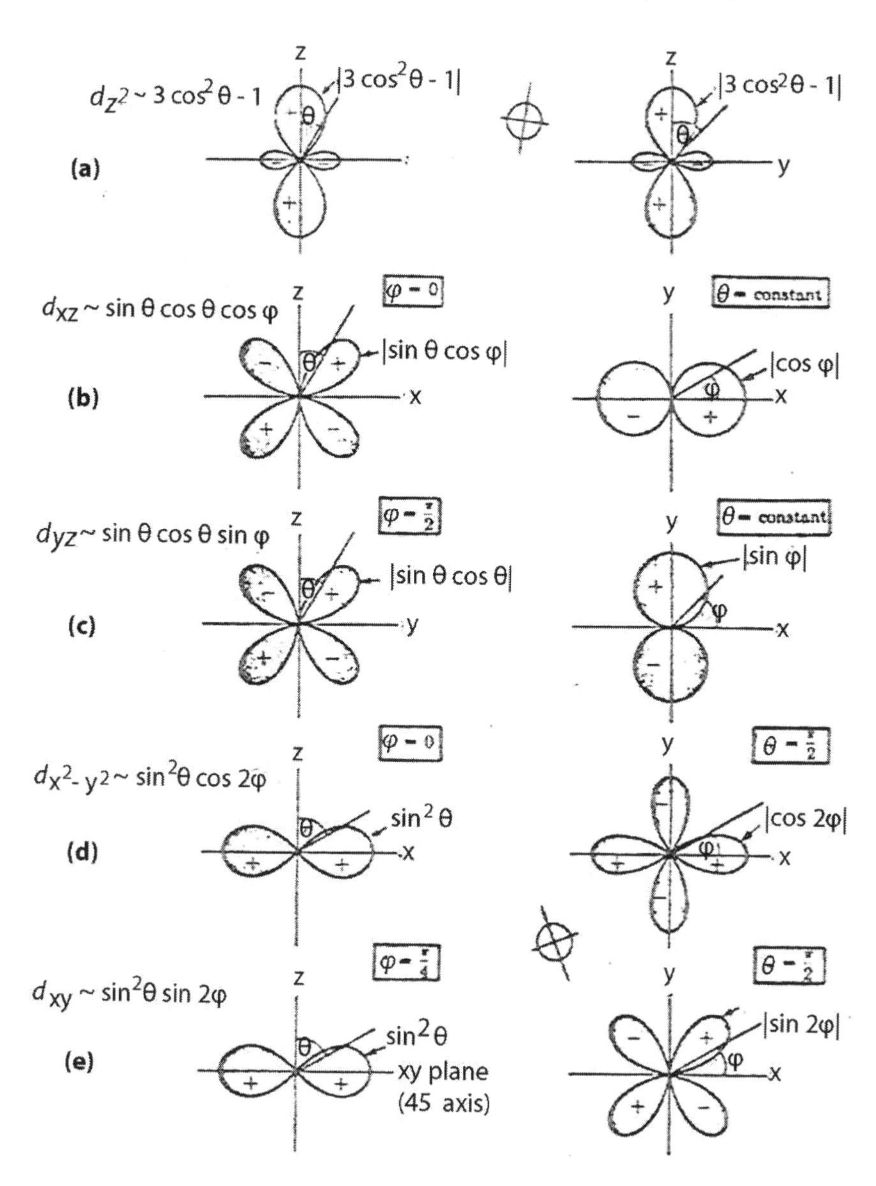

Fig. 21.2. Hydrogen atom orbitals, $\ell = 2$ (after Karplus and Porter [1]).

good as the ones given in Table 21.1, but they may form stronger bonds. We discuss such mixed or hybrid orbitals later, only stating here that they lead to maximal binding in two opposite directions (*dihedral hybridization*), toward the corners of an equilateral triangle (*trigonal hybridization*), or toward the

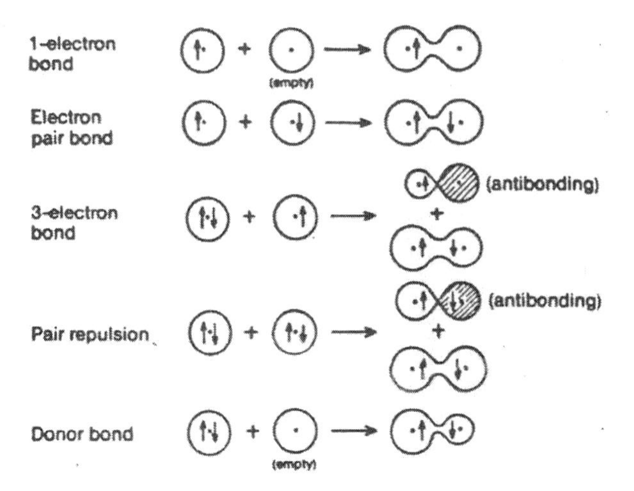

Fig. 21.3. Schematic representation of the possibilities to occupy two overlapping orbitals.

corners of a tetrahedron (*tetrahedral hybridization*). The atoms C, N, and O bind preferentially in trigonal and tetrahedral configurations.

21.3 Molecular Orbitals

Consider a diatomic molecule (Fig. 21.4) consisting of two nuclei A and B of charges (eZ_A) and (eZ_B) a distance R apart and one or more electrons e a distance r_A and r_B from the nuclei. We approximate the Hamiltonian of the system by (1) ignoring all electron-electron interactions and (2) keeping the internuclear distance fixed:

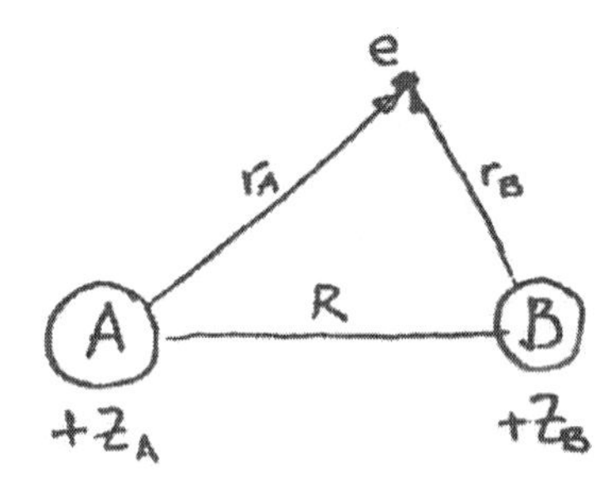

Fig. 21.4. Diatomic molecule.

$$\mathcal{H} = -\frac{\hbar}{2m_e}\nabla^2 - \frac{e^2}{4\pi\epsilon_o}\left[\frac{Z_A}{r_A} + \frac{Z_B}{r_B} - \frac{Z_A Z_B}{R}\right] . \tag{21.7}$$

We thus make the usual independent-particle approximation. Furthermore we use a linear combination of atomic orbitals ϕ_A and ϕ_B centered on A and B as a trial wave function ψ,

$$\psi = c_1\phi_A + c_2\phi_B , \tag{21.8}$$

and we adjust the coefficients c_1 and c_2, which we assume to be real for simplicity, to minimize the expectation value of $\mathcal{H}$. ϕ_A and ϕ_B are normalized, and to normalize ψ we introduce a normalization factor N

$$1 = \int (N\psi)^* N\psi \, d^3x = N^2 \int \left[c_1^2\phi_A^2 + 2c_1c_2\phi_A\phi_B + c_2^2\phi_B^2\right] d^3x$$

$$= N^2(c_1^2 + 2c_1c_2S + c_2^2), \tag{21.9}$$

$$S = \int \phi_A\phi_B \, d^3x, \tag{21.10}$$

where S is the overlap integral. We next calculate the energy expectation value $\langle E\rangle$,

$$\langle E\rangle = \int \psi^* \, \mathcal{H} \, \psi d^3x \Big/ \int \psi^* \, \psi \, d^3x = \frac{c_1^2 H_{AA} + 2c_1c_2H_{AB} + c_2^2 H_{BB}}{c_1^2 + 2c_1c_2S + c_2^2}. \tag{21.11}$$

Here we define

$$H_{k\ell} = \int \phi_k^* \, \mathcal{H} \, \phi_\ell \, d\tau. \tag{21.12}$$

We now determine c_1 and c_2 such that $\langle E\rangle$ is a minimum, that is, by setting $\partial\langle E\rangle/\partial c_1 = \partial\langle E\rangle/\partial c_2 = 0$. This condition leads to

$$c_1(H_{AA} - \langle E\rangle) + c_2(H_{AB} - S\langle E\rangle) = 0,$$

$$c_1(H_{AB} - S\langle E\rangle) + c_2(H_{BB} - \langle E\rangle) = 0.$$

Eliminating c_1 and c_2 gives

$$\begin{vmatrix} H_{AA} - \langle E\rangle & H_{AB} - S\langle E\rangle \\ H_{AB} - S\langle E\rangle & H_{BB} - \langle E\rangle \end{vmatrix} = 0. \tag{21.13}$$

This quadratic equation is easily solved. The physics is most transparent in the simplest case where $H_{AA} = H_{BB}$ (identical nuclei and atomic orbitals, AOs). Under these conditions $c_1 = \pm c_2$, and there are two solutions:

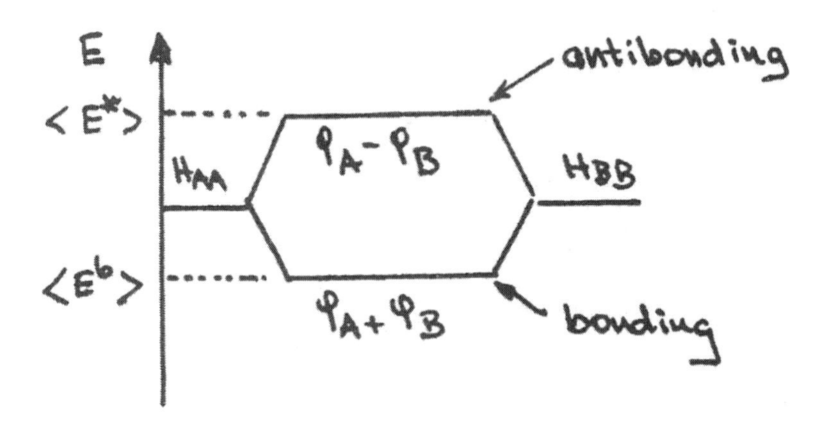

Fig. 21.5. Bonding and antibonding molecular orbitals (MO).

$$\langle E^b \rangle = \frac{H_{AA} + H_{AB}}{1 + S}, \quad \langle E^* \rangle = \frac{H_{AA} - H_{AB}}{1 - S}. \tag{21.14}$$

Energies and wave functions are shown schematically in Fig. 21.5.

Of the two solutions, one is bonding and yields an energy $\langle E^b \rangle$ lower than that of the AO (note that $H_{AA} < 0, H_{AB} < 0$). The bonding MO is the symmetric one, $\psi = \phi_A + \phi_B$, that is, the electron has a high probability of being found between the nuclei. In the antibonding MO, $\psi = \phi_A - \phi_B$, the electron is never found in the symmetry plane between the two nuclei, and the energy is higher than that of the AO.

In Fig. 21.6, the energies of the bonding and the antibonding MOs are plotted as a function of the interatomic distance R. The antibonding orbital is repulsive at all distances R, while the bonding one is attractive at large distances, goes through a minimum at $R_{\min}$ and becomes repulsive at short distances.

The technique used to arrive at the results shown in Figs. 21.5 and 21.6 is called LCAO, for linear combination of atomic orbitals. With this technique, more complicated molecules can be understood. In general, the simplifying assumption $H_{AA} = H_{BB}$ does not hold, since the nuclei A and B and the AOs centered on them are different, and we therefore have to use the general result in Eq. (21.13). We obtain bonding and antibonding orbitals ψ^b and ψ^*, respectively, whenever the integral H_{AB} is nonzero. This case occurs only when the overlap integral S is finite, as a simple calculation shows:

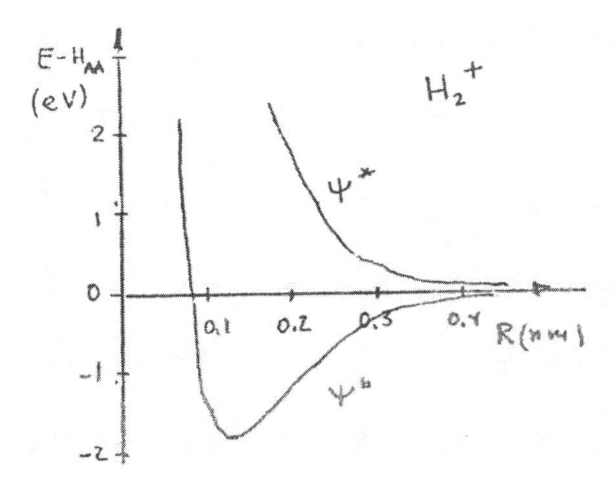

Fig. 21.6. Energies calculated for H_2^+ as a function of the interatomic distance R. A minimum occurs for the bonding orbital at 0.13 nm, with $E - H_{AA} = -1.76\,\text{eV}$.

$$H_{AB} = \int \phi_A^* \left[-\frac{\hbar^2}{2m_e}\nabla^2 - \frac{e^2}{4\pi\epsilon_o}\left(\frac{Z_A}{r_A} + \frac{Z_B}{r_B} - \frac{Z_A Z_B}{R} \right) \right] \phi_B d^3x$$

$$= \int \phi_A^* \left[-\frac{\hbar^2}{2m_e}\nabla^2 - \frac{e^2 Z_A}{4\pi\epsilon_o r_A} \right] \phi_B d^3x + \int \phi_A^* \left[-\frac{\hbar^2}{2m_e}\nabla^2 - \frac{e^2 Z_B}{4\pi\epsilon_o r_B} \right] \phi_B d^3x$$

$$+ \frac{e^2 Z_A Z_B}{4\pi\epsilon_0 R} \int \phi_A^* \phi_B d^3x$$

$$= \left(E_A + E_B + \frac{e^2 Z_A Z_B}{4\pi\epsilon_0 R} \right) S. \tag{21.15}$$

It is useful at this point to consider the symmetry of the AOs ϕ_A and ϕ_B; on the one hand we notice that certain overlap integrals are identically zero, and on the other hand we can estimate the amount of overlap for those that are finite.

The molecular orbital formed from pairs of atomic orbitals are labeled by specifying the AOs, the number of nodal planes containing the nuclear axis (z-axis), and whether the MO is bonding or antibonding, Antibonding orbitals are denoted by an asterisk. The main features are given in Table 21.2 and Fig. 21.7.

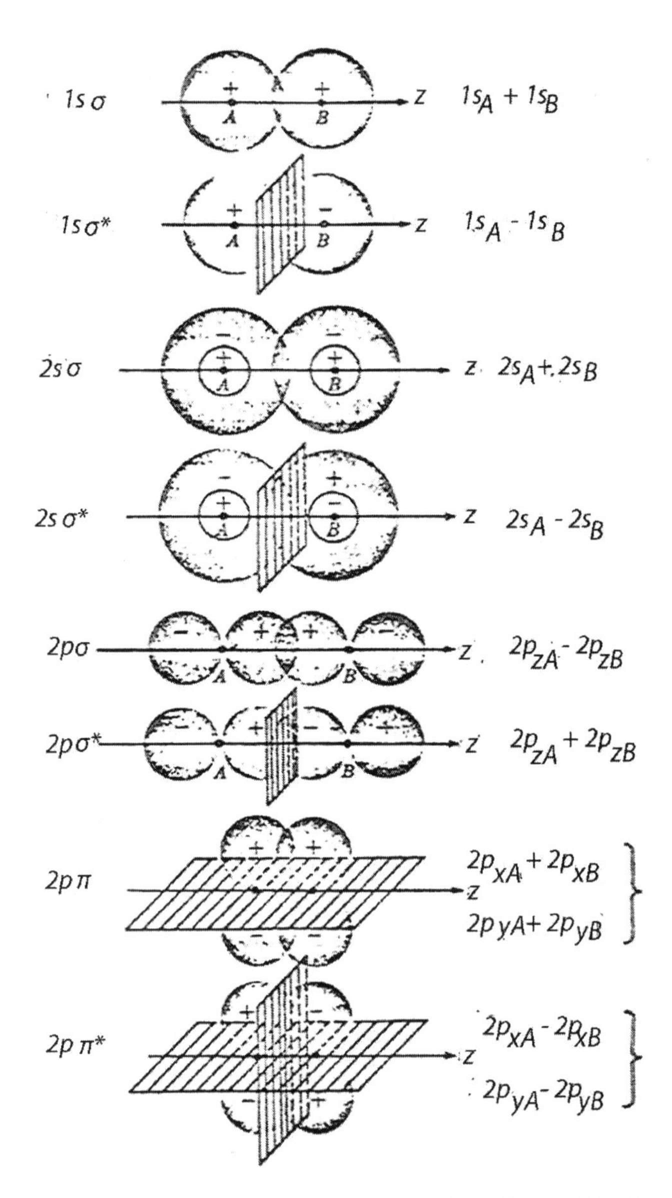

Fig. 21.7. MOs for homonuclear diatomic molecules formed from AO. Nodal planes are shaded (after Karplus and Porter [1]).

Table 21.2. Molecular orbitals.

	Atom	Diatomic Molecule	
ℓ	Designation	Nodal Planes	Designation
0	s	0	σ
1	p	1	π
2	d	2	δ

Figure 21.7 suggests that the σ overlap is larger than the π overlap and that the δ overlap is weakest. This observation is verified by detailed calculations, and the following types of bonds are recognized:

1. *Bonding σ-orbitals* constitute the single bonds between atoms and are very strong. The electrons are almost not delocalized. The electron distribution is cylindrically symmetric about the bond.

2. *Bonding π-orbitals* are found in multiple bonds and are based on a combination of atomic p-orbitals. The electrons are strongly delocalized and interact easily with the environment.

3. *n-orbitals.* If a molecule contains atoms such as oxygen or nitrogen, the occupied orbitals with the highest energy are those of lone pairs, which are not involved in bonds and thus retain their atomic character.

In excited states, two or more types of orbitals are important:

1. *$\sigma*$-orbitals.* These are antibonding, with a node point between the atomic centers.

2. *$\pi*$-orbitals* are delocalized, have a node, and are antibonding.

21.4 Hybridization

The stronger chemical bonds are, the more the orbitals of the two bonding atoms overlap. The bond of two p-orbits shown in Fig. 21.8(a) would be stronger if the electrons could move as shown in Fig. 21.8(b) and increase the overlap. Such a charge distribution appears to violate parity, and it cannot be obtained by using only one AO. However, if an s and a p state are close in energy, a linear combination can lead to lopsided orbitals:

$$\psi = \psi(2s) \pm \lambda\psi(2p). \tag{21.16}$$

The energy of such a hybridized orbital is between that of the $2s$ and the $2p$ AO; it is strongly directional (Fig. 21.9).

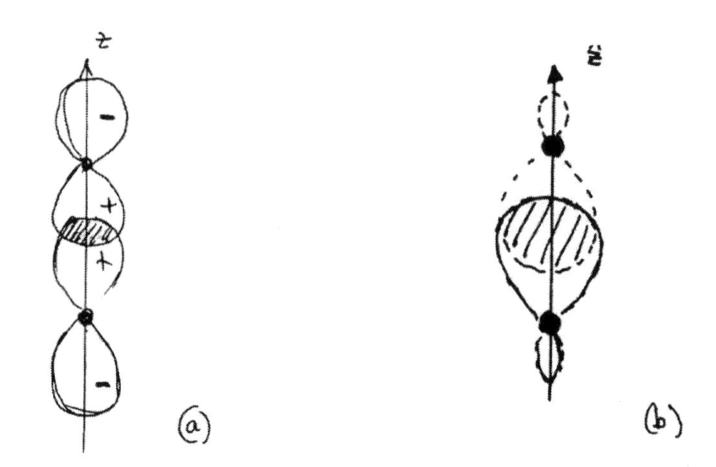

Fig. 21.8. (a) Standard sigma bond, σ. (b) Bond strength is increased if the wave functions can be concentrated.

The ^3P ground state of carbon has the electron configuration $1s^2 2s^2 2p^2$. The two unpaired p electrons account for the C_2 molecule, and they predict a valence of 2. However, they do not explain the observed valence of 4 in molecules such as CH_4 and CCl_4. These compounds must derive from an electronic configuration with four unpaired electrons, or with multiplicity $2S+1 = 5$. Indeed, about 4.2 eV above the ground state of carbon there exists a ^{5}S state, with electron configuration $1s^2 2s 2p^3$ (Fig. 21.10). One s electron from the filled $2s^2$ in the ground state has been promoted to the $2p$ state. Since carbon is nearly always tetravalent, the energy gained by forming four bonds more than offsets the loss stemming from the promotion. With the three unpaired electrons in the p state, and one in an s state, we expect

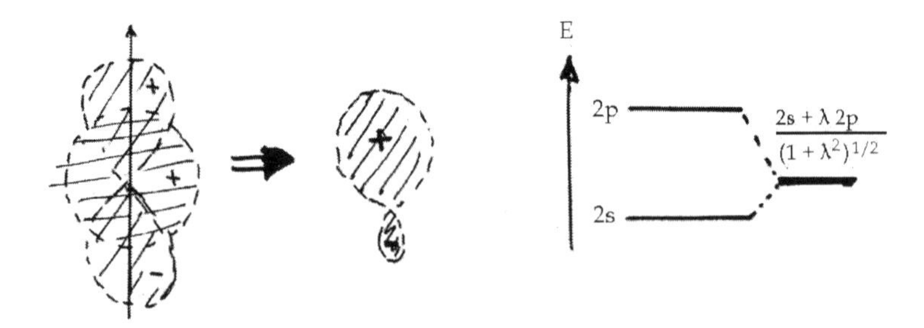

Fig. 21.9. Formation of lopsided orbitals through sp hybridization (left) and the corresponding energy level diagram (right).

Fig. 21.10. Orbital electrons of a carbon atom.

three strong bonds at right angles to each other and one weaker one in another direction. Experimentally, however, all four bonds are equivalent, and the angles between the bonds are tetrahedral [$109°28' = \cos^{-1}(1/3)$]. The puzzle was solved by Slater and Pauling, who pointed out that hybridization can lead to a much higher electron density between the carbon atom and, for instance, the hydrogen atoms in CH_4, thus increasing the bond strength. The possible hybrids of $|s\rangle$ with $|p_x\rangle$, $|p_y\rangle$, and $|p_z\rangle$ are found by writing

$$\psi_i = a_i|s\rangle + b_i|p_x\rangle + c_i|p_y\rangle + d_i|p_z\rangle \tag{21.17}$$

with $i = 1$ to 4.

The states must be orthonormal,

$$a_i a_j + b_i b_j + c_i c_j + d_i d_j = \delta_{ij}, \tag{21.18}$$

where δ_{ij} is the Dirac delta function ($\delta_{ij} = 0$ if i $\neq$ j; $\delta_{ij} = 1$ if $i = j$).

Equation (21.18) represents four normalization and six orthogonality relations; there exist 16 coefficients in Eq. (21.17). Thus there remain six free coefficients. Of these, three describe different orientations of the hybrid orbitals in space, but they do not affect their shape. Thus hybridization sp^3 is specified by three coefficients.

Out of the infinite number of possible hybrid states, some are particularly interesting. The choice

$$a_1 = b_1 = c_1 = d_1 \tag{21.19}$$

leads to

$$|1\rangle = [|s\rangle + |p_x\rangle + |p_y\rangle + |p_z\rangle]/2$$

$$|2\rangle = [|s\rangle + |p_x\rangle - |p_y\rangle - |p_z\rangle]/2$$

$$|3\rangle = [|s\rangle - |p_x\rangle + |p_y\rangle - |p_z\rangle]/2 \tag{21.20}$$

$$|4\rangle = [|s\rangle - |p_x\rangle - |p_y\rangle + |p_z\rangle]/2$$

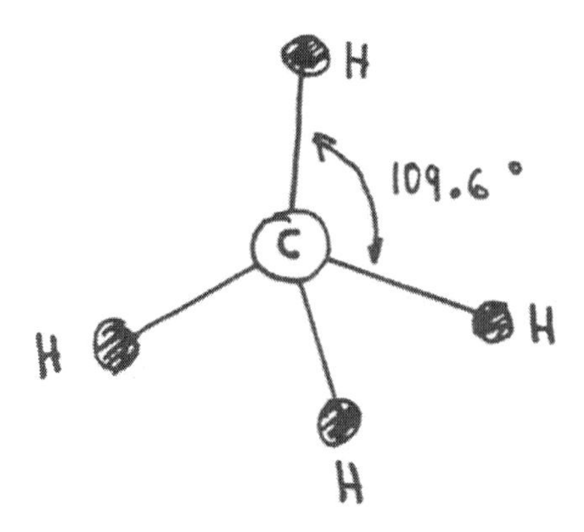

Fig. 21.11. Methane (bond angle = 109° 28').

These four orbitals point in the directions (1,1,1), (1,–1,–1), (–1,1,–1), and (–1,–1,1); they are all equivalent and form tetrahedral angles with each other. This arrangement is just the one found in CH_4, CCl_4, and other saturated organic molecules (Fig. 21.11).

Hybridization is not really a physical effect. Rather, it is a mathematical one, based on the fact that we start with hydrogenlike wave functions to describe the real situation. If the appropriate initial wave functions are chosen, the mathematics may be more complicated, but hybridization does not occur. We are forced to use approximate methods to describe real life, whereas nature "solves the Schrödinger equation exactly."

21.5 Multiple Bonds

The example just given, methane, shows that a particular observed structure can be explained by proper hybridization. To solve a given problem, one must resort to a variational calculation to find the particular hybridization that minimizes the energy of a given molecule. In any case, organic molecules are a good illustration of the connection between geometrical and electronic structure. An understanding, actually far beyond what we are giving here, is important for an unraveling of the spectra and for insight into the construction of biomolecules out of smaller building blocks.

We briefly discuss some examples here.

Ethylene, C_2H_4, has a planar structure (Fig. 21.12).

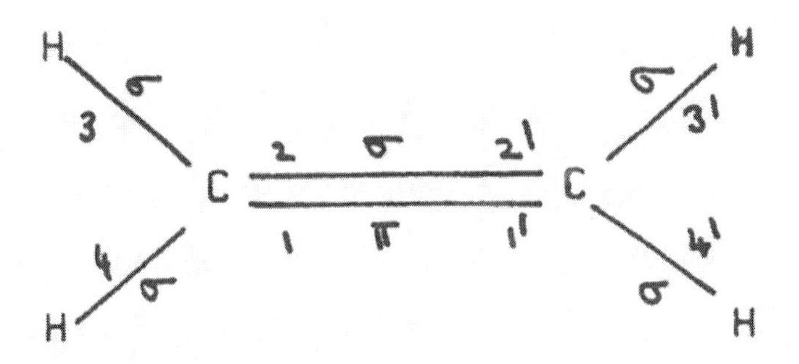

Fig. 21.12. Ethylene.

Such a structure results from the choice (sp^2 hybridization):

$$|1\rangle = |p_z\rangle$$

$$|2\rangle = (1/3)^{1/2}|s\rangle + (2/3)^{1/2}|p_x\rangle$$

$$|3\rangle = (1/3)^{1/2}|s\rangle - (1/6)^{1/2}|p_x\rangle + (1/2)^{12}|p_y\rangle \tag{21.21}$$

$$|4\rangle = (1/3)^{1/2}|s\rangle - (1/6)^{1/2}|p_x\rangle - (1/2)^{1/2}|p_y\rangle .$$

State $|1\rangle$ points along the $\pm z$ direction. The other three orbitals are at 120° in the xy-plane. The situation for the left carbon atom is shown in Fig. 21.13. The orbitals for the carbon atom on the right are mirror images of the first and are denoted by primes. The entire molecule can thus be characterized as follows: The four hydrogen atoms are bound by the four σ bonds $|3\rangle$, $|4\rangle$, $|3'\rangle$, and $|4'\rangle$. The *double bond* between the two carbon atoms consists of two unequal parts. The stronger σ bond is formed by the orbitals $|2\rangle$ and $|2'\rangle$, the weak π bond by $|1\rangle$ and $|1'\rangle$. The orbitals $|1\rangle$ and $|1'\rangle$ must be parallel for maximum overlap and the π bond thus forces ethylene into the planar form.

In *acetylene*, C_2H_2, a *triple bond* turns up. Acetylene is a linear molecule,

$$\mathrm{H - C \equiv C - H}.$$

The $|s\rangle$ state is only mixed with $|p_x\rangle$ (sp hybridization):

$$|1\rangle = [|s\rangle - |p_x\rangle]/2$$

$$|2\rangle = [|s\rangle - |p_x\rangle]/2 \ . \tag{21.22}$$

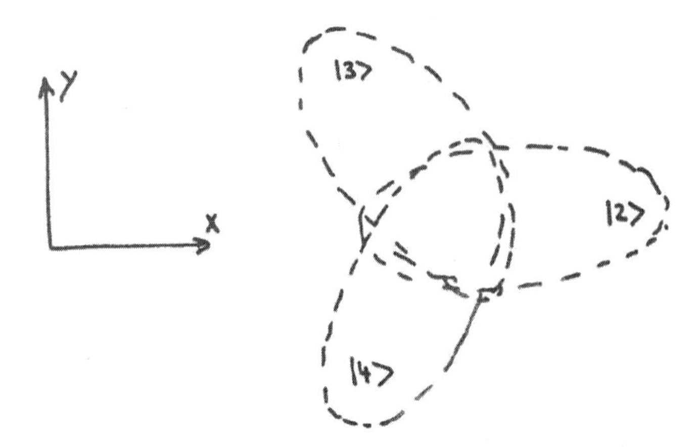

Fig. 21.13. The orbitals $|2\rangle$, $|3\rangle$, and $|4\rangle$ (planar orbitals) of C_2H_4. The orbital $|1\rangle$ points in the $\pm z$-axis at a right angle to the plane of paper (sp^2 hybridization).

These two states, and the corresponding mirror states, extend along the x-axis; they bind the two hydrogen atoms and form a single σ bond between the carbons. The remaining electrons go into linear combinations of $|p_v\rangle$ and $|p_z\rangle$ and form two π bonds between the carbons.

Double bonds and triple bonds are indicated with two or three parallel lines between the atoms. It should be understood that the bonds are not equal; one line describes the stronger σ bond, the other lines the weaker π bonds. The π bonds are, however, responsible for planarity. The measured bond energy of a C=C double bond is 615 kJ/mol = 6.4 eV, that of a C–C single bond is 347 kJ/mol = 3.6 eV. The π bond thus contributes 268 kJ/mol = 2.8 eV. This energy would be required to twist one end of the ethylene molecule by 90°.

21.6 Delocalization and "Resonance"

The molecules that we have described so far have single, double, or triple bonds. More complex situations occur. We first discuss one important case, the carboxyl group, –COO^-. (Do not confuse this group with the molecule, carbon dioxide, CO_2.) The ionization of a carboxyl group removes one electron from the system:

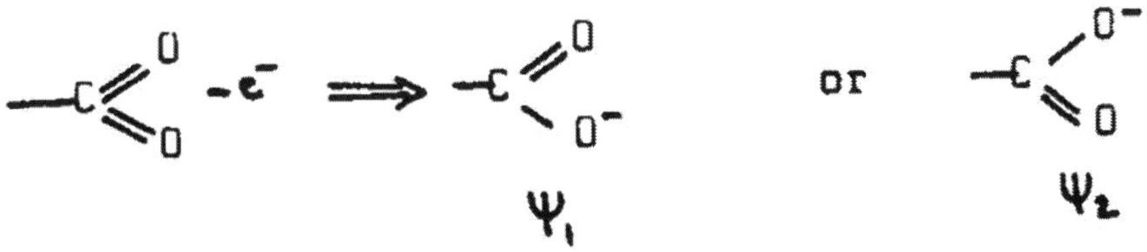

If the electron "hole" remained localized on one oxygen, one double and one single bond would exist, with bond lengths 1.23 Å and 1.36 Å. Not surprisingly, both bonds are equal and of intermediate length, 1.26 Å. Clearly, each wave function ψ_1 and ψ_2 violates symmetry, and the proper solutions to be tried in a variational calculation are

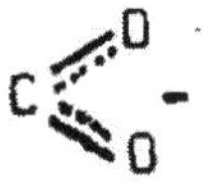

$$\psi = (\psi_1 \pm \psi_2)/\sqrt{2}.$$

The electron wave function of the lone electron spreads over both bonds; each bond has a partial double-bond character, and this structure is more stable than the asymmetric ones by 120 kJ/mol = 1.2 eV. This delocalization is called "resonance." The choice of this word is unfortunate; it implies that resonance is something new. However, the word simply means a nonclassical superposition of states.

Benzene is an outstanding example of a molecule in which electron delocalization occurs. It is also the prototype of the ring structure that occurs in many biomolecules.

From X-ray diffraction and spectroscopic measurements, it is known that benzene is a flat molecule with the six carbon atoms 1.397 Å apart in a hexagonal ring. Six hydrogen atoms radiate out from the ring, one from each carbon atom at a distance of 1.09 Å. As shown in Fig. 21.14, all but one of the valence electrons of each carbon are accounted for as electron-pair or single bonds. The remaining six electrons are indicated as dots. How are they distributed in the ring? The assignment of a double bond to a particular set of states violates symmetry. Delocalizing the electrons around the entire ring lowers the energy because their kinetic energy is decreased. Thus the π electrons running around the ring correspond to a current loop with large area. As a result of the current loop, benzene has a very large diamagnetic susceptibility.

In drawing structures, the π electrons are not shown as distributed, but a single and a double bond are shown and the delocalization "resonance" is understood (Fig. 21.15).

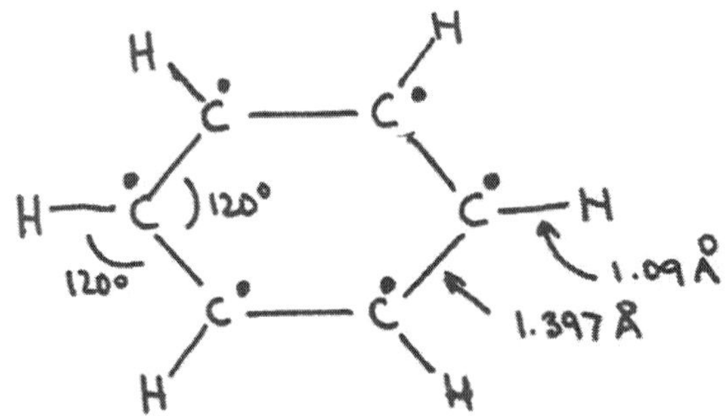

Fig. 21.14. Benzene.

Fig. 21.15. The delocalization of the 6 π electrons in the benzene ring.

21.7 The Peptide Bond

The amino acid residues in a polypeptide chain are connected by peptide bonds (amide links). The carbon atom of one residue links up with the nitrogen atom of the next and a water molecule is released. The bond looks schematically as in Fig. 21.16.

X-ray data indicate that the bond is planar and has the following dimensions: The N–C_α bond length is 1.46 Å, the C–N peptide bond only 1.32 Å (Fig. 21.17). This short bond length compares with a value of 1.25 Å in model compounds with double bonds and is thus taken as a strong indication for a double bond between the N and C. Additional evidence for a double bond comes from the planarity. We saw earlier that the π bond forces planarity on the system (compare with ethylene). The electron configuration of the atom involved gives more information on the binding:

	Ground State 1s 2s 2p			Excited State 2s 2p		
[t] C	2	2	2	1	3	
N	2	2	3	1	4	
O	2	2	4	1	5	(-2)

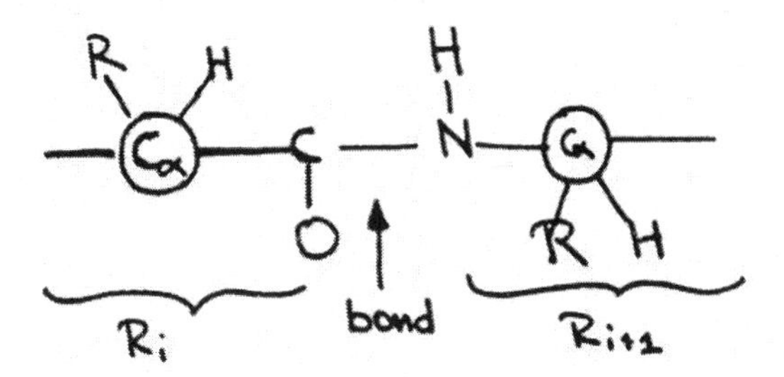

Fig. 21.16. Peptide bond.

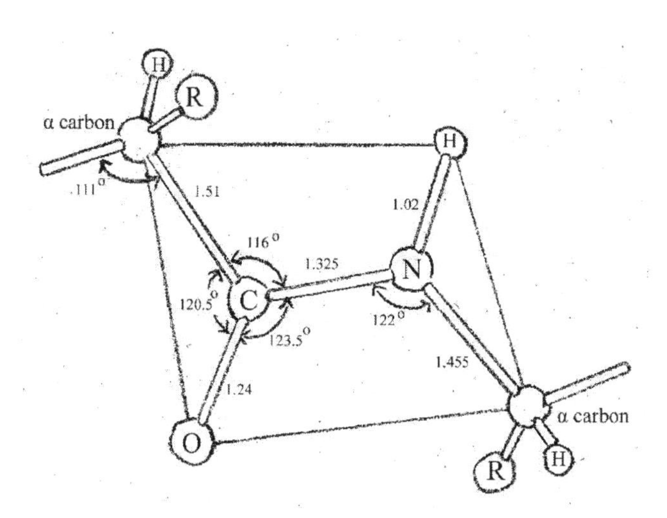

Fig. 21.17. Peptide bond, lengths in Å.

All three atoms are involved in double bonding. The atomic orbitals of C and N have sp^2 hybridization. The nitrogen atom has an electron in each of the three sp^2 orbitals and a lone pair in the p_x orbitals lying perpendicular to the plane of the sp^2 (Fig. 21.18).

Fig. 21.18. Electron occupation in N orbitals: three electrons in sp^2 orbitals and two electrons in a lone pair.

The carbon atom also has three electrons in the sp^2 orbitals and a fourth one in a perpendicular p_x orbital (Fig. 21.12). The oxygen atom has one electron in the bonding sp_z orbital, one in the perpendicular p_x orbital, and two in the nonbonding p_y orbitals (Fig. 21.19). When the three atoms are bonded together, one O(sp_z) and one C(sp^2) overlap to form a single σ bond, as do C(sp^2) and N(sp^2). The three p_x-orbitals are parallel to each other, forming a conjugated (delocalized) π bonding system through which the four π electrons travel: one from oxygen, one from carbon, and two from nitrogen (Fig. 21.20).

After this preparation, we can turn to the electronic spectrum of the peptide bond. The four roving π electrons of the OCN peptide system in the ground state fill the lowest molecular orbitals. The more detailed calculations show that there exist three orbitals, classed as bonding, nonbonding, and antibonding (Fig. 21.21).

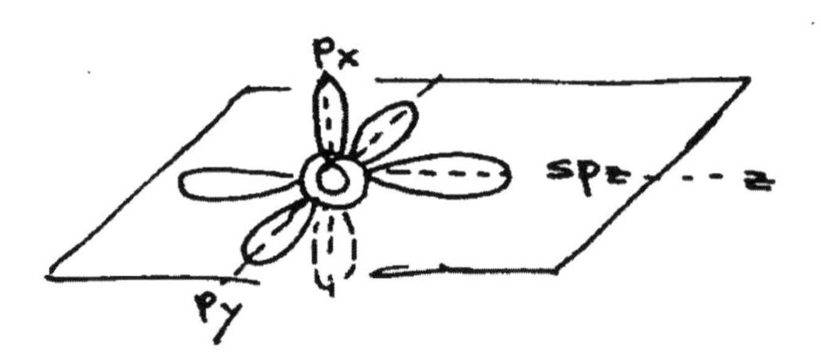

Fig. 21.19. Electron occupation in orbitals.

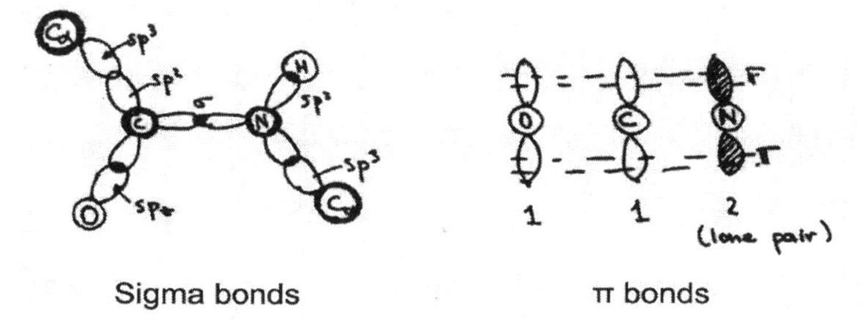

Fig. 21.20. The peptide bond.

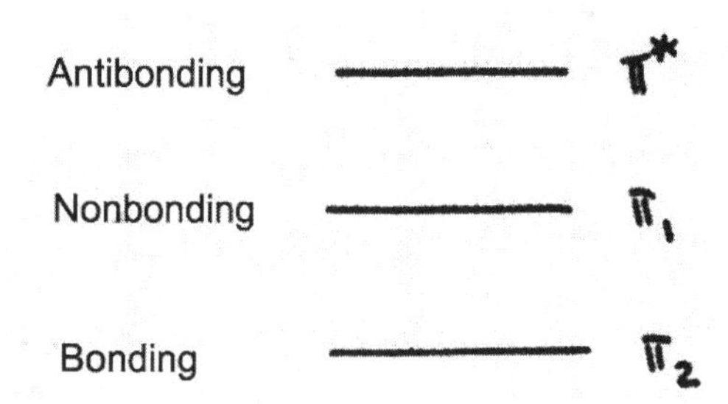

Fig. 21.21. Energy levels of molecular orbitals of π electrons of a peptide system.

Each of these levels can accept two electrons. In the ground state, the bonding and the nonbonding states are filled. The lowest energy electronic transition thus is the one from the nonbonding to the antibonding state, $\pi_1 \rightarrow \pi^*$. The oxygen atom also possesses two nonbonding electron in the atomic p_y orbital. One of these can also be excited to the π^* orbital, giving rise to a so-called $n \rightarrow \pi^*$ transition. The energies and wavelength of the various transitions are shown in Fig. 21.22.

21.8 The Heme Group

Porphyrins play an important, or even crucial, role in many biomolecules [5]–[8]. They are the active centers in molecules involved in oxygen storage and transport, electron transport, and redox reactions. We sketched one special case in Fig. 4.8 and now discuss this molecule in more detail. The basic building block of the porphyrin is the pyrrole ring shown in Fig. 21.23. Its structure is similar to benzene, and a lone pair occurs on the N atom, as in the peptide bond.

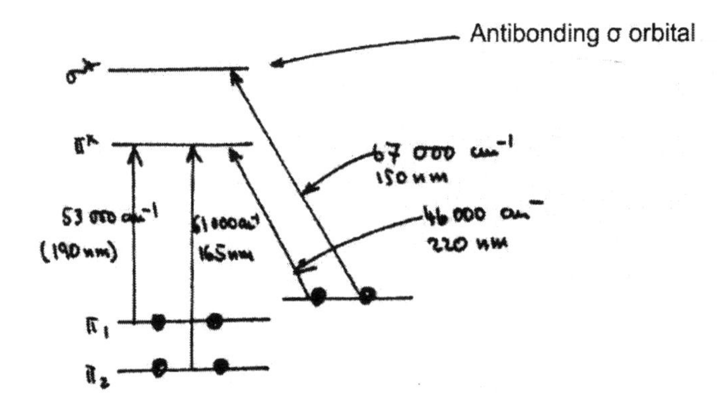

Fig. 21.22. Energy levels and transitions of molecular orbitals of OCN peptide system.

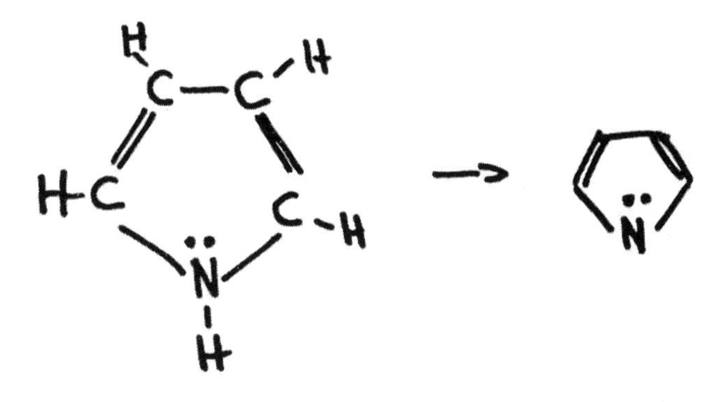

Fig. 21.23. The pyrrole ring.

Four pyrrole rings can link together via methine bridges (= CH–) to form the tetrapyrrole porphyrin (Fig. 21.24). Various porphyrins are obtained formally from porphin by substituting side chains for some or all H. Porphyrins are essentially planar, have a diameter of about 0.85 nm and a thickness of about 0.47 nm, and are intensely colored.

A number of porphyrin derivatives play important roles in biology. We mention three cases here:

1. *Heme.* Protoporphyrin IX is shown in Fig. 21.25. When the center is occupied by iron, it becomes protohemeIX. The heme group appears in

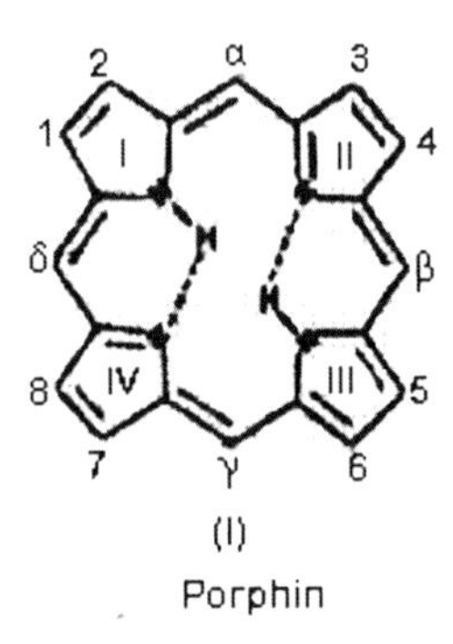

Fig. 21.24. The porphyrin skeleton.

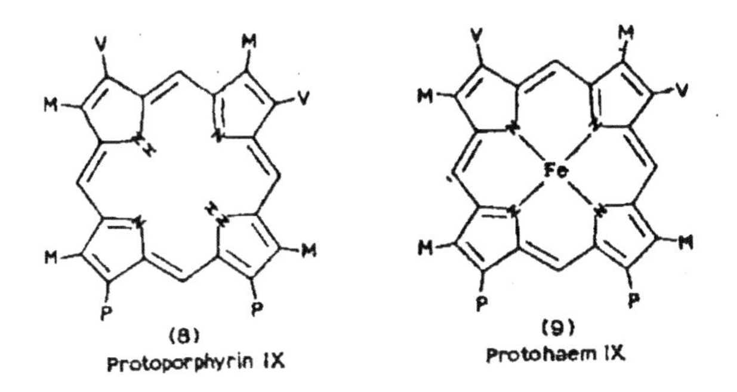

Fig. 21.25. Structures of (a) protoporphyrin IX and (b) protoheme IX.

hemoglobin, myoglobin, in cytochromes, and in some other enzymes. It thus does a variety of jobs well, from oxygen storage and transport to electron transport, to oxydation-reduction reactions.

2. *Chlorophylls.* With Mg instead of Fe, and with some changes in the side groups of a porphyrin ring, an essential part of the photosynthetic system is obtained as shown in Fig. 21.26.

3. *Vitamin* B_{12}. With Fe replaced by CO, and with two of the pyrrole rings joined directly rather than through methene bridges, vitamin B_{12} results.

The most primitive approach to the electronic structure of a porphyrin molecule consists in considering the π electrons in the inner ring free electrons, moving along a one-dimensional path—"one-dimensional electron gas" [9, 10]. The path is shown dashed in Fig. 21.27; it includes 16 atoms: 12 C and 4 N. Each N contributes, as in Fig. 21.18, an electron pair; each C contributes, as in Fig. 21.20, one electron. There are thus 20 electrons ready to move in the π-orbit. However, the central metal atom is usually not neutral; if it contributes two electrons to the N-metal bond, these will distribute over the four bonds

(12) Chlorophyll a (* Phytol)

(13) Chlorophyll b

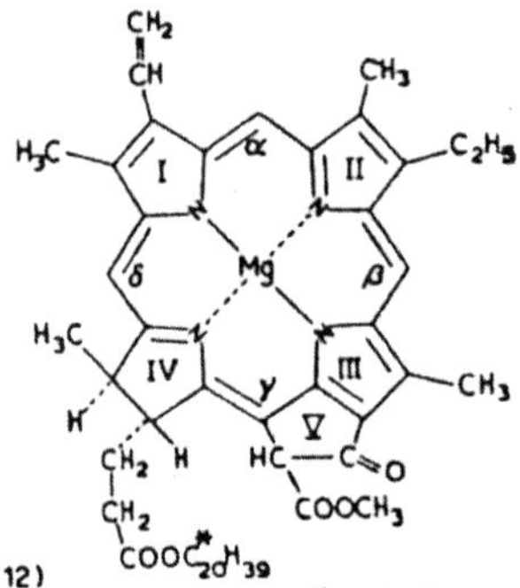

Fig. 21.26. Two types of chlorophylls.

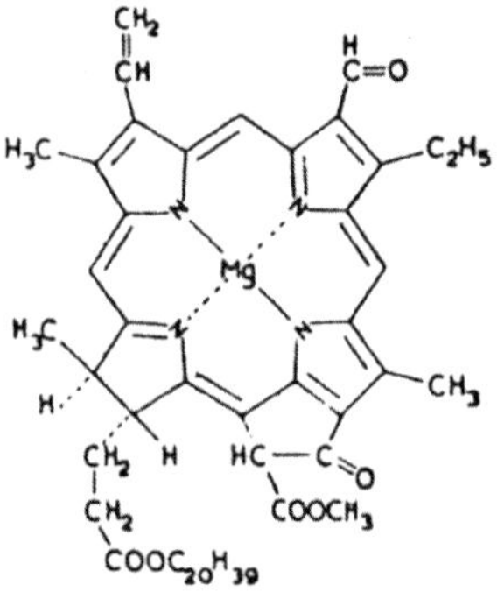

and remove two electrons from the ring. There will thus be 18 electrons in the ring.

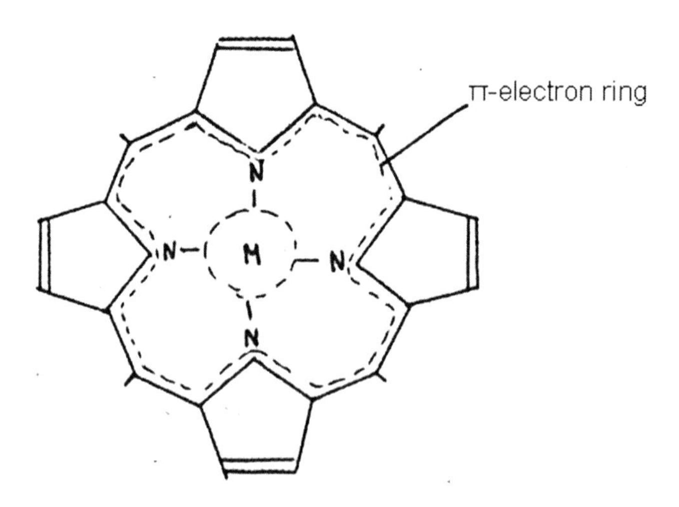

Fig. 21.27. The π-electrons in the "inner ring" of a porphyrin can be treated as a "free electron gas."

To find the energy levels of the free electron gas, we assume that the electrons are uniformly distributed over the one-dimensional chain of length L. The Schrödinger equation then is

$$-\frac{\hbar^2}{2m_e}\frac{d^2\psi}{ds^2} = E\psi\,, \tag{21.23}$$

where s is the coordinate along the ring. The wave function must satisfy the periodic boundary condition

$$\psi(s) = \psi(s+L). \tag{21.24}$$

The solutions of Eq. (21.23) satisfying the boundary conditions Eq. (21.24) are

$$\psi = N\exp(2\pi i\, m\, s/L)\ , \tag{21.25}$$

where the quantum number m can assume the values

$$m = 0, \pm 1, \pm 2, \pm 3, \ldots\ . \tag{21.26}$$

Inserting Eq. (21.25) into (21.23) gives the energy eigenvalues

$$E_m = \frac{2\pi^2\hbar^2}{m_e L^2}\, m^2. \tag{21.27}$$

The energy level diagram for the "one-dimensional free electron gas" is given in Fig. 21.28. In the ground state, the 18 electrons in the ring fill all states up to $m = \pm 4$. The lowest unfilled states thus are the one with $m = \pm 5$. In the first excited state, one electron is promoted to the state with $m = 5$.

The wavelength corresponding to a transition $m = 4$ to $m = 5$ is given by

$$\lambda = 2\pi\hbar c/(E_5 - E_4) = \frac{m_e c L^2}{9\pi\hbar}. \tag{21.28}$$

With proper bond lengths we find for L : $L = 8 \times 0.14\,\text{nm} + 8 \times 0.13\,\text{nm} = 2.2\,\text{nm}$. The electron Compton wavelength is $\hbar/m_e c = 386$ fm and we find

$$\lambda \approx 440\,\text{nm}. \tag{21.29}$$

The absorption line corresponding to the excitation to the first excited state of the π-electron ring should be at about 440 nm. Indeed, such a line exists; it is called the *Soret band.*

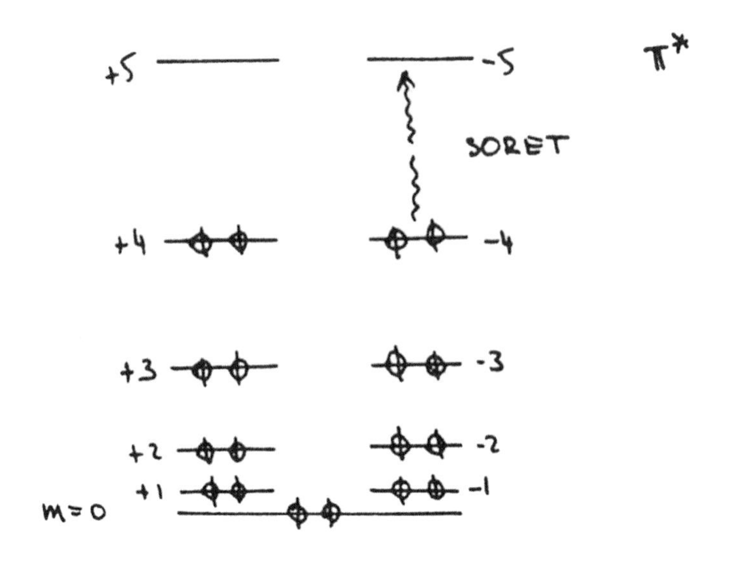

Fig. 21.28. Energy levels for the π-electrons in heme.

The free-electron model provides a simple insight into the electronic spectrum of the porphyrin compounds but neglects all finer details. For a more complete understanding and in particular a discussion of the strength of the absorption lines, more sophisticated MO calculations are required. A vast amount of literature covering these aspects exists [8, 11]–[14]. We state only some ideas here. Molecular orbits ϕ_i, with energies $\sum_i$, are expressed as linear combinations of atomic orbitals (LCAO) χ_p,

$$\phi_i = \sum_p C_{pi}\,\chi_p\ . \tag{21.30}$$

The atomic orbitals come from three sources, the atoms in the porphyrin, the central metal atom, and ligands below and above the metal atom in Fig. 21.29. The eigenvalue equation

$$H_{eff}\ \phi = \sum_i \phi_i \tag{21.31}$$

is then approximately solved. Some of the MO turn out to be concentrated in the central atom, some to be predominantly formed by the π electrons of the porphyrin ring, and some are strongly mixed.

In iron porphyrin, three different types of transitions are important, d–d, π–π^*, and charge transfer. In d–d, transition, an electron is excited from one iron d orbital to another; such transitions are very weak. The very strong

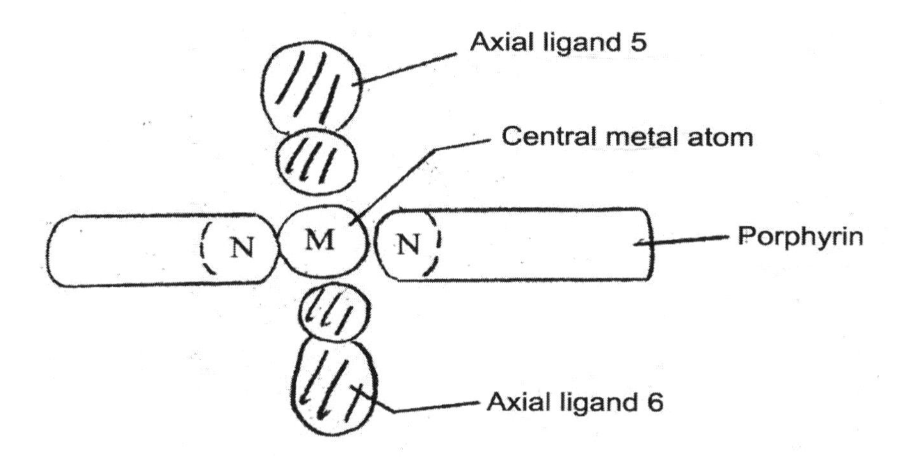

Fig. 21.29. Contribution to molecular orbitals come from the porphyrin, the central metal atom, and axial ligands.

π–$\pi*$ transitions occur between orbitals that are dominated by the π-electron ring. Charge transfer transitions take place between orbitals that are in one state concentrated on the central metal atom, and in the other in the porphyrin. They thus involve porphyrin $\leftrightarrow$ metal, and the axial ligands may also contribute. The salient features are summarized in Fig. 21.30.

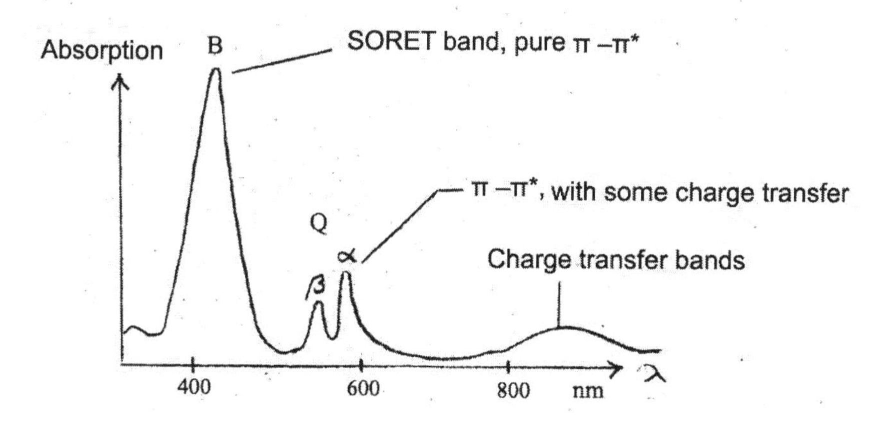

Fig. 21.30. Main features of the optical absorption of iron porphyrin.

References

1. M. Karplus and R. N. Porter. *Atoms and Molecules*. Benjamin, New York, 1970.
2. W. H. Flygare. *Molecular Structure and Dynamics*. Prentice-Hall, New York, 1978.
3. R. S. Berry, S. A. Rice, and J. Ross. *Physical Chemistry*. Wiley, New York, 1980.
4. G. L. Hofacker. Intra- and intermolecular interactions in biophysics. In W. Hoppe et al., editors. *Biophysics*. Springer Verlag, Berlin, 1983, Section 4.1, pp. 235–237.
5. J. E. Falk. *Porphyrins and Metalloporphyrins*. Elsevier, New York, 1964.
6. K. M. Smith, editor. *Porphyrins and Metalloporphyrins*. Elsevier, New York, 1975.
7. K. M. Smith. Protoporphyrin IX: Some recent research. *Accts. Chem. Res.*, 12:374–81, 1979.
8. A. B. P. Lever and H. B. Gray, editors. *Iron Porphyrins*. Addison-Wesley, Reading, MA, 1983. 2 vols.
9. H. Kuhn. A Quantum-mechanical theory of light absorption of organic dyes and similar compounds. *J. Chem. Phys.*, 17:1198–1212, 1949.
10. W. T. Simpson. On the theory of the $\alpha-$electron system in porphines. *J. Chem. Phys.*, 17:1218-1221, 1949.
11. M. Zerner, M. Gouterman, and H. Kobayashi. Extended Huckel calculations on iron complexes. *Theor. Chim. Acta*, 6:363–400, 1966.
12. M. Gouterman. Excited states of porphyrins and related ring systems. In C. W. Shoppee, editor, *Excited States of Matter* (Texas Tech University graduate stueis, no. 2). Texas Tech Press, Lubbock, 1973, pp. 63–101.
13. J. Hofrichter and W. A. Eaton. Linear dichroism of biological chromophores. *Ann. Rev. Biophys. Bioeng.*, 5:511–60, 1976.
14. F. Adar. Electronic absorption spectra of hemes and hemoproteins. In D. Dolphin, editor, *The Porphyrins, Vol. III*. Academic Press, New York, 1978, pp. 167–209.

22

Energy Levels from Nuclei to Proteins

Consider the energy levels of a number of systems. *Nuclei*, with a characteristic length of a few fm, possess excitation energies of the order of keV to MeV. A typical example is given in Fig. 22.1. The first excited state of the nucleide ^{57}Fe is at 14.4 keV; it has a mean life of 1.4×10^{-7} sec. The spin of the ground state is 1/2, of the first excited state 3/2. In a magnetic field, both ground state and excited state show a Zeeman splitting. In an electric field, the ground state remains degenerate, the excited state splits into two substates. The dominant excitation in *atoms* is electronic and too well known to be discussed here.

Diatomic molecules possess three types of excitation, electronic, vibrational, and rotational. The essential features are shown in Fig. 22.2.

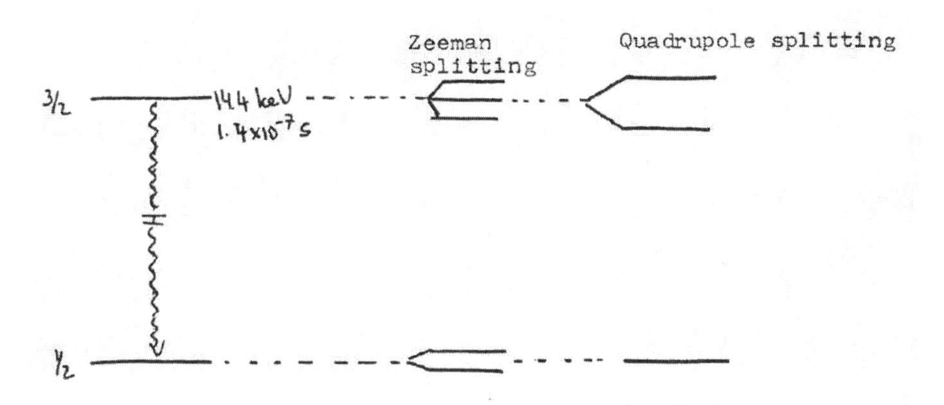

Fig. 22.1. Ground state and first excited state of the nucleide ^{57}Fe.

H. Frauenfelder, *The Physics of Proteins*, Biological and Medical Physics, Biomedical Engineering, DOI 10.1007/978-1-4419-1044-8_22,

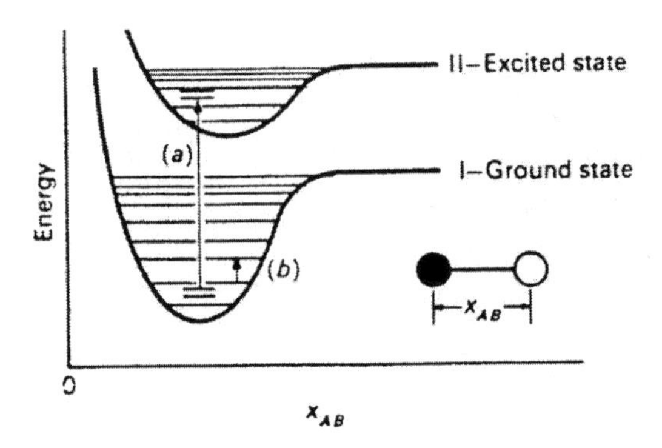

Fig. 22.2. The energy levels of a simple diatomic molecule. The potential energy (heavy curve) is a function of the internuclear distance (X_{AB}). Two electronic states are shown (I and II). They differ in their potential energy and in the position of the minimum, the equilibrium internuclear distance. For each electronic state there are different possible levels of vibrational energy (long, thin lines) and rotational energy (short lines, shown only for two vibrational levels). A possible electronic transition is shown by (a) and a vibrational transition in the electronic ground state by (b).

In *polyatomic molecules*, more complex motions can occur. As an important new feature, we mention the existence of tunnel states in the ammonia molecule [1]. The NH_3 is sketched in Fig. 22.3.

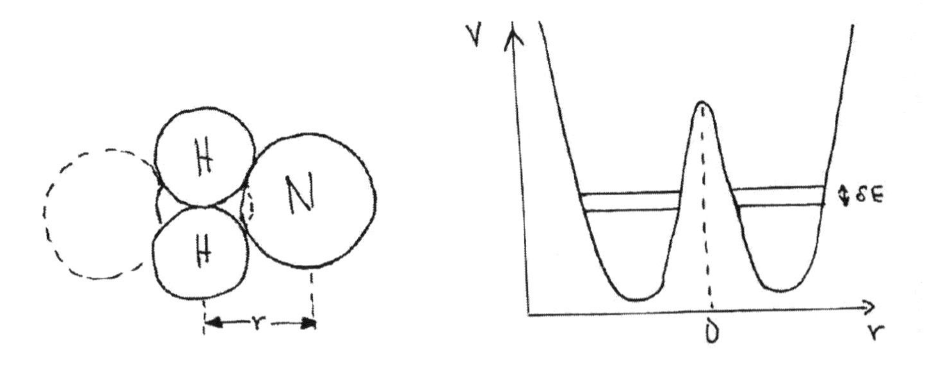

Fig. 22.3. The two substates of the ammonia molecule. The ground state is split into two tunnel states.

It can exist in two different arrangements (or substates in our terminology). The degeneracy leads to a splitting of the energy levels. The splitting is caused by the fact that the N atoms can tunnel through the barrier formed by the three H atoms. The tunnel splitting is given by

$$\delta \mathrm{E} = 10^{-4}\mathrm{eV} = 24\,\mathrm{GHz}.$$

In more complex molecules, we can expect more conformational substates, as we have already discussed for the case of heme in Section 11.2.

Solids also show features that are importance to biomolecules. The specific heat, Eq. (20.26), is classically expected to be independent of temperature. Experimentally, it goes to zero for $T \rightarrow 0$. Einstein first explained this observation with discrete energy levels in solids, as indicated in Fig. 22.4(a) [2]. A typical value of the Einstein energy is 0.1 eV. Debye generalized Einstein's spectrum, and a typical Debye spectrum is shown in Fig. 22.4(b). A realistic spectrum is shown in Fig. 22.4(c). Note that (b) and (c) use a different representation from (a).

The vibrations in a solid are quantized and are called *phonons*. Phonons are not the only elementary excitations in solids. Plasmons, magnons, polarons, and excitons can also exist.

In *proteins*, we expect to see most or all of the various excitations and energy levels, from tunnel states to electronic states. A treatment of all possible spectroscopies needed to obtain a complete picture of the energy levels and the excitations in a protein would far exceed the space and time available here.

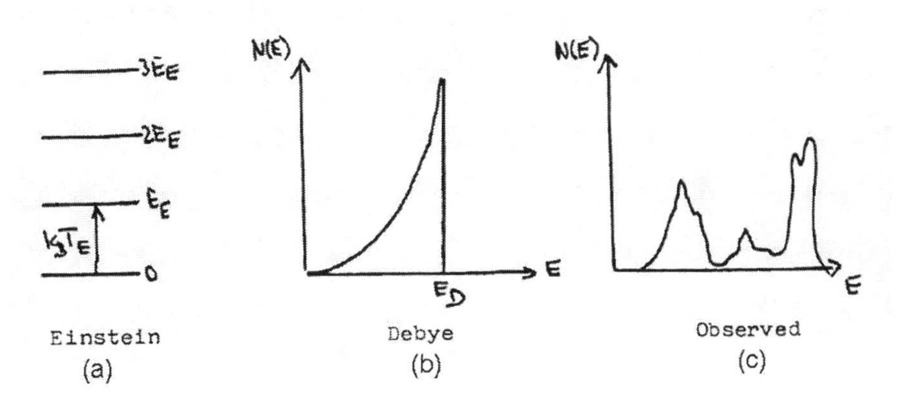

Fig. 22.4. Vibrational spectra in solids. (a) Einstein's energy levels, (b) Debye spectrum, and (c) observed spectrum.

The following chapters are consequently not exhaustive but should provide a glimpse of the richness of protein spectroscopy. Most of the time we use heme proteins as an example. As indicated in Fig. 22.5, all components contribute to the excitations.

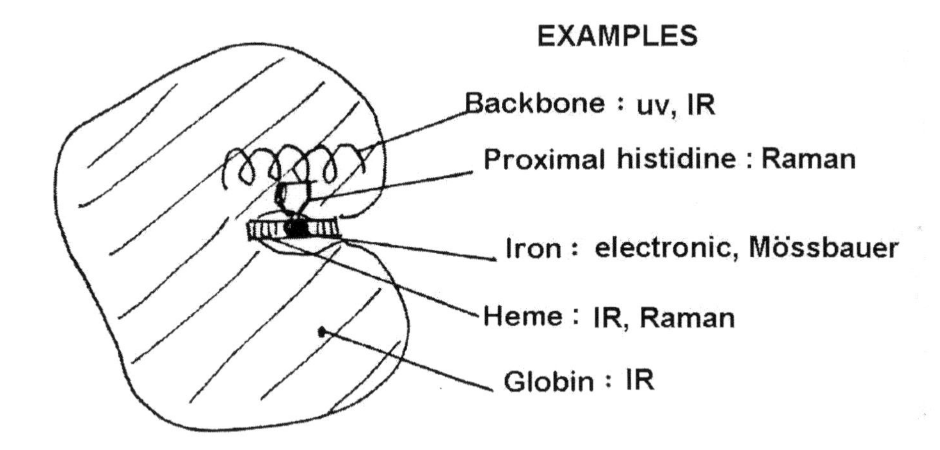

Fig. 22.5. Excitations in a heme protein.

References

1. R. P. Feynman, R. B. Leighton, and M. Sands. *The Feynman Lectures on Physics. Vol. III.* Addison-Wesley, New York, 1965. Chapter 9.
2. A. Pais. *Subtle Is the Lord ..., The Science and the Life of Albert Einstein.* Oxford Univ. Press, Oxford, 1982. Chapter. 20.

23

Interaction of Radiation with Molecules

We now discuss the most important features of the interactions. We only present the main ideas and results, without derivations. Derivations involve quantum electrodynamics—since the photon always moves with the velocity of light (it is light!), no nonrelativistic theory of light exists. The techniques and ideas used in nonrelativistic quantum mechanics thus must be extended and generalized—a task that leads too far astray here. For details of the aspects that we introduce here we therefore refer to the relevant texts.

23.1 Absorption and Emission

Consider a molecule as in Fig. 23.1 that has only two energy levels [1]–[4]. Assume that radiation of frequency ω impinges on the molecule.

Three processes can then happen:

1. *Spontaneous emission.* An atom in the excited state can decay spontaneously to the ground state with emission of a photon of energy $E = \hbar\omega_{ij}$. We denote by A_{ij} the corresponding transition probability.

Fig. 23.1. Emission and absorption of radiation.

H. Frauenfelder, *The Physics of Proteins*, Biological and Medical Physics, Biomedical Engineering, DOI 10.1007/978-1-4419-1044-8_23,

2. *Induced absorption.* The incident radiation, of frequency $\hbar\omega_{ij}$, can excite the system by moving it from the lower to the higher state. We denote the transition probability for induced absorption by B_{ij}.

3. *Induced emission.* An atom in the excited state can be forced to the ground state by the incident radiation; the corresponding transition probability is called B_{ji}.

While the first two processes are easily understood, the third requires some adjustment. The fact that incident radiation can lead to the additional emission of photons is not immediately obvious. However, as we will see, all three processes are necessary for a complete picture.

The following approach is due to Einstein, and the three coefficients A_{ji}, B_{ij}, B_{ji} are called Einstein coefficients. To find a relation among these coefficients, we note that these equations should be satisfied:

$$B_{ij} = B_{ji}, \tag{23.1}$$

$$\frac{n_j}{n_i} = \exp(-E/k_B T), \tag{23.2}$$

$$\rho(\omega_{ij}) = \frac{\hbar\omega_{ij}^3}{\pi^2 c^3} \frac{1}{e^{\hbar\omega_{ij}/k_B T} - 1}. \tag{23.3}$$

The first relation indicates the symmetry between the two states, and it follows from detailed balance or microscopic reversibility. The second relation is the Boltzmann equation. The third relation is the well-known Planck expression for the density of light energy per unit frequency at temperature T [5].

If the molecular system is in equilibrium with the radiation field, as many transitions ij occur as ji. If n_i and n_j denote the number of atoms that are in states i and j, respectively, the equilibrium condition reads, with Eq. (23.1),

$$\rho(\omega_{ij}) B_{ji}(n_i - n_j) - A_{ji} n_j = 0.$$

With Eqs. (23.2) and (23.3) and $E = \hbar\omega_{ij}$, this relation leads to

$$A_{ji} = \frac{\hbar\omega_{ij}^3}{\pi^2 c^3} B_{ij}. \tag{23.4}$$

All three Einstein coefficients are thus related. Equation (23.4) also shows that spontaneous emission dominates if E is large, induced if E is small.

A characteristic difference between induced and spontaneous emission lies in the directional dependence. The direction of emission in induced emission is the same as that of the incident radiation; in spontaneous emission, the direction is determined by the orientation of the molecule.

23.2 Line Width

The energy levels in Fig. 23.1 are shown infinitely sharp. In reality, however, the energy levels will have finite width. Assume, for example, that at time $t = 0$ all molecules are in the excited state, and that no external radiation field is present. The molecules will then decay to the ground state according to the "radioactive decay law,"

$$n_j(t) = j_j(0)e^{-A_{ij}\,t}. \tag{23.5}$$

The lifetime τ (mean life) of the state is given by

$$\tau = 1/A_{ji}. \tag{23.6}$$

The Heisenberg uncertainty relation connects the mean life and the energy uncertainty Γ:

$$\Gamma\tau = \hbar. \tag{23.7}$$

The excited state will therefore have an uncertainty in energy, or a width given by [6]

$$\Gamma = \hbar\, A_{ji}. \tag{23.8}$$

The complete quantum-mechanical treatment [7] indicates that the line indeed has such a width, but it also yields the line shape. The line is a Lorentzian, with distribution

$$I(\omega) \propto \frac{1}{(\omega - \omega_{ij})^2 + \frac{1}{4}\Gamma^2}. \tag{23.9}$$

If the lower state is not stable but also has a finite lifetime, its width also enters Eq. (23.9).

If the molecular system is not isolated, other factors can broaden the line. One particular broadening influence comes from the motion of the molecule, the Doppler broadening. If the excited state can decay into more than one way, the rates for all these channels add:

$$\Gamma^{\text{total}} = \Gamma_1 + \Gamma_2 + \Gamma_3 \ldots\ . \tag{23.10}$$

The line can therefore be much broader than the natural line width.

23.3 Transition Rates

Classically, charged particles radiate when they are accelerated. The strength of radiation is proportional to the square of the electric dipole moment, $e\mathbf{r}$. The quantum mechanical calculation gives for the transition probability for spontaneous emission [1]–[4]

$$A_{ji} = \frac{4}{3} \frac{e^2\omega_{ji}^3}{\hbar c^3} |\langle j|\mathbf{r}|i\rangle|^2 \tag{23.11}$$

where the matrix element $\langle j|\mathbf{r}|i\rangle$ is defined by

$$\langle j|\mathbf{r}|i\rangle = \int d^3r \; \psi_j^* \; \mathbf{r}\, \psi_i . \tag{23.12}$$

It is customary to write Eq. (23.11) in terms of the dimensionless *oscillator strength* f, defined by

$$f_{ji}^x = \frac{2m\omega_{ji}}{\hbar} |\langle j|x|i\rangle|^2 , \tag{23.13}$$

$$f_{ji} = f_{ji}^x + f_{ji}^y + f_{ji}^z . \tag{23.14}$$

The transition probability becomes

$$A_{ji} = \frac{2}{3} \frac{e^2}{mc^3} \omega_{ji}^2 f_{ji} . \tag{23.15}$$

The expression given so far describes the radiation from an electric dipole, and it is therefore called *electric dipole radiation*. Whenever it is allowed, it is fast and dominates. Sometimes selection rules prohibit the emission of electric dipole radiation, and other radiations then become observable, for instance, magnetic dipole or electric quadrupole radiation. In nuclear physics, these are extremely important. In molecular physics, however, electric dipole radiation dominates.

23.3.1 Selection and Sum Rules

Transitions between two states can occur via electric dipole radiation only if certain selection rules are obeyed. Consider Eq. (23.12). We first investigate the behavior of the matrix element under the parity operation,

$$P \; : \; \text{parity operation} \qquad \mathbf{r} \rightarrow -\mathbf{r} \tag{23.16}$$

Careful experiments have shown that the electromagnetic interaction is invariant under the parity operation. The matrix element Eq. (23.12) therefore should be invariant under the parity operation. However, the operation $\mathbf{r}$ changes sign according to Eq. (23.16). The parity of the two wave functions

ψ_j and ψ_i must therefore be opposite; otherwise the transition is forbidden by electric dipole radiation. Symbolically, we write this condition in the form

$$P_j P_i = -1. \tag{23.17}$$

A second selection rule also follows from Eq. (23.12): The orbital angular momentum ℓ of the two states j and i can differ by at most one unit:

$$\Delta\ell = 0, \pm 1. \tag{23.18}$$

With no spin change,

$$\Delta S = 0.$$

Finally it is also easy to see that

$$\text{No } J_j = 0 \;\rightarrow\; J_i = 0. \tag{23.19}$$

As a generalization of the two-state system shown in Fig. 23.1, we consider a system with many excited states j that all can decay to the final state i by electric dipole radiation. Can the radiation emitted be arbitrarily strong? The dipole- or Thomas-Reiche-Kuhn sum rule, which is easily derived, states that the sum of all oscillator strengths leading to state i is fixed,

$$\sum_j f_{ji} = 3. \tag{23.20}$$

The emitted radiation is consequently limited; it cannot be arbitrarily strong.

The discussion given here of the interaction of radiation with matter has been cursory, but far greater details and depth can be found in the cited books.

References

1. G. Baym. *Lectures on Quantum Mechanics.* W. A. Benjamin, New York, 1969.
2. H. A. Bethe and R. Jackiw. *Intermediate Quantum Mechanics*, 2nd edition. W. A. Benjamin, New York, 1968.
3. J. I. Steinfeld. *Molecules and Radiation*, 2nd edition. MIT Press, Cambridge, MA, 1985.
4. W. H. Flygare. *Molecular Structure and Dynamics.* Prentice-Hall, New York, 1978.
5. A beautiful derivation of Eq. (23.3) is due to S. Bose; a translation of his original derivation, The beginning of quantum statistics, was published in *Amer. J. Phys.*, 44:1056-7, 1976.
6. M. Moshinsky. Diffraction in time and the time-energy uncertainty relation. *Amer. J. Phys.*, 44:1037–42, 1976.
7. V. F. Weisskopf and E. P. Wigner. Berechnung der naturlichen Linienbreite auf Grun der Diracschen Lichttheorie. *Z. Physik*, 63:54–73, 1930 and 65:18, 1930.

24

Water (R. H. Austin[1])

We didn't know where to put this chapter. Water is *the* solvent of biology yet it is a very complex liquid. Certainly some discussion of water belongs in a book on the physics of proteins, but in order to get at the complexity of the liquid, one has to work through all the fundamentals that we have covered up to this point in the appendices. So, here is water.

What do we mean when we say water in its liquid form is a very complex liquid? Is it really complicated in comparison to say, liquid helium 3? The answer seems to be "yes," but in a different sense from the mathematical and quantum mechanical complexities of helium. Water is a liquid with very strong many-body effects, but it is hot compared to helium 3, and many of the problems are almost classical in nature.

Ordinarily we are trained to run from such complexities, but if you want to understand biomolecules you must understand water to some degree, since it is the primary solvent for biomolecules. This isn't too surprising, since as Franks points out in his wonderful little introductory volume [1], water is the only naturally occurring inorganic liquid on the surface of the earth so life on earth has to play the hand it was dealt. However, it seems that as we study the physics of biological molecules it isn't simply a matter of taking advantage of what was around, rather water's unique properties seem quite important in terms of allowing life to exist at all. It is altogether wonderful that the only liquid present in abundance is also one of the most complex and useful in its properties. People have speculated in the past that on other planets silicon may substitute for carbon chemistry, but they also should consider the primary role that water plays in biology. It may well be the truly crucial ingredient that cannot be replaced. We seek simplicity, yet we find that even the solvent for biomolecules is very complex. In keeping with our goals, we will keep the discussion in this chapter aimed toward the major points of interest, in particular those of biological interest. Perhaps this will be a cautionary tale for those who would do biological physics.

[1] Department of Physics, Princeton University, Princeton, NJ 08544, USA.

H. Frauenfelder, *The Physics of Proteins*, Biological and Medical Physics, Biomedical Engineering, DOI 10.1007/978-1-4419-1044-8_24,

We would be amiss if we didn't mention one of the truly bizarre chapters that has occurred in the history of science, namely, the polywater craze. We will see that water is a highly correlated liquid, held together by a network of hydrogen bonds. It was proposed that under certain conditions water could "polymerize" into a highly viscous form called "polywater." Those of you familiar with Kurt Vonnegut's book *Cat's Cradle* will have heard this before (there really is an Ice 9 by the way). Obligingly, an obscure Russian chemist found that by exhaustive distillation of distilled water a highly viscous "form" of water could be isolated. Pandemonium broke loose among the physical chemistry community, and there were warnings not to let polywater loose into the environment or all the oceans might polymerize. Probably if this hysteria had occurred in 2009, water would be labeled a pollutant and a tax would be put on water producers.

Unfortunately Princeton University was a hotbed of polywater research. Franks has documented this whole amazing story in his wonderful little book *polywater*! [2], must-reading for the experimentalist. Of course, ultimately the whole bubble burst as it was revealed that the isolators of polywater were purifying either highly concentrated salt solutions (known as sweat outside the Ivy League) or silica gels derived from the powerful solvent capability of water. The point to this **is** that water is so unusual that it isn't too hard to ascribe truly fantastic properties to it.

24.1 Clues to the Complexity of Water

Is water really unusual? The best place to find a quick answer to this question that I have found is Franks' *Water* [1]. This book is not to be confused with the far more comprehensive seven-volume (!) tome called *Water: A Comprehensive Treatise*, also by Franks [3]. A volume in between these two extremes is the classic volume by Eisenberg and Kauzmann, *The Structure and Properties of Water* [4].

The first clue to the unusual nature of water can be found in the melting and boiling points of water. Table 24.1, taken from [1], reveals that the melting and boiling points of water are significantly higher than would be expected for a liquid made from a molecule as light as H_2O. Clearly, there are strong associative forces between water molecules. However, many molecules have strong associative forces. There is another aspect to water that is quite unusual, and it can be found in some numbers that we dutifully feed out in our introductory physics courses, namely, the latent heats of fusion and evaporation for water and the specific heat of water. Other entries in Table 24.1 illustrate the problem. The latent heat of evaporation for liquid water, 40.5 kJ/mol, is indeed large (as we would expect), but it is not amazingly large: For example, liquid sodium at 107 kJ/mol is considerably larger. What is strange is the amazingly *small* molar latent heat of fusion of water, which is often misrepresented as being very large! Note that water's molar latent heat of fusion, 5.98 kJ/mol,

Table 24.1. Physical Properties of Some Liquids.

Property	Argon	Benzene	Sodium	Water
Latent heat of fusion, kJ/mol	7.86	34.7	109.5	**5.98!**
Latent heat of evaporation, kJ/mol	6.69	2.5	107.0	**40.5!**
Heat capacity, J/(mol· K), (solid)	25.9	11.3	28.4	37.6
Heat capacity, J/(mol· K), (liquid)	22.6	13.0	32.3	**75.2!**
Melting point, K	84.1	278.8	371.1	273.2

is very small, in fact smaller than the molar latent heat of fusion of liquid argon, an inert element! One can easily conclude that either liquid water is very highly structured, rather like ice, or that ice is rather highly disordered, like water. In reality, both viewpoints have elements of truth to them. This is clear from the unusually large specific heat of liquid water, 75.2 J/(mol· K), which is far larger than similar materials, and twice the heat capacity of ice. Appearently there is much structure in "liquid" water, which is being broken down with increasing temperature.

Other physical parameters of water are not so amazing. For example, the viscosity of water, while high, is not outrageous, nor are the self-diffusion constants of water highly unusual. However, as we probe water in other sections on a microscopic scale we find that other unusual properties of water will manifest themselves on a molecular level. The parameter that probably best characterizes the interaction of water molecules with each other is the latent heat of evaporation. To cast that into more usable units, it is 40.5 kJ/mol, or about 0.5 eV/mol, which is a decent-sized number. But, the mysteries of water do not lie in this number.

24.2 Basic Structure of the Water Molecule

Can we understand some of the unusual properties of water from an examination of the molecule itself? We apply here many of the principles outlined in Chapter 21, at the risk of some redundancy but with the hope that this will re-enforce some of the basic physics of that chapter.

The oxygen atom $^{16}O_8$ has the following atomic orbitals:: 2(1s), 2(2s), 4(2p). The s orbitals are, of course, spherically symmetric, while the p orbitals are of the form:

$$p_x = \sin\theta\cos\varphi;\quad p_y = \sin\theta\sin\varphi;\quad p_z = \cos\theta \tag{24.1}$$

Since another two electrons from two hydrogens will fill the 2p shell, you would guess that H_2O would be a happy molecule and quite inert. However, simple valence counting doesn't go very far in predicting chemical bonding.

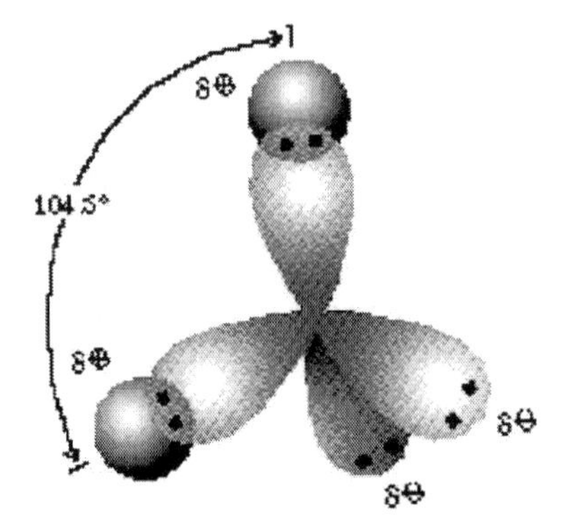

Fig. 24.1. The basic hydridized orbitals for the water moelcule.

In the simplest approximation, bonding occurs via delocalization of atomic orbitals between two sites. A gross approximation is to simply take a linear combination of atomic orbitals to describe the actual complex molecular orbital that forms as was discussed in Chapter 21, The construction of the water orbitals follows very closely the construction of the carbon bonding orbitals in that chapter.

In the case of oxygen it is thus feasible to consider *hybridizing* the oxygen orbitals by mixing in 2s oxygen orbitals with the 2p oxygen orbitals. A little trigonometry will reveal that these orbitals form a tetrahedron that points along the 1,1,1 and similar combination directions. Of these four vertices, two are taken up by hydrogen atoms and have a net positive character, while two other orbitals (lone pairs) are unbonded and have a net negative character. The expected angle from this model between the hydrogen atoms is 109.6°, the actual value is 104°. Figure 24.1 shows roughly what we might expect the electron distribution around a water molecule to look like.

If we consider one of the two mirror equivalent configurations of the water molecule, we can guess that the molecule probably has a large dipole moment. In fact, from vapor phase measurements the value of the electric dipole moment μ is 1.86 x 10^{-18} esu-cm for each mirror symmetric molecule. Since an electron has charge of 4.8x10^{-10} esu, there is a strong ionic character to the bond between oxygen and hydrogen, on the order of 0.5e over a 1-Å distance. The water molecule is **very** polar, and gets even more polar in the condensed phase. This is where water becomes really interesting.

24.3 Condensed Water: Intermolecular Forces

Now we need to understand what happens when water molecules get close to one another. Although in this section we concentrate on water, of course in the long run we have to understand these forces if we have a hope of understanding protein structure and action. There are basically four kinds of "long-range" interactions that are of importance in biological systems, as we discussed in Chapter 18: Van der Walls (dispersion) forces between neutral atoms, dipole-dipole forces between polar molecules, hydrogen bonding, and finally coulombic (charged ions). We'll ignore the coulombic part for now. We will discuss dispersion forces and dipole-dipole coupling, then concentrate on hydrogen bonding if we manage to convince you that the latter is the dominant part of the attractive forces holding water together. A great textbook that describes in detail the physics of the four forces we talk about in passing here is *Molecular Driving Forces*, by Dill and Bromberg [5].

24.3.1 Dispersion Forces

Dispersion forces result in an attractive force felt between electrically neutral atoms. A clear description of the dispersion force can be found in the book *Dispersion Forces* by Mahanty and Ninham [6] or in *Intermolecular and Surface Forces* by Israelachvilli [7]. Dispersion forces are strictly quantum mechanical in nature and exist between **all** atoms and molecules, independent of their charge state or the presence of a large dipole moment. In the case of two metal plates separated in a vacuum, this force is known as the Casimir force. The force that allows argon to form the solid we mentioned in Table 24.1 is due to dispersion forces. The name "dispersion force" arises from the frequency dependence of the force, which has a maximum value in the UV-to-optical-spectral range, where the "dispersion" (the derivative of the dielectric constant with frequency) is greatest. Intuitively, dispersion forces arise from the time-dependent dipole moments arising from the movement of an electron around a nucleus. This flickering dipole moment induces a dipole moment in a molecule nearby, and thus a force exists between the two molecules.

This is clearly an enormously complex interaction, but following Israelachvilli we can roughly estimate (in a Bohr-like mix of classical and quantum mechanics) the strength and frequency dependence of this force. Consider an electron orbiting a proton, as in hydrogen. Let the electron orbit be at the Bohr radius R_o. The radius R_o of the orbit has an energy E(R) given by:

$$E(R) = \frac{e^2}{2R_o} = hf, \tag{24.2}$$

which is equal to the energy of a photon of frequency f that can ionize this state. The orbiting electron has a (instantaneous) dipole moment $\mu = eR_o$. This time-dependent dipole moment will polarize a neutral atom that is a distance r away, giving rise to a net attractive force and potential energy of interaction $w(r)$.

To compute $w(r)$ we need to know the electronic polarizability of a neutral atom, α. The polarizability of an atom is defined as the coefficient between the induced electric dipole p in response to an applied field E_{ext}:

$$\mu = \alpha E_{\text{ext}}. \tag{24.3}$$

In the case of an electron orbiting a proton we can, as per Bohr, calculate that the center of the orbit of the electron gets shifted a distance x because of the applied field E_{ext}. The distance $x = R_o \sin\theta$, where R_o is the original radius of the orbit. Balancing the applied and coulombic forces we get:

$$eE_{ext} - \frac{e^2}{R_o^2}\sin\theta = 0, \tag{24.4}$$

where we take the projection of the coulombic force in the direction of the deformation. We can rewrite the right-hand side as:

$$\frac{e^2 x}{R_o^3} = \frac{e\mu}{R_o^3}, \tag{24.5}$$

where μ is the induced electric dipole moment. Thus:

$$\frac{e\mu}{R_o^3} = eE_{\text{ext}}, \tag{24.6}$$

Then we finally get that the polarizability α using Eq. (24.3) is simply:

$$\alpha = R_o^3. \tag{24.7}$$

Perhaps you can see now why argon actually has a decent latent heat of vaporization; the bigger the atom the more polarizable it is.

To get the distance dependence of the dispersion force takes a bit more work. The electric field $E(\theta, r)$ created by an electric dipole μ at an angle θ and distance r in a medium of dielectric constant ϵ is given by:

$$E_{ext} = \frac{\mu(1 + 3\cos^2\theta)^{1/2}}{\epsilon r^3}. \tag{24.8}$$

The angle average interaction energy $w(r)$ is then:

$$w(r) = < \frac{1}{2}\mu E_{\text{ext}} > = \frac{1}{2}\alpha < E_{\text{ext}}^2 > = \frac{\alpha\mu^2}{\epsilon^2 r^6}. \tag{24.9}$$

Going back to our dispersion interaction calculation, we can use Eq. (24.9) to compute the dispersion force strength, assuming that $\mu \sim eR_o$ (this is a cheat of course and misses the whole point of the self-consistency of the dispersion force, but the units are right!). Since $\alpha = R_o^3$ we have:

$$w(r) = -\frac{\alpha\mu^2}{\epsilon^2 r^6} = \frac{R_o^3 e^2 R_o^2}{\epsilon^2 r^6} = \frac{e^2 R_o^6}{r^6 R_o}. \tag{24.10}$$

We finally get, using $R_o = \frac{e^2}{2fh}$,

$$w(r) = -\frac{\alpha^2 hf}{\epsilon^2 r^6}. \tag{24.11}$$

Although as we derived this expression f was a fixed frequency, we could imagine that for a real molecule there might be a number of resonances and in the continuum limit the frequency becomes a continuous variable. Then the dispersion interaction *increases* with frequency until the polarizability rolls off in the ultraviolet, we can see the origin of the expression "dispersion."

The dispersion interaction strength is very dependent on the separation between the atoms. In the case of atoms of radius R_o =1 Å= 10^{-8} cm, the neutral atom polarizability α is 10^{-24} cm^3. If the atoms are roughly "in contact" then the ratio α^2/r^6 is about 0.1 and thus the dispersion energy of interaction can easily be several kT! It can't explain the *unique* properties of water because there is nothing remarkable about the packing density of water, but it serves as a sobering reminder of how complex and innately quantum mechanical are the forces by which the various parts of a protein interact with each other.

24.3.2 Dipole-Dipole Coupling

But water is polar, as we noted earlier. Thus, it has an electric dipole moment, which one would guess would give rise to a stronger interaction than the induced dipoles governed by the dispersion forces. How can we estimate the strength of static dipole-dipole coupling? We can (and will) do a simple dipole-dipole coupling calculation, but note that for a molecule with a large dipole moment like water the effect of the electric field of the dipole moment on neighboring molecules leads to substantial changes in the polarization of the original molecule; in fact at some critical value the system can undergo a *ferroelectric transition* that leads to effects totally not predicted by a simple isolated dipole-dipole calculation.

If we have simple isolated dipoles the calculation is very simple. Let molecule 1 have electric dipole moment μ and be a distance r from a second molecule of the same dipole moment. The electric field at a distance r from a dipole (this of course is the same as Eq. (24.8), but is written in a different way, to confuse the student as much as possible) is:

$$\mathbf{E}(r) = \frac{\mu \cos\theta}{R^3}\mathbf{r} + \frac{\mu \sin\theta}{r^3}\theta, \tag{24.12}$$

and the potential energy of a dipole in the presence of such an electric field is simply:

$$U(r) = \mu \bullet \mathbf{E}(\mathbf{r}). \tag{24.13}$$

At a finite temperature T we can calculate the net alignment that one dipole will achieve in the presence of a second dipole, and from that we can guess what the rough average interaction is between two dipoles.

Briefly, we calculate the average dipole moment using the Boltzmann factor $\exp(-U/kT)$:

$$< \mu >= \int \exp[-[\mu E \cos\theta/kT]\mu \cos\theta d\Omega \tag{24.14}$$

A little work assuming that the energy is small compared to kT yields:

$$< \mu >= \frac{\mu^2}{3kT} E. \tag{24.15}$$

Thus, the average interaction energy is:

$$U(R) = \frac{\mu^4}{3kTr^6}. \tag{24.16}$$

We find an interaction that falls off as r^6 like the dispersion force, but now we have temperature dependence. For water molecules at 300K, separated by approximately 2.8 Å, with dipole moments of 1.86 D (1 D $= 1\times10^{-18}$ esu-cm), the average interaction energy is 0.1 eV/mol, too small again to account for the observed anomalies in water, but getting bigger.

24.3.3 Ferroelectric Ice?

So we see that in the case of an isolated pair of dipoles, the cohesive energy of water cannot be explained. The next step is to consider the effect of the dipole moments on each other for the case of many dipoles, not just one. Peter Debye's classic book *Polar Molecules* [8] is still one of the best places to begin to understand how polar molecules interact. The basic problem is to include the effect that the local field makes on the overall polarization. Our goal here is to find the net dielectric constant that we would expect water to have given the dipole moment of 1.83 D that we know from the previous section. If the dielectric constant comes out to be significantly higher than this, then we know that significant corrections need to be made.

We need to calculate the local electric field **F** that the dipoles feel, since that is the electric field that aligns a given dipole. As we all know, if a dielectric is placed between the plates of a capacitor with charge density $\pm\sigma$ on the plates there is a displacement field **D** that is determined by the free charge alone, and a bulk field **E** within the dielectric that is less than D because of the induced polarization **I** within the dielectric. None of these, however, is the local field **F** that aligns the dipole! The relationship between **D**, **E**, and **I** is:

$$\mathbf{D} = \mathbf{E} + 4\pi\mathbf{I}. \tag{24.17}$$

The dielectric constant ϵ arises from the assumption that there is a constant relating D and E:

$$D = \epsilon E. \tag{24.18}$$

Now, **I** is related to the dipole moment of the polar molecule, μ, since the internal field in the dielectric wants to align the dipole moment, as we just saw. But the field that actually aligns the dipole is not the average macroscopic field **E** in the dielectric but instead the local field **F** that the dipole feels. It is confusing to consider yet another field, yet the crux of the problem is that the fields **D**, **E**, and **I** are all macroscopic fields that are in effect an average over space and never deal with the microscopic and atomic nature of the real polar material.

In fact, the calculation of **F** is an extremely difficult problem. We will first follow Debye's simple calculation that ignores correlations at a local level, and then do a very simple mean field calculation that attempts to take local correlations into effect. Professor Roberto Car of Princeton University is actively engaged in this problem [9].

The idea here is to carve out a small sphere ("small" means small compared to the macroscopic dimensions of the capacitor but large compared to the atomic parts of the dielectric). We do this in the hope that the dielectric has a small enough aligment that only large numbers of molecules summed together will give rise to an appreciable field. We can split the local field F into three parts (groan!): F_1 is due to the actual charge on the plates, F_2 is due to the polarization of the charge on the dielectric facing the plates, and F_3 is the local field due to the exposed charges on the interior surface of the little sphere that we carved out (this is the new part).

We already know that

$$F_1 = 4\pi\sigma \tag{24.19}$$

and

$$F_2 = -4\pi I + \frac{4\pi}{3}D. \tag{24.20}$$

The calculation of the internal field F_3 inside a sphere is best left to reading the classic textbook by Jackson [10]. If the field with the dielectric is E in a medium of dielectric constant ϵ, one gets:

$$F_3 = \frac{3\epsilon}{2\epsilon + 1}E. \tag{24.21}$$

Note that the inner field adds to the external field and hence serves to increase the polarization of anything within the carved-out sphere, an important point. Adding everything together we can finally arrive at the renowned equation of Clausius and Mosotti:

$$\frac{\epsilon - 1}{\epsilon + 2} = \frac{4\pi}{3}n\alpha, \tag{24.22}$$

where α is the polarizability of the molecule within the sphere. Well, we know what α is since we calculated it earlier, namely,

$$n < \mu >= I = n\alpha F = n\frac{\mu^2}{3kT}F, \tag{24.23}$$

so we have

$$\alpha = \frac{\mu^2}{3kT} \tag{24.24}$$

and thus the C-M relation can be rewritten as:

$$\epsilon = 2\frac{4\pi n\mu^2}{9kT+1}(1 - \frac{4\pi n\mu^2}{9kT}). \tag{24.25}$$

Well! Let's plug in and see how well we can calculate the dielectric constant of water. The answer is:

$$\epsilon = -3.2. \tag{24.26}$$

Ugh! A negative number.This means that the large dipole moment of water results in a huge self-interaction and that it should be, by our analysis, a ferroelectric. Looks like our attempt to rationally explain water is all wet, which shouldn't be too surprising since the C-M relationship is a mean-field theory that will break down for dense materials, especially for water, which is 55 molar and has a huge dipole moment! The C-M relation has a divergence built into it: Since the term

$$\frac{\epsilon - 1}{\epsilon + 2} \tag{24.27}$$

cannot exceed unity, there is a divergence in the dielectric constant for a very finite value of the dipole moment of the molecule, and in fact water is well past that limit, hence the amazing result that the predicted dipole moment of water from the known dielectric constant is a negative number. In fact, neither water nor ice is ferroelectric. The basic problem here is that we have ignored the local bond entropy of the water molecule due to the four valence orbitals. This so-called Onsager entropy kills the ferroelectric transition.

24.4 The Ice Problem and Entropy

We saw in our sp hybridization scheme that we had four bonding orbitals forming a tetrahedral symmetry around the oxygen atom. There is a partial charge water molecule model called the Bjerrum model that is based on the model of bond hybridization. Given the dipole moment of the water molecule, the angle of 105° between orbitals, and the bond length of 1 Å, one quickly finds that each arm contains a partial charge of 0.2e. Industries have arisen trying to calculate this number to ever higher precision, but for now this will suffice. Now, we let other water molecules form a tetrahedral environment around the water. A strange kind of lattice is formed. Any two of the four lobes contain positively charged hydrogen atoms, while the other two lobes contain negatively charged lobes of excess electron density. Note that there is an intrinsic amount of disorder contained in such a lattice. Charge neutrality requires that any oxygen atom can have at most two hydrogen atoms near it. The lattice that forms has a residual amount of entropy due to the possible ways of arranging the hydrogen atoms and still obey the so-called ice rule: *For the four nearest-neighbor hydrogens surrounding the oxygen atom, two are close to it and two are removed from it.* A particularly lucid discussion of this problem can be found in the beautiful book by Baxter, *Exactly Solved Models in Statistical Mechanics* [11]. The residual entropy of ice is quite substantial:

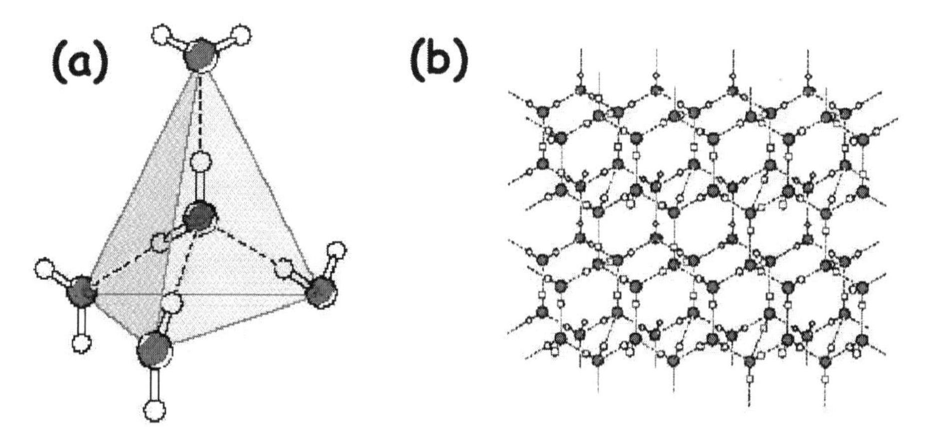

Fig. 24.2. (a) The basic unit cell of the water molecule in ice. (b) A possible ice lattice pattern of hydrogen bonds.

S/k_B is 0.4 extrapolated to 0 K! Thus, in the case of ice it isn't true that as T goes to 0 that the entropy S goes to 0.

In Fig. 24.2, we show some of the possible patterns that are possible in an ice lattice that satisfy the ice rules. Note that the hydrogen atoms are not at all free to individually move back and forth between the two equivalent positions that can be seen to exist between the the oxygens, but rather the motions in the ice lattice are by necessity of a collective nature. In some respects, this lattice shows many aspects of frustration that play such a predominant role in spin-glass systems. In this model, there are no charged defects.

In reality, the ice lattice we presented cannot represent the true lattice that is formed by water, since such a lattice has ferroelectric transitions that we know do not occur in a real ice or water system. There must be considerable dynamics in this lattice, highly coupled as it must be, to explain the lack of dipole ordering transitions.

There is a significant point to be made here, however. The ice model has built into it a signficant amount of entropy due to the empty orbitals that allow various bonding patterns to form. If defects are put into this lattice that disrupt the bonding patterns, they can actually *decrease* the entropy of the system since the bonds can no longer jump among a collection of degenerate states. Thus, the defects via a pure entropic effect can raise the net free energy of the system. This increase in free energy due to decrease in entropy is called the *hydrophobic* effect. It makes it energetically unfavorable for the nonpolar amino acids to expose themselves to bulk water, and thus contributes substantially to the free energy of various protein structures.

24.4.1 The Hydrophobic Effect: Entropy at Work

As we have seen, there is evidence that liquid water is a highly ordered liquid that strangely has a large amount of internal disorder due to its hybridization scheme. When something is introduced into the water that cannot form hydrogen bonds, it forces the water to form a cage around the molecule that has a lower entropy than the bulk liquid itself, which we discussed in Chapter 18. The free energy change is predominantly due to entropic rather than internal energy changes. This negative entropy change associated with ordering water is the hydrophobic effect. There is a masterly book on the subject by Tanford (*The Hydrophobic Effect*) [12] and a great deal of theoretical activity.

The hydrophobic effect can be seen in the fact that many aliphatic molecules are quite soluble in alcohols and other reasonably polar molecules but very sparingly soluble in water. Further, the solubility of aliphatic molecules in water takes on a characteristic temperature shape. If we consider the transfer of a benzene molecule (for example) from a pure solution to a water environment, the free energy change ΔG can be written as:

$$\Delta G = \Delta H - T\Delta S. \tag{24.28}$$

The enthalpy change ΔH of bringing benzene into a highly polar environment is actually likely to be negative: All of the dispersion effects we considered make it better for the benzene to be in water. However, the fact that benzene cannot form hydrogen bonds means that the entropy change upon entering water ΔS is negative: The entropy is smaller for benzene in water and overwhelms the negative enthalpy. In fact, solubility plots indicate that benzene is more soluble at low temperatures than at high temperatures (up to a point), as you would expect for free energy change with rather small negative ΔH and large negative ΔS. However, beyond a point, the effect of increasing temperature is to break down the lattice of hydrogen bonds formed in water and the effective ΔS decreases with increasing temperature.

There is a simple demonstration (we think) of this effect, which you may have already seen. In a bag is a very concentrated solution of sodium acetate, a molecule that is incredibly soluble in water (1 gram dissolves in 0.8 ml of water!). This high solubility means that it likes to form hydrogen bonds with water. When it crystallizes, the water is forced to form a shell around the incipient crystal, and the resulting negative entropy change is quite unfavorable; on the other hand there is a latent heat of crystallization due to the packing of sodium and carboxylic groups that is quite favorable. This term is related to the volume. Suppose that a microcrystal of radius R forms. Then, in general, the net free energy change is:

$$\Delta G = -AR^3 + T \times BR^2, \tag{24.29}$$

If $B \times T$ is considerably greater than A, then crystals up to some radius $R_o = 3A/2BT$ actually raise the free energy and do not form. That is, there is a minimum nucleation size below which the crystals are actually unstable. If the B term is big enough, the barrier can actually be high enough to allow the liquid to drastically supercool. If you press the magic button and allow crystals above this critical radius to form, they nucleate the crystallization of the entire bag with much release of heat.

24.4.2 Hydrogen Bonds

We are left with only Coulomb interaction as the last force that can contribute to the strong coupling of water molecules to each other. When the hydrogen atom approaches the oxygen atom, some fraction δe of the charge on the hydrogen atom is transfered over to the oxygen atom. This gives rise to an electrostatic interaction. It would appear that the positively charged hydrogen atom at any one site is not involved in a covalent bond but more electrostatic in nature. As we have presented it here, it is strictly electrostatic in nature. From the known crystal structure of ice, it is easy to guess what the size of this attractive potential energy will be. We can guess that in ice the distance between lone pair orbital and the hydrogen atom is roughly 2 Å, therefore the energy is trivially:

$$U = \frac{\delta^2 e^2}{r} = 0.1 \text{ eV}. \tag{24.30}$$

At this point we have been able to understand some of the consequences of the high density and large dipole moment of the water molecule, and we have been able to do this by doing a strictly classical electrostatic calculation. However, the strictly electrostatic picture of water self-interactions is incomplete, and some form of covalent bonding (delocalization) of the hydrogen atom in condensed phases must occur. This semidelocalized hydrogen state is the true hydrogen bond, and it is extremely important in biology, as we already saw when we considered base-pairing in DNA. A very readable book on the hydrogen bond can be found in the volume by Pimental and McClellan [13].

You can do some simple calculations concerning what such bonds must look like. At the simplest level the hydrogen atom can be in one of two (degenerate) sites. For example, in water, it can either be 1 Å away from an oxygen atom at site A or site B, and the two sites are separated by about 3 Å. Now, we have noted the hydrogen atoms are not free to make arbitrary transitions between these two equivalent sites, yet it is amusing to calculate the tunneling rate that you would expect in the no correlation limit for hydrogen atom transfer. Note that hydrogen atoms will have the maximal tunneling rates between the two equivalent sites, and so the disorder produced by tunneling will be maximized in the case of the hydrogen bonded system. Replacement of the hydrogen with deuterium will decrease the tunneling rate.

A simple estimate of the tunneling rate can be made by appealing to the relatively simple double harmonic oscillator problem. The Schrödinger equation is:

$$\frac{-\hbar^2}{2m}\frac{d^2\psi(x)}{dx^2} + 1/2(k \mid x \mid -a)^2\psi(x) = E\psi(x) \tag{24.31}$$

where 2a is the separation between the minima of the two parabolic potential surfaces.

Solving this is, of course, not trivial, since any realistic problem in physics is usually not solvable. If we define

$$\omega = \sqrt{k/m} \tag{24.32}$$

and make the well barrier very high compared to the ground state energy, that is,

$$V_o = 1/2ka^2 \gg 1/4\hbar\omega, \tag{24.33}$$

we can find approximate wave functions that have even and odd parity. The even parity states clearly are the equivalent of our binding orbitals that we derived in our LCAO talks, while the odd parity states are the antibonding orbitals. The energy splitting ΔE between these two states is

$$\Delta E = 2\hbar\omega\sqrt{\frac{2V_o}{\hbar\omega\pi}}exp(-\frac{2V_o}{\hbar\omega}), \tag{24.34}$$

and of course the *tunneling frequency* between these two states is

$$\omega' = \frac{\Delta E}{\hbar}. \tag{24.35}$$

We can guess what this tunneling frequency might be. The infrared absorption spectrum of water has a very strong feature at about 3 microns that is due to the hydrogen vibrating against the oxygen atom. Figure 24.3 gives an incredibly wide view of the absorbance of water versus wavelength. This graph reveals that the first excited vibrational state of the hydrogen atom has a very large energy of about 0.3 eV, so we can guess that the barrier height V_o in Eq. (24.33) is on the order of 0.5 eV or so. Our tunneling calculation gives a tunneling rate of on the order of 10^{13} sec^{-1}, which would give rise to a far-infrared absorbance in water, which is clearly visible. This calculation and the high residual entropy of water points to a picture of rapidly tunneling hydrogens in the ice network.

There have been some direct measurements of the tunneling rate. Oppenlander et al. [14] used a hole-burning technique to actually measure the tunneling rate of the hydrogens in a hydrogen-bonded crystal (benzoic acid) and found a tunneling rate of approaximately 10^{10} Hz, which is very fast.

This result has several significant consequences. First, since the proton can rapidly tunnel between sites one expects that the protons should be highly *delocalized* in ice. Now, we have to be rather careful here since: (1) the ice

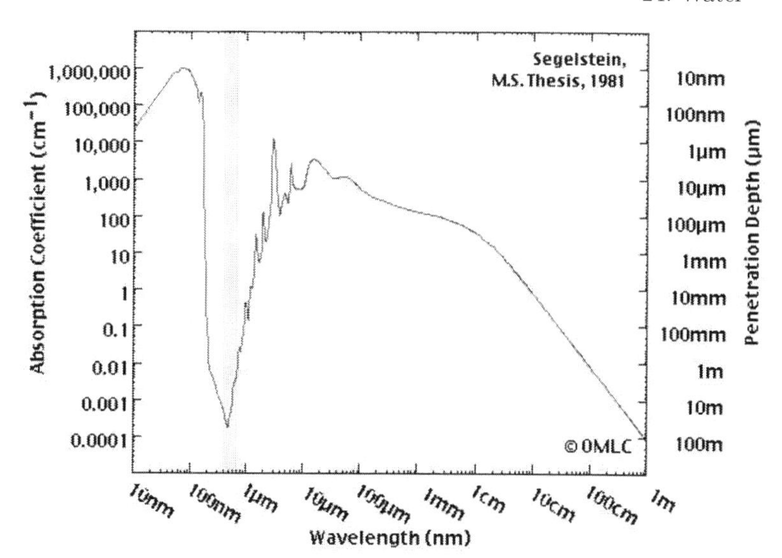

Fig. 24.3. The complete absorbance spectra of water. The gray band is the visible optical window.

rules allow substantial disorder even in the absence of tunneling and (2) the ice rules force significant correlations in the tunneling process, so that the effect mass of the tunneling state must be substantially higher than the bare proton mass. However, it is true that X-ray diffraction of ice Ih reveals complete disorder in the protons and hence ice Ih is termed a *proton glass* with significant residual entropy at $T = 0\,\mathrm{K}$, as we mentioned. Second, since the hydrogen bond is quite "soft" and in fact is best approximated as a bistable double minimum we would expect that this bond is *extremely* nonlinear, that is, very non-hookean in restoring force versus displacement, particularly at large amplitudes of displacement. In fact, it is exactly this nonlinearity in the displacement that Davydov has used in his theory of dynamic soliton propagation in hydrogen bonded systems, which we discuss in Chapter 16.

There is another interesting aspect to this problem, and that is the role of percolation in understanding the phase transitions of the ice lattice, and possibly proteins. If you look at the ice model you might become worried about the stability of the structure, assuming that it is the hydrogen bonds that hold the whole thing together. Only two of the four possible links can be filled at any one time. You might try to construct a tinker-toy kind of object with four holes and only two links allowed per hole: Is such a structure stable or not?

Thus, we have a system held together in a rather fragile way and for which simple rotations allow the scaffold to be cut out. The problem is to basically trace a path of connected hydrogen bonds from one side of the bulk material to the other. This so-called percolation of bonds determines the rigidity of the object. The classic example, discussed beautifully in the book *The Physics of Amorphous Solids* by Zallen [15], is the so-called vandalized grid, where a disgruntled telephone employee cuts links at random in a two-dimensional resistor net. The resistance of such a net is a surprisingly nonlinear function of ρ, the fraction of uncut bonds, where we can have a "valency" Z_c of resistors (or wires) per site. Again, surprisingly, in an infinite lattice if ρ is below some critical number ρ_c there is absolutely **no** current flow in the net, and the transition is quite accurately a second-order phase transition. In fact, and we do not know how to prove this, for a 2-D lattice with $Z = 4$ the critical percolation threshold is 0.5. Again, surprisingly, for dimensions greater than two there are no analytical ways to find the critical threshold. There have been numerical values found by computers for several different types of lattices. Water with a $Z = 4$ diamond lattice in three dimensions has a ρ_c (bond concentration) of .388, so ice seems safely solid since it has a bond concentration of 0.50.

The whole subject of bond percolation and the related issue of rigidity transitions is a fascinating field that links directly with many aspects of biology and networks. Polymers typically can undergo rigidity transitions as the number of cross-links approaches a critical number per node. There is probably **no** algorithm known that can predict the rigidity of a network in three dimensions, and rigidity is an inherently long-range interaction.

There are many amusing examples of percolation problems. One interesting toy connected to protein dynamics are the "happy and unhappy balls," which can be bought from Arbor Scientific (http://www.arborsci.com/). These are black polymer spheres that when squeezed seem to have identical static elastic constants. Yet, if you drop the balls you will find that one of the balls has almost no rebounding ability while the other is quite resilient. This is an example of a system where the dynamic behavior of the ball is quite different from the static behavior. We think it is due in the case of the balls to a rigidity phase transition in these polymer balls. We probably have a glass-rubber phase transition, which is the next step down from a gel. In other words, in a gel you have a simple rigidity percolation that gives the solid a finite shear modulus, but one can quite easily have rotational and translational freedom on a local scale, which can make the object quite "soft." If there is cross-coupling between the changes, as there always is and as we now know how to calculate, then yet another phase transition can occur, which results in a glass state, which we discussed in Chapter 3 and has played a major role throughout the development of the physics of proteins.

References

1. F. Franks. *Water.* The Royal Society of Chemistry, Letchworth, UK, 1983.
2. F. Franks. *Polywater!* The MIT Press, Cambridge, MA, 1981.
3. F. Franks. *Water, A Comprehensive Treatise.* Plenum Publishing Corp, New York, 1982.
4. D. Eisenberg and W. Kauzmann. *The Structure and Properties of Water.* Clarendon Press, Oxford, 1969.
5. K. A. Dill and S. Bromberg. *Molecular Driving Forces.* Garland Science, New York, 2002.
6. J. Mahanty and B. W. Ninham. *Dispersion Forces.* Academic Press, London, 1976.
7. J. Israelachvilli. *Intermolecular and Surface Forces.* Academic Press, San Diego, 1985.
8. P. Debye. *Polar Molecules.* Chemical Catalog Company, Lancaster, PA, 1929.
9. M. Sharma, Y. D. Wu, and R. Car. Ab initio molecular dynamics with maximally localized Wannier functions. *Int. J. Quant. Chem.*, 95(6):821–9, 2003.
10. J. D. Jackson. *Classical Electrodynamics.* John Wiley and Sons, New York, 1999.
11. R. Baxter. *Exactly Solved Models in Statistical Mechanics.* Academic Press, London, 1982.
12. C. Tanford. *The Hydrophobic Effect: Formation of Micelles and Biological Membranes.* Krieger Publishing Company, Malabar, FL, 1992.
13. G. Pimentel and A. McClellan. *The Hydrogen Bond.* W.H. Freeman, San Francisco, CA, 1960.
14. A. Oppenlander, C. Rambaud, H. P. Trommsdorff, and J. C. Vial. Translational tunneling of protons in benzoic-acid crystals. *Phys. Rev. Lett.*, 63(13):1432–5, 1989.
15. R. Zallen. *The Physics of Amorphous Solids.* John Wiley and Sons, New York, 1983.

25

Scattering of Photons: X-Ray Diffraction

Studying biomolecules of unknown structure is like walking in an unknown forest at night without a map or a light. Major progress in understanding biomolecules came when John Kendrew and Max Perutz and colleagues used X-ray diffraction to determine the structures of myoglobin [1] and hemoglobin [2]. Since these pioneering studies, X-ray diffraction has continued to lead advances in biology and biological physics. In the present chapter, we describe some of the main aspects of X-ray diffraction. Updated information on protein structures can be found at http://www.nigms.nih.gov.

X-ray diffraction has been of the utmost importance for the study of biomolecules. Within the span of about 10 years, between 1949 and 1959, biomolecules emerged from darkness and their structure no longer had to be guessed. The enormous difficulties that had to be overcome can only be partially appreciated by an outsider. Some measure of understanding can be obtained by realizing that von Laue, who knew the field of X-ray diffraction as well as anyone, wrote in 1948 [3], "From the start it is hopeless to determine the electron distribution in proteins." The following sketch can only give some idea of how this gigantic task was solved [4]. Bernal and Crowfoot observed the first X-ray diffraction pattern from a protein in 1934. In 1938, Max Perutz started working in Bragg's laboratory. Bragg likened the research to an Everest ascent in which successive camps are established at higher and higher altitudes. The final dash to the summit is taken from the highest camp. In Bragg's words, the first camp was reached in 1949, when some initial "Patterson" maps were determined. The next major step became possible when it was found that heavy atoms could be built into hemoglobin, to yield reference phases. The final success came in Kendrew's investigation of the simpler myoglobin (Mb). In 1959, the structure of Mb to a resolution of 0.6 nm (6 Å) was established and the result is shown in Fig. 25.1.

Since 1959, progress in determining the structure of biomolecules has steadily accelerated. At present, new structures, with ever-increasing resolution, are being published at a rate unimaginable in 1959. This avalanche of information is due to improved X-ray sources, detectors, data-evaluation

H. Frauenfelder, *The Physics of Proteins*, Biological and Medical Physics, Biomedical Engineering, DOI 10.1007/978-1-4419-1044-8_25,

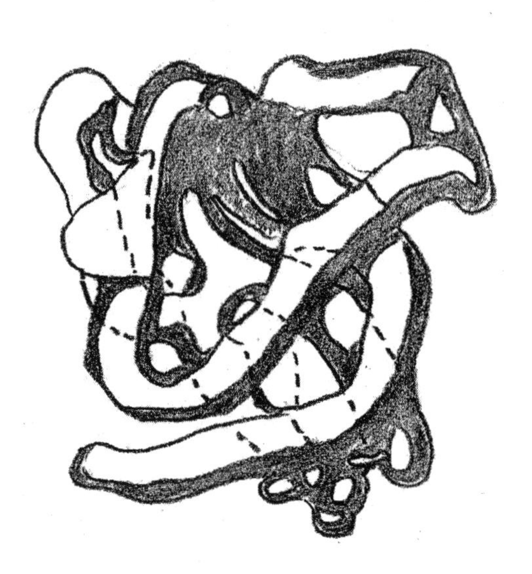

Fig. 25.1. Myoglobin, in 1959.

techniques, and single-crystal production. The X-ray work is described in a large number of classic and modern books. We give a short list: The references [3]–[5] are classics; useful information can be found in [6]–[10].

25.1 The Principles of Structure Determination

The primary goal of X-ray crystallography is the determination of the charge distribution of proteins. "Static structure" and charge distribution are synonymous. The use of the word "crystallography" suggests that crystals are necessary in principle and that the desired information is obtained from the distribution of the Laue spots in the diffraction pattern. This suggestion is wrong, as can be easily understood. Consider a single protein molecule, fixed to the tip of a holder as indicated in Fig. 25.2. Scattering of the incident X-rays then gives an intensity pattern, $I(\theta, \phi)$. From $I(\theta, \phi)$, the charge distribution in the plane perpendicular to the incident beam can be found. Measurement of $I(\theta, \phi)$ for a number of orientations of the protein then yields, with sufficient accuracy and computer power, the charge distribution of the entire protein. No single crystal is involved.

The practical problem in trying to perform the experiment in Fig. 25.2 is threefold. It is impossible to handle and orient a single protein. Even if we could, the X-ray beam would destroy the protein. Even if the protein survived, scattering from one protein is too weak to give good data. Protein crystals overcome all three problems: The proteins in a crystal are oriented,

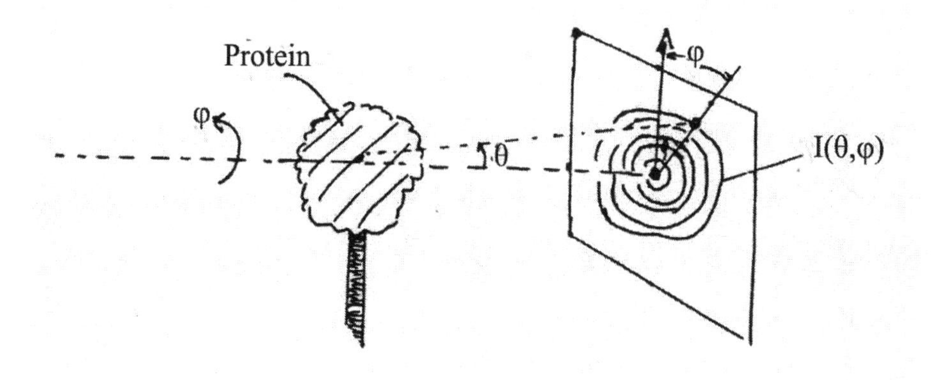

Fig. 25.2. Scattering from a single oriented protein molecule.

the destruction of some molecules does not affect the data much, and the intensity is (within limits) proportional to the square of the number of protein molecules.

The essential argument given earlier, however, still applies. The charge distribution is deduced from $I(\theta, \phi)$, the intensity of the diffraction spots, and a number of different crystal orientations must be used to obtain the full charge distribution.

25.2 Interaction of X-Rays with Atoms

X-ray diffraction is based on the interaction of X-rays with atoms, and so is EXAFS (extended absorption fine structure). Knowledge of this interaction is therefore necessary. We give here some of the basic facts but refer for all details to the literature [11, 12]. Two processes are of interest here: elastic X-ray scattering and photo effect. In the relevant energy range, a Compton effect also occurs. In elastic scattering, X-rays are scattered coherently by the electrons of the atom or molecule under consideration; the atom or molecule as a whole takes up the recoil momentum and the scattered X-rays have essentially the same energy as the incident ones. In photo effect, the X-ray gives all its energy to an atomic electron. The electron is then ejected from the atom with an energy $E_{e\ell} = E_X - B$, where B is the binding energy of the electron. In Compton effect, the photon scatters from an electron and ejects the elctron. In the final state, a scattered photon and a free electron are present. Some properties of these three processes are summarized in Table 25.1.

Table 25.1. The Important X-Ray Processes.

Process	Interaction	Approximate Z Dependence
Photo effect	Photon gives the entire energy to a bound electron, ejects electron	Z^5
Elastic scattering	X-ray scatters coherently from all electrons of system	Z^2 small angles Z^3 large angles
Compton effect	Photon scatters from a quasi-free electron, ejects electron.	Z

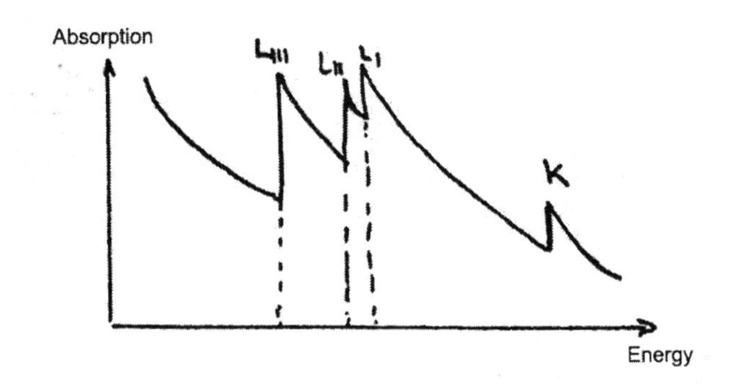

Fig. 25.3. The absorption coefficient as a function of photon energy.

X-ray diffraction is based on elastic scattering, EXAFS on the photo effect. For later reference, we sketch in Fig. 25.3 the energy dependence of photo absorption. Four "edges" are shown; they can be explained easily. The L shell in atoms has three different sublevels with different energies. As the energy of an incident photon is increased past the binding energy of the weakest bound electron in the L shell (L_{III}), the electron will be ejected and the absorption coefficient increases abruptly. The process continues as the energy is further increased.

25.3 Intensity and Geometry

We summarize in Fig. 25.4 some of the essential facts about the information that can be obtained from X-ray scattering and diffraction. For a preliminary discussion of the four basic phenomena shown in Fig. 25.4, we assume that

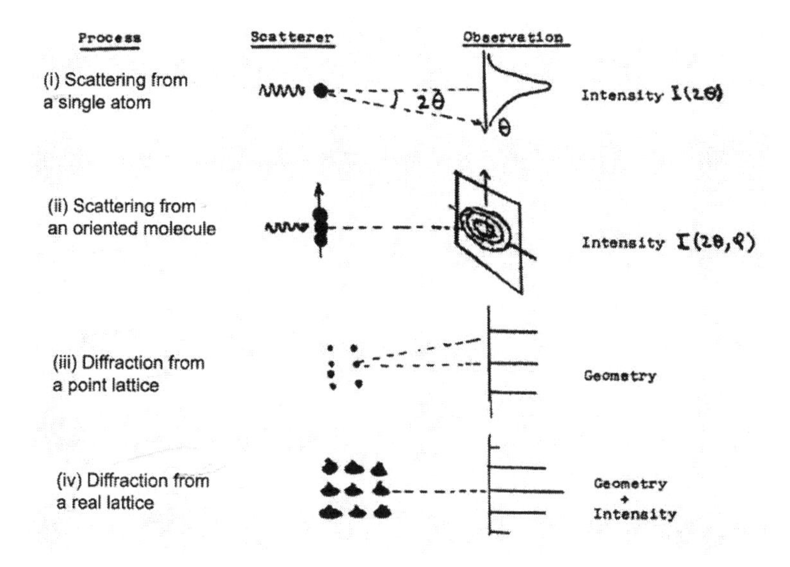

Fig. 25.4. The basic processes in scattering and diffraciton—geometry and intensity.

the scatterers are infinitely heavy so that thermal motion can be neglected and no recoil occurs. The four processess then can be sketched as follows:

1. A single atom is spherically symmetric. Observation of the intensity $I(\theta)$ for any one orientation of the scattering atom with sufficient resolution provides the information to deduce the radial charge distribution of the atom.

2. Assume that we can orient a single molecule and that we can determine the scattered intensity $I(\theta, \phi)$ for all orientations. We can then compute the charge distribution and consequently the structure of the molecule.

3. Diffraction from a regular lattice with point scatterers is entirely determined by geometry. Maxima result where wavelets from the individual scatterers arrive in phase. All maxima are equally intense. The position of the maxima gives information about the arrangement (geometry) of the crystal. The intensities do not yield any information. The situation is similar to diffraction of light from a grating with infinitely narrow slits.

4. If the scatterers possess structure, then intensity and geometry provide information. Each individual scatterer contributes with an angular distribution given essentially by (2). However, the intensities from the individual scatterers add up to a diffraction spot only when the different wavelets are in phase. Thus, the geometry of the crystal lattice determines the angles at which spots occur, the charge distribution of the unit cell governs the intensity of the spots. The individual molecules at the lattice sites are oriented, and the orientation of the overall crystal therefore permits changing the orientation of the molecules.

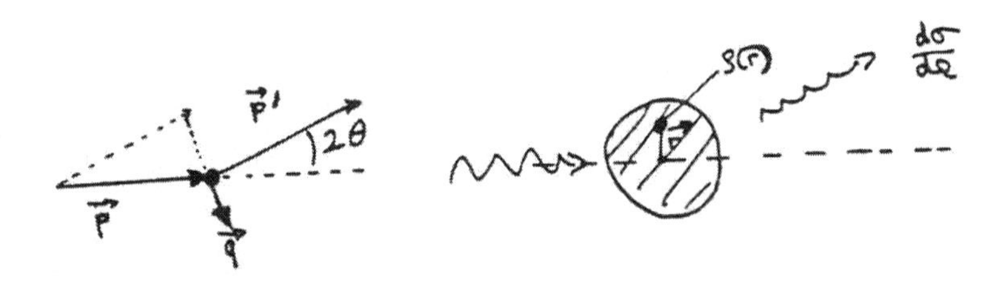

Fig. 25.5. Scattering from an atom.

To summarize: The crystal lattice determines the position of the diffraction spots, the electron distribution of the individual molecule(s) (unit cell) the intensity.

25.4 Scattering from Single Atoms and Molecules

The differential cross section for scattering from a *single atom* can be calculated in a straightforward way by using the first Born approximation. Consider the situation shown in Fig. 25.5. The differential cross section into the direction (**p**) is given by

$$\frac{d\sigma}{d\Omega} = |f(\mathbf{q})|^2, \tag{25.1}$$

where $f(\mathbf{q})$ is the scattering amplitude and $\mathbf{q}$ the momentum transfer (elastic scattering):

$$\mathbf{q} = \mathbf{p} - \mathbf{p}', \quad q = 2p\sin\theta \,. \tag{25.2}$$

$\mathbf{p}$ is the momentum of the incident, $\mathbf{p}'$ that of the scattered photon.[1] Denote by $\rho(r)$ the charge density at the point $\mathbf{r}$. The cross section can then be calculated, with the result [13]

$$\frac{d\sigma}{d\Omega} = \left(\frac{d\sigma}{d\Omega}\right)_{pt} |f(\mathbf{q})|^2 \tag{25.3}$$

where

$$f(\mathbf{q}) = \int d^3r \; \rho(\mathbf{r}) e^{i\mathbf{q}\cdot\mathbf{r}/\hbar} \tag{25.4}$$

is called the *atomic form factor* or *scattering factor*. $(d\sigma/d\Omega)_{pt}$ is the scattering from a point atom with unit charge. In the forward direction, $\mathbf{q} = 0$, and the atomic scattering factor is

$$f(q = 0) = \int d^3\, r \; \rho(r) = Z, \tag{25.5}$$

[1] In biological texts, $\mathbf{q}$ is usually replaced by the scattering vector $\mathbf{s}$ with $\mathbf{q} = 2\pi\hbar\,\mathbf{s}$.

where Z is the atomic charge. At other angles, scattering factors have been computed [14]. Equation (25.4) shows that $f(\mathbf{q})$ for a spherically symmetric atom depends only on the magnitude of $\mathbf{q}$, $|\mathbf{q}|$. With Eq. (25.2) and $p = h/\lambda$, we see that it is a function of $x = 1/(2\lambda)\sin(\theta)$, where λ is the X-ray wavelength. Equation (25.4) shows that $f(\mathbf{q})$ is the Fourier transform of the charge density. The integral can be inverted to give

$$\rho(\mathbf{r}) = \frac{1}{(2\pi\hbar)^3} \int d^3q\, f(\mathbf{q}) \exp\left[-i\mathbf{q}\cdot\mathbf{r}/\hbar\right]. \tag{25.6}$$

The charge density thus could be found if $f(\mathbf{q})$ is known for all values of the momentum transfer $\mathbf{q}$. The famous phase problem shows up, however, and makes life difficult! Equation (25.3) demonstrates that the experiment yields $|f|^2$, not f. Since f is in general complex, the inversion to get the charge density cannot be performed. We will discuss in Section 25.8 how the phase problem is solved. Before continuing to molecules, we note that f for atoms can be calculated in a very good approximation.

Next, consider X-ray diffraction from a molecule, as in Fig. 25.6. Each atom in the molecule contributes the amplitude $\mathbf{f_i}$. We write $\mathbf{f_i}$ as a vector because it is in general a complex function. The total scattering from the molecule then is given by

$$\mathbf{G}(\mathbf{q}) = \sum_{i=1}^{N} \mathbf{f_i}. \tag{25.7}$$

From $\mathbf{G}(\mathbf{q})$ for all values of $\mathbf{q}$, the charge density of the molecules can be obtained by a Fourier transform as in Eq. (25.6). However, since the molecule is usually not spherically symmetric, one orientation alone is not enough. Scattering has to be observed with many different orientations.

In principle, scattering from one large molecule is enough to find $\mathbf{G}(\mathbf{q})$. As pointed out in Section 25.2, this approach is impossible. Molecules cannot be oriented singly, the intensity from one scattering center is not sufficient,

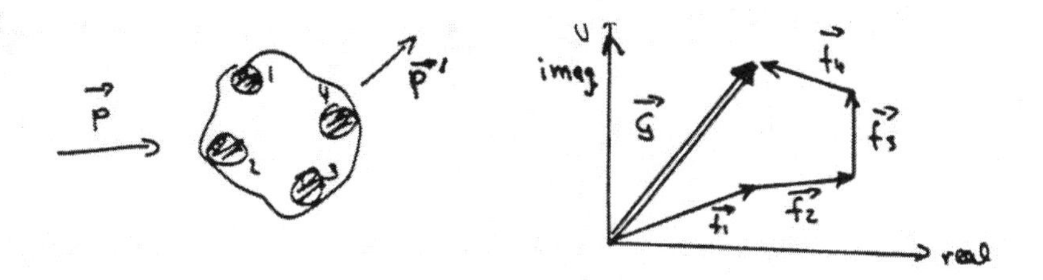

Fig. 25.6. Scattering from a molecule. The scattering factors add vectorially.

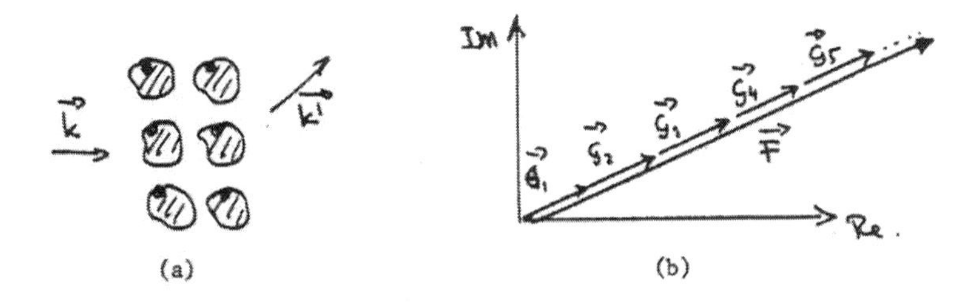

Fig. 25.7. (a) X-ray diffraction from protein crystal. (b) Constructive interference.

and a single protein would be quickly destroyed by the X-rays. The problems of orientation, intensity, and radiation damage are overcome by using *protein crystals.* In the situation shown in Fig. 25.7(a), the individual protein molecules are arranged in a regular array, a crystal. The vectors $\mathbf{G}(\mathbf{q})$ from each protein then add. In most directions the result is small. In some directions, however, the phase differences between any two scattering amplitudes are integral multiples of 2π and the **G**s add as shown in Fig. 25.7(b). The gain in intensity is equal to the square of the number of interferring proteins. The crystal thus acts as a low-noise high-gain amplifier. From the diffraction pattern, the charge density is obtained by Fourier transform.

25.5 Bragg's Law

When do all molecules interfere constructively? This problem was solved in different forms by Bragg and by von Laue. We consider first Bragg's approach.

In 1913, W. L. Bragg found that a very simple model could explain the *position* of the diffracted beam spots. He assumed that the X-rays are reflected from the various lattice planes; spots are produced where the beams from various parallel planes interfere constructively. Consider a series of atomic planes separated by equal distances d. The path difference for rays reflected from adjacent planes is $2d \sin\theta$ (Fig. 25.8). Constructive interference occurs if the path difference from successive planes is an integral number, n, of wavelengths. This condition yields the Bragg's Law,

$$2d \sin\theta = n\lambda \; . \tag{25.8}$$

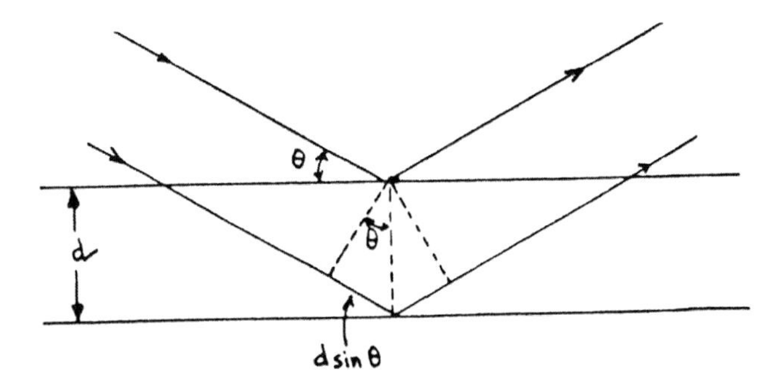

Fig. 25.8. Primitive derivation of the Bragg condition.

25.6 The Laue Diffraction Equations

Consider the scattering from two lattice points, separated by the vector $\mathbf{r}$. $\mathbf{p}$ and $\mathbf{p}'$ denote the momenta of the incident and the scattered photons, $\hat{p}$ and $\hat{p}'$ are the corresponding unit vectors. As seen in Fig. 25.9, the path difference between the two scattered waves is

$$\Delta s = \mathbf{r} \cdot (\hat{p} - \hat{p}') = \mathbf{r} \cdot \mathbf{q}/p. \tag{25.9}$$

Here $\mathbf{p}$ is the momentum transfer defined in Eq. (25.2), and we have used the fact that scattering is elastic so that $|\mathbf{p}| = |\mathbf{p}'| \equiv p$. The momentum transfer vector is normal to the reflecting plane, the plane that reflects the incident into the scattered direction.

Constructive interference occurs if the path difference Δs is an integral multiple of the wavelength λ. We denote the three basis vectors defining the unit cell by $\mathbf{a}$, $\mathbf{b}$, and $\mathbf{c}$, as shown in Fig. 25.10. A diffraction maximum occurs if the path difference in all three directions is an integral multiple of λ [7],

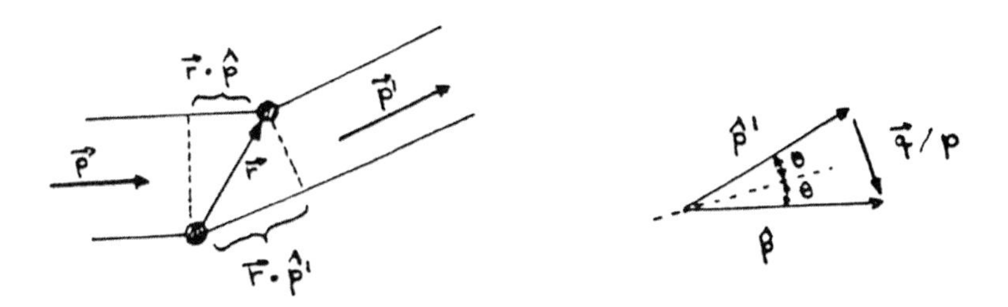

Fig. 25.9. Path difference in scattering.

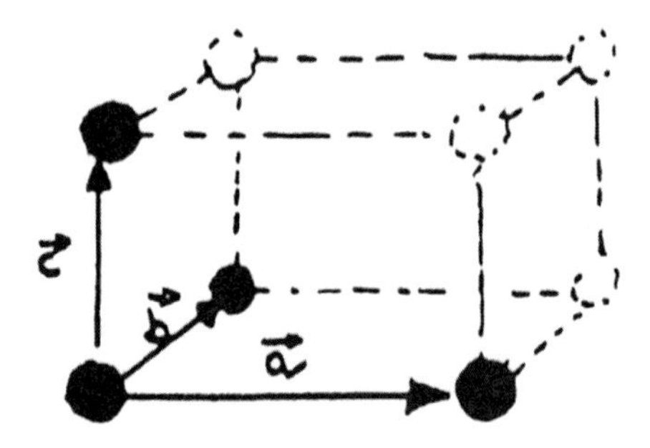

Fig. 25.10. Unit cell and basis vectors **b**, **c**, and **a**.

$$\begin{aligned} \Delta s_a &= \mathbf{a} \cdot \mathbf{q}/p = h\lambda, \\ \Delta s_b &= \mathbf{b} \cdot \mathbf{q}/p = k\lambda, \\ \Delta s_c &= \mathbf{c} \cdot \mathbf{q}/p = \ell\lambda, \end{aligned} \tag{25.10}$$

where h, k, and ℓ are integers. These relations were first given by Max von Laue and are called the Laue equations. They contain all geometrical information. For a photon, momentum and wavelength are connected by

$$p = 2\pi\hbar/\lambda,$$

where $2\pi\hbar$ is Planck's constant. The Laue equations can therefore be rewritten as

$$\begin{aligned} \mathbf{a} \cdot \mathbf{q}/\hbar &= 2\pi h, \\ \mathbf{b} \cdot \mathbf{q}/\hbar &= 2\pi k, \\ \mathbf{c} \cdot \mathbf{q}/\hbar &= 2\pi \ell. \end{aligned} \tag{25.11}$$

(Do not confuse the Planck constant and the integer h.)

25.7 Crystal Diffraction

Consider finally scattering from a complete crystal, as already sketched in Fig. 25.7 and redrawn in Fig. 25.11.

The incoming direction and the crystal are assumed to be fixed, and we are interested in the intensity of the scattered wave as a function of the direction $\mathbf{p}' = \mathbf{p} - \mathbf{q}$. Each protein molecule will, according to Fig. 25.6, contribute the amplitude $\mathbf{G}(\mathbf{q})$. The total scattering amplitude is therefore given by

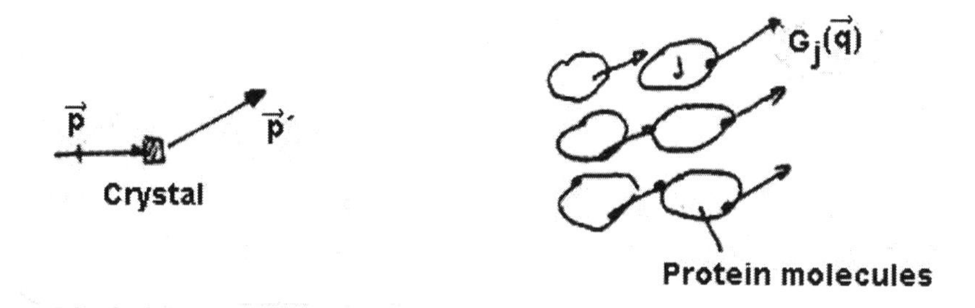

Fig. 25.11. Scattering from a crystal.

$$F(\mathbf{q}) \equiv \sum \mathbf{G}_j(\mathbf{q}) e^{i\phi_j} \tag{25.12}$$

where ϕ_j is the phase of molecule j with respect to a certain origin. To find the phase we note that the path difference between two points connected by the vector $\mathbf{r}$ is given by Eq. (25.9) as $\Delta s = \mathbf{r} \cdot \mathbf{q}/p$. The corresponding phase difference is, with $p = 2\pi\hbar/\lambda$,

$$\phi_j = \frac{2\pi}{\lambda}\Delta s \approx \mathbf{r}_j \cdot \mathbf{q}/\hbar. \tag{25.13}$$

In general, most of the interference will be destructive and the intensity into the arbitrary direction $\mathbf{p}$ will be small. However, for a given crystal orientation there will be some directions in which the situation of Fig. 25.7 is realized and where constructive interference occurs. To determine these directions, we write the position of atom j in terms of the basis vectors $\mathbf{a}$, $\mathbf{b}$, and $\mathbf{c}$ as

$$\mathbf{r}_j = u_j\mathbf{a} + v_j\mathbf{b} + w_j\mathbf{c}. \tag{25.14}$$

With Eq. (25.13), the phase then becomes

$$\phi_j = \mathbf{r}_j \cdot \mathbf{q}/\hbar = u_j\mathbf{a} \cdot \mathbf{q}/\hbar + v_j\mathbf{b} \cdot \mathbf{q}/\hbar + w_j\mathbf{c} \cdot \mathbf{q}/\hbar.$$

Constructive interference occurs where the Laue equations hold. With these relations in the form of Eq. (25.11), the phase becomes

$$\phi_j = 2\pi(u_jh + v_jk + w_j\ell)\ . \tag{25.15}$$

Inserting Eqs. (25.7) and (25.15) into (25.12) finally yields the *structure factor equation*

$$F(hk\ell) = \sum_i f_i \exp[2\pi(u_ih + v_ik + w_i\ell)]\ . \tag{25.16}$$

The sum is extended over all atoms in the crystal. The structure factor $F(hk\ell)$ is in general a complex quantity; its modulus $|F(hk\ell)|$ is called the structure amplitude.[2] The intensity of a reflection is proportional to $|F|^2$

$$\text{Intensity} \propto F^*(hk\ell) \cdot F(hk\ell). \tag{25.17}$$

In Eq. (25.17), the atoms are taken as discrete units, each scattering with strength f_i. We can also consider the electron density at each point, and the structure factor becomes

$$F(hk\ell) = \int d^3r \; \rho(\mathbf{r}) \; \exp[2\pi i(hu + kv + lw)], \tag{25.18}$$

where $\mathbf{r} = (u, v, w)$. Fourier inversion gives the charge density in the unit cell,

$$\rho(u, v, w) = \text{const} \sum_{h=-\infty}^{\infty} \sum_{k} \sum_{\ell} F(hkl) \exp[-2\pi i(hu + kv + lw)]. \tag{25.19}$$

If the structure of a protein and protein crystal are known, f_i or equivalently $\rho(\mathbf{r})$ is known. Equations (25.16) or (25.18) can then be used to find the diffraction pattern. The direction of each reflection (spot) is determined by $hk\ell$. The inverse problem, finding the structure from observed patterns, is far more difficult and will be considered in the following section.

25.8 The Phase Problem

We have pointed out that the experiment yields F*F; phase information is lost. For the inverse Fourier transform, the phase information is needed, however. There are four methods to overcome the phase problem: the Patterson map [15], use of the fact that the electron density has to be positive everywhere [16, 17], the isomorphous replacement method, and anomalous scattering. We discuss the third method here and make some remarks about the fourth next.

Under certain circumstances it is possible to combine the native biomolecule (protein) with a small molecule containing a heavy atom and to crystallize the complex in a form isomorphous with the native protein crystal. The heavy atom complex probably occupies space that is filled by liquid of crystallization in the native crystal.

The first problem is the determination of the position of the heavy atom(s), usually with the Patterson method. Once the position of the heavy atom is known, its (complex) structure factor $\mathbf{f_H}$ can be computed.

[2] Note the discrepancy in terminology between quantum mechanics and X-ray work. In quantum mechanics, the scattering amplitude is a complex quantity, the amplitude here is real.

With the structure factor $\mathbf{f_H}$ known, the unknown phases can be determined for each spot $hk\ell$. The intensity of a particular spot of the native protein is given by Eq. (25.17) as

$$I(h,k,\ell) \propto |F_{hk\ell}|^2 \equiv |F_{\rm P}|^2,$$

where F_P indicates the structure factor for the native protein. The scattering of the biomolecule after insertion of the heavy atom is given by

$$I'(h,k,\ell) \propto |\mathbf{F_P} + \mathbf{f_H}|^2.$$

We have written $\mathbf{F_P}$ and $\mathbf{f_H}$ as vectors to stress the fact that they are complex quantities. We denote the phase angle between the complex vectors $\mathbf{F_P}$ and $\mathbf{f_H}$ by ϵ. Neglecting the term in $|\mathbf{f_H}|^2$, the change in intensity after insertion of the heavy atom becomes

$$\Delta I(h,k,\ell) = I' - I \propto 2|F_P||f_H|\cos\epsilon. \tag{25.20}$$

Since $|\mathbf{F}_P|$ and $|\mathbf{f_H}|$ are known, the change in intensity permits determination of $\cos\epsilon$ and hence of the magnitude of the angle ϵ between $\mathbf{f_H}$ and $\mathbf{F_P}$. The direction of $\mathbf{f_H}$ is known and two possible directions of the structure factor $\mathbf{F_P}$ result. To resolve the ambiguity, the entire procedure is repeated with a heavy atom inserted at another site in the biomolecule. Again two possible directions for $\mathbf{F_P}$ are obtained. If the measurements are accurate enough, one of the new directions will coincide with one of the two choices from the first substitution and the sign ambiguity is resolved.

The procedure just described for the determination of the phase of $F_P \equiv F_{hk\ell}$ must be repeated for *every value* of h, k, and ℓ! Once all the structure factors F_P are known, with phase and magnitude, the Fourier inversion Eq. (25.19) gives the probability density $\rho(u,v,w)$.

Two remarks can be added at this point.

1. Once the structure has been obtained as described in the previous steps, it can be *refined*. The initial values of the coordinates of the atoms are adjusted until the calculated values of the X-ray reflections match the observed values as nearly as possible. Before the existence of powerful computers, the determination of protein structures was very slow and painful. At present, however, the determination of protein structures has been automated, and different programs exist that make the life of the protein crystallographer easier. (e.g. [18, 19]).

2. The isomorphous replacement method can be avoided in some cases by using X-rays with adjustable wavelengths [20]. Assume that the biomolecule contains a naturally occurring heavy atom that can be seen well in a Patterson map. The structure can now be determined with two different X-ray energies, just below and above an absorption edge, as indicated in Fig. 25.12. The diffraction pattern will be different for the two energies and the method can proceed as described earlier.

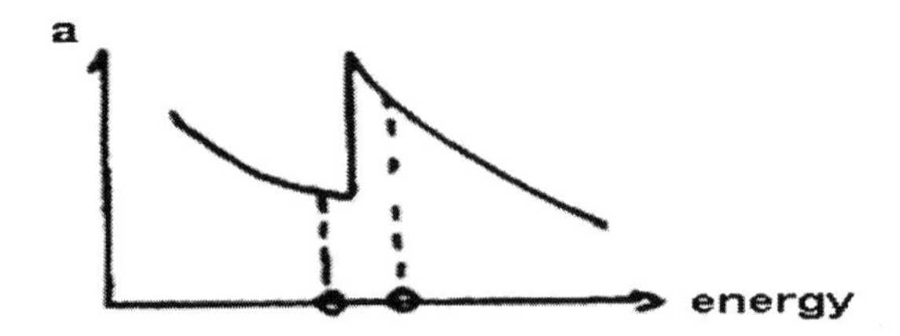

Fig. 25.12. Using X-rays with energies below and above the absorption edge of a heavy atom represents a type of isomorphous replacement.

25.9 Data Collection

In order to apply the previous considerations to a real protein, a protein crystal must be grown, and a diffraction pattern must be produced, recorded, and evaluated. As pointed out in Section 25.3, both geometry and intensity must be obtained. The technical details are very well understood, and the techniques are far advanced. We only sketch some of the simplest ideas here.

The Laue equations (or the Bragg's Law) indicate that an arbitrary orientation of a crystal plane does not necessarily produce a diffraction maximum. Only if the three Laue equations are satisfied does a reflection occur. One way to achieve reflection is to use a powder sample, where some planes always satisfy the Bragg's Law. This approach works well for some problems, but is far too crude for the study of biomolecules. Two different methods have yielded seminal information. The two methods can be called monochromatic and polychromatic (Laue). In the *monochromatic* method, an X-ray with a given energy is scattered by a protein crystal in a given orientation. A given crystal orientation will, however, only provide part of the desired information. To obtain all the information, many different crystal orientations must be used. In practice, the approach is as sketched in Fig. 25.13. In the rotating crystal method, the proper orientation for a reflection is obtained by using a single crystal (of the order of 1 mm linear dimensions), orienting it so that one axis is perpendicular to the incident X-ray pencil, and then rotating it about this axis, as sketched in Fig. 25.13.

In the example shown in Fig. 25.13, a cubic crystal rotates around the **c**-axis. Reflections from the shaded plane and all planes parallel to it occur when these planes make an angle θ, given the Bragg equation, Eq. (25.8), with the direction of the incident X-ray beam. A spot will appear on the photographic plate. Other spots will appear on the same line, coming from other planes parallel to the **c**-axis and from higher-order reflections. Planes not parallel to **c** will produce spots that are higher or lower. For a complete study, rotations about **a** and **b** must also be performed.

The spots on the film will appear as in Fig. 25.14(a). To label the spots with indices, h, k, and ℓ, consider Eqs. (25.2) and (25.10). For the line shown

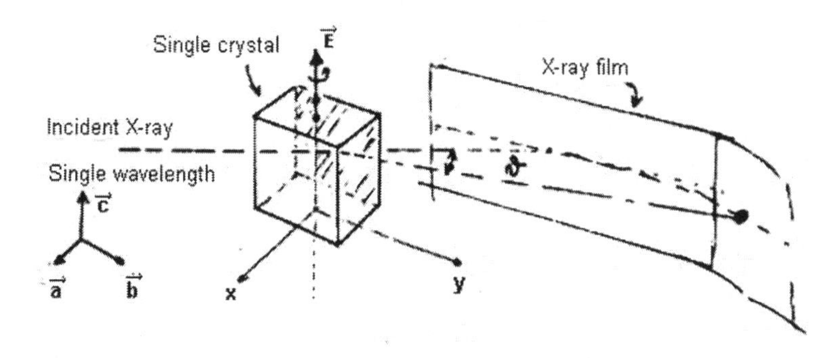

Fig. 25.13. Diffraction pattern obtained from a rotating oriented single crystal.

in Fig. 25.13, $\mathbf{c} \cdot \mathbf{q} = 0$ and hence $\ell = 0$. To see how h and k are found, assume that the crystal axes **a**, **b**, and **c** are known and that the geometry is in Fig. 25.14(b). At this particular angle, $\mathbf{a} \cdot \mathbf{q} = 0$ and hence $h = 0$. Furthermore, $\mathbf{b} \cdot \mathbf{q} = bq$ and hence from Eqs. (25.2) and (25.8) $2b \sin\theta = k\lambda$. From the measured spot position, it is thus possible to find k if b is known. The generalization to other positions can be written without much difficulty. The opposite problem, finding the vectors **a**, **b**, and **c** from the observed spots, is more difficult and we refer to the books listed earlier for all details. It is clear, however, that the spacing of the spots gives the dimensions of the unit cell.

To find the electron density once h, k, and ℓ for each individual reflection are known requires a careful measurement of the intensity of each reflection.

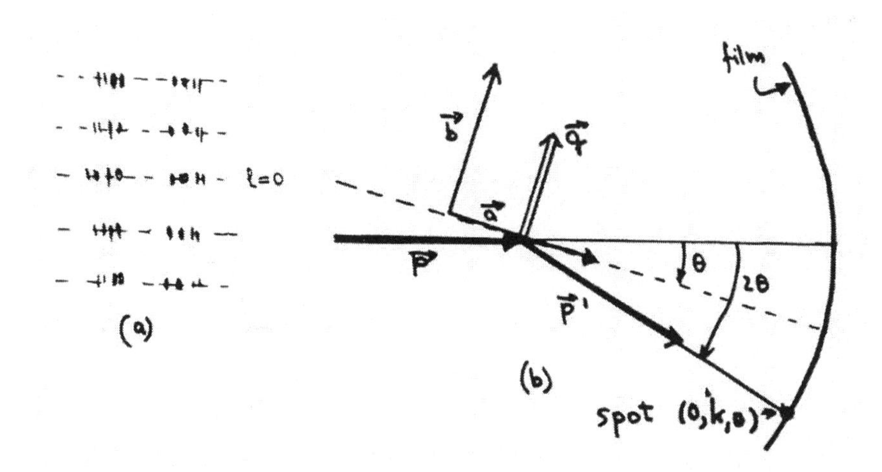

Fig. 25.14. (a) Spots (reflections) produced by rotating the crystal. (b) Diffraction, **c** perpendicular to plane.

Equation (25.17) then shows that the intensity yields the magnitude of the structure factor $F(hk\ell)$, the structure amplitude $|F(hk\ell)|$.

In the *polychromatic (Laue)* method a stationary crystal is illuminated by a polychromatic beam of X-rays [21]–[25]. Diffraction then occurs from all crystal planes for which the Bragg condition Eq. (25.8) is satisfied. If the range of wavelengths of the polychromatic beam is large enough, the information obtained with one crystal orientation suffices to yield the protein structure.

The monochromatic and the polychromatic methods both have advantages and disadvantages [24]. Rotating a crystal takes time, whereas a Laue diffraction pattern can be obtained rapidly if the X-ray beam is intense. The Laue technique is consequently best adapted to study time-dependent processes. On the other hand, the large X-ray intensity necessary for Laue diffraction may affect the crystal, and the data evaluation is also more complex than in the monochromatic method. Both methods, however, are essential for the study of biomolecules.

In Fig. 25.3, we show an X-ray film. This technique has been superseded. In fact, modern protein crystallography has advanced by many orders of magnitude because of techology. The old-fashioned X-ray discharge tubes have been replaced by very powerful synchrotrons and free-electron lasers, as described in Section 25.10. The photographic film has been replaced by charge-coupled devices (CCD) [26, 27]. The hand-cranked Marchand calculator evolved into the Blue Gene [28]. The advances, largely due to physics, have changed the discovery of protein structures from a science to an industry.

25.10 X-Ray Sources

So far we have talked about the use of monochromatic X-rays without saying where they come from. For many years, the only feasible source was based on the discovery of Roentgen [29, 30], and Roentgen or X-ray tubes could be found in every crystallography lab. They are based on ionization of inner shells by electrons: Electrons of sufficient energy hit an anode and ionize the inner shells of the atoms. X-rays are emitted when the holes are filled.

While Coolidge tubes, in which the electrons are produced by an electron gun, are used everywhere and will not soon be replaced completely, they have disadvantages. The energy of the X-rays is given by the atomic structure of the anode, and only a limited number of "characteristic lines" are available. The intensity is also limited, essentially by the power that can be dissipated in the anode. We therefore will not discuss ordinary X-ray tubes further, but turn to synchrotron radiation and free-electron lasers (FEL).

Just as X-rays were not found in the search for a diagnostic tool for broken legs, synchrotron radiation sources were not developed to help biology, but for pure high-energy research. Consider a particle of charge e that moves with velocity $v = \beta c$ on a circular path of radius R. Since it is continuously accelerated, it radiates with a power [31]

$$P = \frac{2e^2c}{3R^2}\frac{\beta^4}{(1-\beta^2)^2}. \tag{25.21}$$

For relativistic particles, where $\beta = v/c = pc/E \approx 1$, the factor $\beta^4/(1-\beta^2)^2$ tends to $(E/mc^2)^4$ and the power becomes

$$P = \frac{2e^2c}{3R^2}\left(\frac{E}{mc^2}\right)^4. \tag{25.22}$$

The time T for one revolution is given by $2\pi R/v \approx 2\pi R/c$ so that the energy lost per particle per revolution is

$$-\Delta E = PT = \frac{4\pi e^2}{3R}\left(\frac{E}{mc^2}\right)^4. \tag{25.23}$$

For a given particle, say an electron, the energy loss increases with the fourth power of the energy and, with the inverse of the radius. In high-energy accelerators, the energy loss is undesirable and the radius is made as large as feasible. If an electron accelerator is to be used as a *dedicated light source*, the radius is made as small as possible.

The lost energy is emitted in the form of Bremsstrahlung or, as it is called here, synchrotron radiation [31, 32]. Light is emitted tangentially to the electron orbit in a narrow cone. The light is polarized, with the electric vector in the plane of the electron orbit (Fig. 25.15). The radiation spectrum is characterized by the critical wavelength

$$\lambda_c = (4/3)\pi R(mc^2/E)^3, \tag{25.24}$$

which indicates approximately where the spectrum has its maximum. Equations (25.21) to (25.24) are in Gaussian units [31].

Originally, electron synchrotrons were built for high-energy research. However, synchrotron radiation turns out to be such a useful tool in many parts of physics and biochemistry that they can be built as light sources. In this case there is no need to plow the accelerated electron beam into a target. A simple storage ring is sufficient. In such a device, the electron beam circulates for days, and only enough energy is provided to make up for the radiation. Dedicated light sources thus have multiplied like mushrooms after a rain. At last count, there are at least 40 synchrotron light sources in operation, from Beijing to Siberia.

Energy and intensity of the synchrotron's radiation can be enhanced by using wigglers and undulators [33]–[35]. The concept of these insertion devices is simple: Consider an array of magnets, separated by a small distance d, and with the direction of the magnetic field changing up-down-up-down designed so that the field varies approximately sinusoidally. An electron beam then moves in a sinusoidal path through the array, continuously radiating light

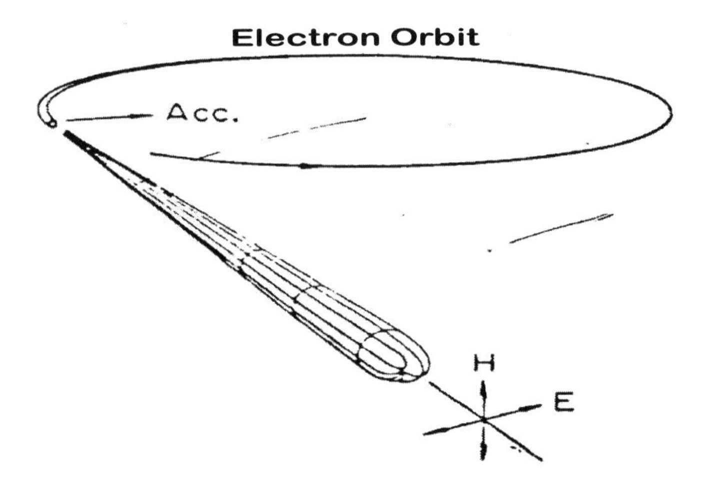

Fig. 25.15. The radiation pattern for synchrotron radiation.

according to Eq. (25.22). The intensity and the spectral shape of the radiation emerging from such an array depends on the incident electron energy and intensity, the field strength, the number N_m of magnets, and the distance d between the magnets. In a wiggler, the parameters are chosen so that radiations emanating from the electrons in each magnet are independent; the intensity of the radiation then is proportional to N_m. In an undulator, the parameters are chosen so that the radiations from successive wiggles interfere and add constructively. The intensity of the emerging beam then is proportional to ${N_m}^2$ and concentrated in the forward direction. Wigglers and undulators thus increase the X-ray intensities available from synchrotrons and also change the spectral shape of the radiation. Undulators, however, are also essential in free-electron lasers.

An FEL converts the kinetic energy of a relativistic electron beam into coherent radiation [36, 37]. The conversion is achieved by letting a short burst of monochromatic relativistic electrons pass through an undulator. With proper choice of the electron energy, the magnetic field strength, and the separation d, the emerging X-ray photons will be coherent. Their intensity will be proportional to $(N_m N_e)^2$, where N_e is the number of electrons in the incident electron bunch.

At present, three very-high-energy FEL are under construction or in the planning stage. Their main characteristics are given in Table 25.2 [38].

The peak brightness (1032 photons/s mm^2 mrad^2) of the X-ray FEL will be about 10^{10} times larger than existing sources. Past experience shows that it is difficult to predict what new results will come from such powerful X-ray sources. It is likely, however, that they will provide new insight into biological systems.

Table 25.2. X-Ray Free-Electron Lasers.

FEL	LCLS (Stanford, US)	DESY XFEL (Europe)	SCSS (Japan)
Electron beam energy (GeV)	4–4	≤ 20	≤ 8
Wavelength (nm)	0.1–6	0.1–1.5	0.1– 5
Pulse duration (fs)	< 230	100	80
Repetition rate (Hz)	120	10	60
Estimated start date	2009	2012	2010

25.11 Limitations of X-Ray Structures

The vast majority of known protein structures have been determined using X-ray diffraction. The X-ray technique has, however, a number of serious limitations:

1. The technique requires protein crystals. Some proteins are difficult to crystallize.

2. Crystallization may force the protein into a particular conformation and thus may lead to wrong conclusions concerning the connection between structure and function.

3. The intensity of the scattering of X-rays from particular atoms is proportional to their electron number. X-rays therefore have difficulties seeing hydrogen atoms, which are very important for biological functions.

4. Protein functions in a liquid environment, and not in crystals. X-rays may therefore sometimes provide misleading information. Structure determinations using neutron diffraction and NMR (nuclear magnetic resonance) provide ways around the limitations of X-rays. These two techniques will be sketched in chapters 29 and 30.

References

1. J. C. Kendrew, R. E. Dickerson, B. E. Strandberg, R. G. Hart, D. R. Davies, D. C. Phillips, and V. C. Shore. Three-dimensional Fourier synthesis at 2Å resolution. *Nature*, 185:422–7, 1960.
2. M. F. Perutz, M. G. Fossmann, A. F. Cullis, H. Muirhead, and G. Will. Structure of haemoglobin: A three dimensional Fourier synthesis at 5.5–Å. *Nature*, 185:416–22, 1960.
3. M. von Laue. *Röntgenstrahl-interferenzen.* Akademische Verlagsgesellschaft, Leipzig, 1948.
4. L. Bragg. First stages in the X-ray analysis of proteins. *Rept. Prog. Phys.*, 28(1):1–14, 1965. Reprinted In: W. Fuller, *Biophysics*, W. A. Benjamin, New York, 1969.
5. A. H. Compton and S. K. Allison. *X-Rays in Theory and Experiment.* D. Van Nostrand Co., Princeton, NJ, 1926.
6. G. H. Stout and L. H. Jensen. *X-Ray Structure Determination: A Practical Guide*, 2nd edition. McMillan, New York, 1989.
7. M. M. Woolfson. *An Introduction to X-Ray Crystallography.* Cambridge Univ. Press, Cambridge, 1997.
8. G. Rhodes. *Crystallography Made Crystal Clear.* Academic Press, New York, 2001.
9. J. Als-Nielsen and D. McMorrow. *Elements of Modern X-Ray Physics.* Wiley, New York, 2001.
10. A. McPherson. *Introduction to Macromolecular Crystallography.* Wiley-Liss, New York, 2002.
11. C. M. Davisson. Interaction of γ-radiation and matter. In K. Siegbahn, editor, *Alpha- Beta- and Gamma Ray Spectroscopy, Vol. 1.* North-Holland, Amsterdam, 1965. pp. 37–78.
12. H. A. Bethe and J. Ashkin. Passage of radiation through matter. In E. Segre, editor, *Experimental Nuclear Physics, Vol. 1.* Wiley, New York, 195, pp. 166–357.
13. See nearly any text on quantum mechanics or scattering theory or Hans Frauenfelder and Ernest M. Henley. *Subatomic Physics*, 2nd edition. Prentice-Hall, New York, 1991.
14. *International Tables for Crystallography.* Springer, Berlin, 1984.

15. A. L. Patterson. A Fourier series method for the determination of the components of interatomic distances in crystals. *Phys. Rev.*, 46:372–6, 1934.
16. J. Karle. Recovering phase information from intensity data. *Science*, 232:837–43, 1986.
17. H. Hauptman. The direct methods of X-ray crystallography. *Science*, 233:178–83, 1986.
18. T. C. Terwilliger and J. Berendzen. Automated MAD and MIR structure solution. *Acta Cryst.*, D 55:849–61, 1999.
19. T. C. Terwilliger. SOLVE and RESOLVE: Automated structure solution and density modification. *Methods Enzymol.*, 374:22–37, 2003.
20. W. A. Hendrickson. Determination of macromolecular structure from anomalous diffraction of synchrotron radiation. *Science*, 254:51–8, 1991.
21. J. L. Amoros, M. J. Buerger, and M. C. Amoros. *The Laue Method.* Academic Press, New York, 1975.
22. J. R. Helliwell. *Macromolecular Crystallography with Synchrotron Radiation.* Cambridge Univ. Press, Cambridge, 1992.
23. K. Moffat, D. Szebenyi, and D. Bilderback. X-ray Laue diffraction from protein crystals. *Science*, 223:1423–5, 1984.
24. K. Moffat. Laue diffraction. *Methods Enzymol.*, 277:433–47, 1997.
25. F. Schoot et al. In D. Mills, editor, *Third-Generation Synchrotron Radiation Sources.* Wiley, New York, 2002.
26. S. M. Gruner, M. W. Tate, and E. F. Eikenberry. Charge-coupled device area X-ray detectors. *Rev. Sci. Instr.*, 73:2815–42, 2002.
27. P. Falus, M. A. Borthwick, and S. G. J. Mochrie. Fast CCD camera for X-ray photon correlation spectroscopy and time-resolved x-ray scattering and imaging. *Rev. Sci. Instr.*, 75:4383–4400, 2004.
28. F. Allen et al. Blue Gene: A vision for protein science petaflop supercomputer. *IBM Systems J.*, 40:310–27, 2001.
29. A. Haase, G. Landwehr, and E. Umbach, editors. *Röntgen Centennial X-Rays in Natural and Life Sciences.* World Scientific, Singapore, 1997.
30. W. C.Röntgen. *Sitzungsber. der Würzburger.* Physik. Mediz. Gesellschaft, 1895.
31. J. D. Jackson. *Classical Electrodynamics*, 3rd edition. Wiley, New York, 1999.
32. J. L. Bohn. Coherent synchrotron radiation: Theory and experiments. In *Advanced Accelerator Concepts*, AIP Conference Proceedings, Vol. 647, pp. 81–95, Melville, NY, 2002.
33. H. Winick, G. Brown, K. Halbach, and J. Harris. Wiggler and undulator magnets. *Phys. Today*, 34(5):50–63, 1981.
34. G. Brown, K. Halbach, J. Harris, and H. Winick. Wiggler and undulator magnets – a review. *Nuclear Inst. Meth.*, 208:67–77, 1983.
35. J. R. Helliwell. *Macromolecular Crystallography with Synchrotron Radiation.* Cambridge Univ. Press, Cambridge, 1992.
36. H. Motz, W. Thorn, and R. N. Whitehurst. Experiments on radiation by fast electron beams. *J. Applied Phys.*, 24:826–33, 1954.
37. E. Prosnitz, A. Szoke, and V. K. Neil. High-gain, free-electron laser amplifiers: Design considerations and simulation. *Phys. Rev. A.*, 24:1436–51, 1981.
38. T. Feder. Accelerator labs regroups as photo science surges. *Phys. Today*, 58(5):26–8, 2005.

26

Electronic Excitations

One of the first measurements performed on a newly discovered or isolated biomolecule is the optical absorption as a function of wavelength. An absorption spectrum provides a first clue as to the nature of the beast. In this chapter, we discuss some of the important features. More information can be found in [1].

26.1 Electronic and Vibrational Transitions

From an experimental point of view, emission and absorption lines are usually denoted not according to their origin, but according to their wavelength. The following names are customary:

10–185 nm	Vacuum ultraviolet
185–200 nm	Far UV
200–300 nm	Short UV
300–400 nm	Near UV
400–700 nm	Visible
0.7–2.5 μm	Medium IR
15 μm–0.1 mm	Far IR

The division is useful because techniques differ in the various regions. Electronic transitions lie in the UV and visible ranges, vibrational ones in the IR.

H. Frauenfelder, *The Physics of Proteins*, Biological and Medical Physics, Biomedical Engineering, DOI 10.1007/978-1-4419-1044-8_26,

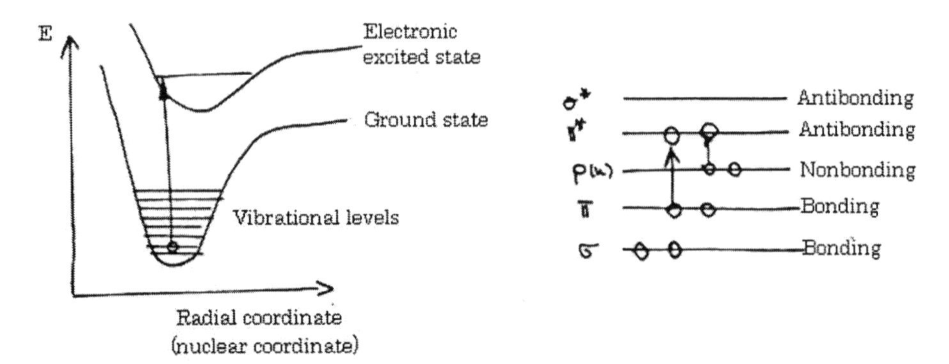

Fig. 26.1. (a) Electronic and vibrational levels; (b) the main transitions.

Consider Fig. 26.1, where two different electronic levels are shown, together with the corresponding vibrational levels.

Two general remarks apply to these transitions:

1. The energy E of the electronic transition depends on the characteristic size of the chromophore. The wavelength λ at which the absorption occurs is proportional to L^2, the square of the linear dimension (see Eq. (21.28)). This result is general; a large molecule will give rise to longer wavelengths.

2. The electronic transition occurs with very small change in the nuclear coordinate (Franck-Condon principle). The transition in Fig. 26.1 thus is shown "straight up." This effect arises from the fact that electronic absorption takes about 10^{-15}s, nuclear motion about 10^{-13}s, and rotational motion about 10^{-10}s.

In contrast to atomic transitions, the lines are very wide, because vibrational and conformational sublevels are involved. The main absorption lines (or bands) arise as follows:

Visible (1) Orbitals concentrated on metal ions, mainly transition metals, embedded in the biomolecule.
Visible (2) Large aromatic structures and conjugated double-bond systems.
Near UV (3) Small conjugated ring systems, aromatic and heterocyclic. (Benzene is an example of an aromatic ring, pyrrole one of a heterocyclic.)
Far UV (4) In the far UV, nearly everything absorbs (amino acid side chains, peptide backbone).

26.2 Electronic Spectra—Experimental

Electronic spectra are usually observed in absorption; a basic setup is simple (Fig. 26.2): A monochromator selects the desired wavelength from a light

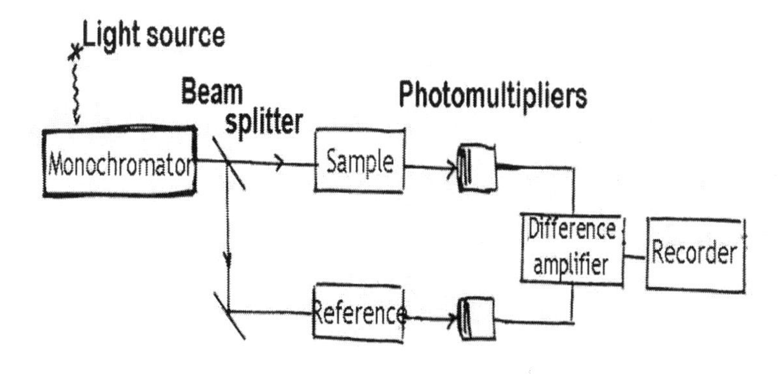

Fig. 26.2. Absorption spectroscopy.

source. The beam is then split. One beam passes through the cell with the sample to be studied; the other goes through a reference cell. The reference cell contains the same solvent as the sample cell, but no biomolecules. Each beam is detected with a photomultiplier, and the difference of the output currents is recorded. Such systems are available commercially; they record automatically over a predetermined wavelength interval.
Usually, the absorption in a sample will be exponential, so that the intensity after passing through a cell of length ℓ is given by

$$I(\ell) = I(0)e^{-\epsilon' \ell c} = I(0)10^{-\epsilon \ell c}. \tag{26.1}$$

Here c is the concentration of the solution in moles/liter. The quantity ϵ,

$$\epsilon = \epsilon' / \ln 10 = 0.434\, \epsilon', \tag{26.2}$$

is called the extinction coefficient, with units liter/mol·cm. Data are frequently reported in *absorbance*,

$$A = \log[I(0)/I(\ell)] = \epsilon\, \ell\, c. \tag{26.3}$$

A is dimensionless and usually quoted as, for instance, 2 OD, meaning $A = 2$, or a decrease by a factor 100 in intensity. Commercial spectrometers can be found in every chemistry lab; in physics labs they are usually older models! Modern types are coupled to a minicomputer and they digitize the data.

The system shown in Fig. 26.2 records at a given instant only one wavelength. This disadvantage is overcome with *optical multichannel analyzers* (OMA). The basic idea is sketched in Fig. 26.3. White light traverses the sample cell; dispersion occurs after the cell and the dispersed beam fall on a vidicon array detector, where the intensity is recorded simultaneously at many wavelengths.

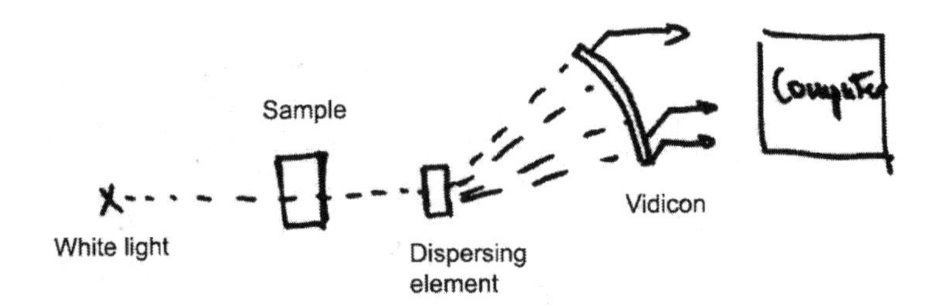

Fig. 26.3. Optical multichannel analyzer.

26.3 Electronic Transitions

In Fig. 26.1(b), we indicated the main transitions between the most important molecular orbits. In reality, of course, absorption energies are determined with the ground state at "energy 0." The transition energy diagram (Jablonski diagram) then appears as shown in Fig. 26.4.

The transitions in Fig. 26.4 are labeled by the initial and final orbitals. The electrons in the bonding orbitals are normally paired, with spins antiparallel, and are consequently denoted as single states, S_1 and S_2. In an antibonding orbital, the electrons may have parallel spins, giving rise to triplet states, T_1, $T_2, \ldots$. The triplet state associated with a singlet state always has a somewhat lower energy.

Absorption and emission of photons have already been discussed. In *internal conversion*, the excitation energy is removed without radiation through relaxation via vibrational states. In other words, the energy is removed by

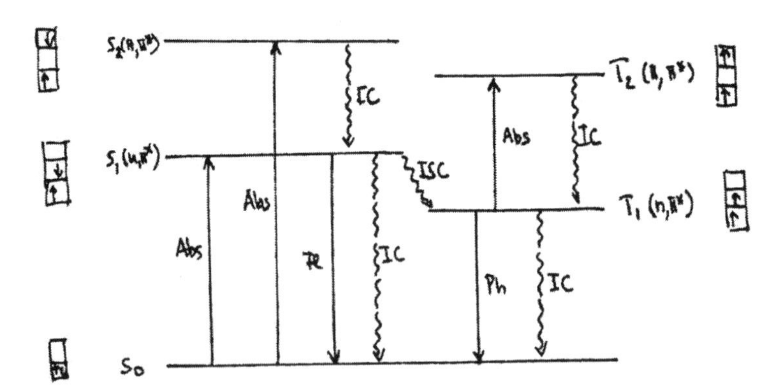

Fig. 26.4. Transition energy diagram. The spin state in each molecular orbital is indicated by arrows. The transitions are indicated as: → radiative transition, ⇝ nonradiative transition, Abs: absorption, Fl: fluorescence, Ph: phosphorescence, IC: Interval conversion, and ISC: intersystem crossing.

phonons. In *resonance fluorescence*, the energy of the reemitted photon is the same as that of the absorbed photon. In ordinary *fluorescence*, the emitted photon has a smaller energy; the remaining energy goes into phonons. In *intersystem crossing*, the system moves via a radiationless transition to the triplet state. Slower photon emission then is called *phosphorescence*.

26.4 Chromophores and Their Spectra

Atoms or groups of atoms containing the electrons that participate in an excitation are called *chromophores*. For transitions in the range from 200 to 800 nm, chromophores must contain loosely bound (n or π) electrons. Three classes of chromophores occur in proteins: the peptide bond, amino acid side chains, and prosthetic groups (like the heme group). We noted earlier that the wavelength of an absorption band is proportional to L^2, the square of the linear dimension of the radiating system (Eq. (21.28)). The peptide bond, the smallest of the three chromophores, has an absorption maximum at 190 nm. The relevant wavelengths and the absorption coefficients for a number of chromophores are given in Table 26.1.

Spectra of the side chains of the residues that give rise to the strongest electronic absorption are shown in Fig. 26.5. The absorption is again caused by delocalized electrons.

As a typical prosthetic transition group, we consider protoheme IX, shown in Fig. 21.25. Protoheme IX is a transition-metal complex and represents a large group of important cases [2]–[4]. The molecular orbitals involved in the

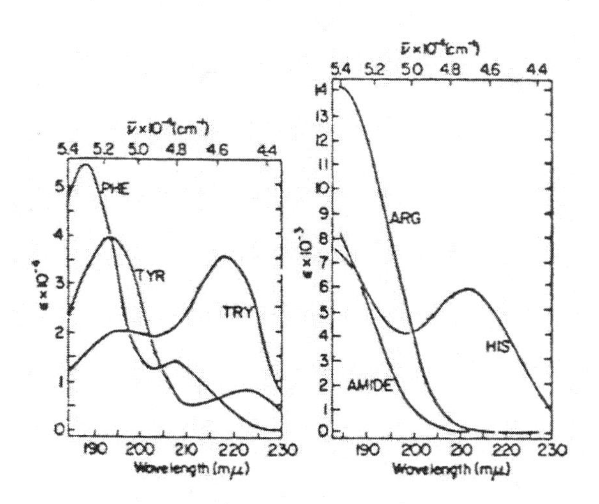

Fig. 26.5. UV absorption spectra of some side chains.

Table 26.1. Wavelength maxima of some chromophores (after Hoppe et al.[5]).

Chromophore	λ_{max}[nm]	ε[M^{-1} cm^{-1}]	Transition
—COO—R	205 165	50 $4 \cdot 10^3$	$n \to \pi^*$ $\pi \to \pi^*$
>C=O	280 190 150	20 $2 \cdot 10^3$	$n \to \pi^*$ $n \to \sigma^*$ $\pi \to \pi^*$
>C=S	500 240	10 $9 \cdot 10^3$	$n \to \pi^*$ $\pi \to \pi^*$
—S—S—	250-330	10^3	$n \to \sigma^*$
>C=C<	190	$9 \cdot 10^3$	$\pi \to \pi^*$
—C≡C—	175	$8 \cdot 10^3$	$\pi \to \pi^*$
(pyrimidine ring, N N)	300 245	325 $2 \cdot 10^3$	$n \to \pi^*$ $\pi \to \pi^*$
(purine ring, N N N N)	220 265	$3 \cdot 10^3$ $8 \cdot 10^3$	$\pi \to \pi^*$ $\pi \to \pi^*$

electronic transitions are schematically indicated in Fig. 21.29—contributions come from the porphyrin, the iron atom, and the axial ligand(s). Three types of transitions are important:

1. *Porphyrin $\pi - \pi^*$ transitions.* These transitions were discussed in Section 21.8. The characteristic features are a very intense band near 25,000 cm^{-1} (400 nm) called the Soret band (B); two bands near 18,000 cm^{-1} (550 nm), called Q_O and Q_V (or α and β); and the almost complete absence of vibronic fine structure. The free-electron model treated in Section 21.8 gives the correct order of magnitude for the wavelength of the Soret band B. If the model is extended to include Coulomb interactions, the splitting between the Q and B bands, shown in Fig. 21.30, is also reproduced [6].

2. *Ligand field transitions.* These transitions occur between molecular orbitals that are mainly localized on the central iron atom. They are transitions between d orbitals of the iron that have maintained their metal character in spite of the bonding of the iron to the heme group. In general, the transitions are spin- and symmetry-forbidden and are very weak.

3. *Charge transfer transitions* [7]. These are transitions between orbitals mainly localized on the iron and orbitals localized on the ligands (heme). Ligand → metal and metal → ligand transitions are distinguished depending

on whether the excitation lifts an electron out of a heme orbital into an iron orbital or vice versa.

We can now ask: What can we learn about biomolecules from the electronic spectra? It is still hopeless to calculate the electronic spectra from first principles. Thus, a semiempirical approach is indicated, and two main possibilities exist.

1. In cases where the complete composition of a biomolecule is not known, the characteristic features of certain groups permit conclusions as to their existence.

2. More important, when the biomolecule is subjected to some kind of perturbation, the resulting spectral change provides information. Since groups in different environments react differently to an external change, conclusions concerning the positions of the groups can be drawn. The following perturbations give useful information:

Ionization. Spectral changes occur when the biomolecule is subjected to high pH. The resulting ionization can lead to large red shifts. Groups that are exposed to the solvent ionize rapidly and reversibly, while groups that are buried deep inside usually ionize slowly and irreversibly.

Chemical modification. Sometimes a chemical modification affects one particular group. The corresponding change in the spectrum of this group can be observed.

Solvent influence. The electronic spectrum of a protein does not unambiguously show which chromophores are directly exposed to the solvent. Organic solvents, such as ethylene glycol, produce a red shift in the absorption spectrum. If such a solvent is added to the protein solution, then only the spectra of the exposed chromophores shift.

Conformational changes. Proteins and nucleic acids undergo conformational changes. These changes can be followed by observing shifts in the electronic spectra. We discussed some examples in detail in Part IV, where it became clear that spectral lines can serve as sensitive markers.

Unfolding [8]. A globular protein can be forced to unfold partially or completely, for instance, by changing the temperature or by proper choice of the solvent. When unfolded, essentially all groups become exposed to the solvent and the spectrum changes correspondingly. Complete unfolding should produce the same spectrum from all identical groups. The spectrum can thus be used to assess the extent of residual structure.

Ligand binding. Absorption spectroscopy is often used to follow the binding of various ligand molecules to the protein. Two cases can be distinguished: The ligand is chromophoric and the binding is followed by the change in its absorption spectrum or the change in the protein's spectrum upon binding is monitored.

26.5 Fluorescence

So far, we have restricted our discussion to the position of electronic levels. Valuable information can also be obtained from the time dependence of emission processes. We return to Fig. 26.4 to discuss what happens after a photon has been absorbed. Usually, the excited state gives off its excess energy in the form of thermal vibrations (radiationless transitions), but certain molecules have a high probability for radiative decay: They show fluorescence or phosphorescence [9]. Fluorescence involves an allowed singlet-singlet transition and has characteristic lifetimes of the order of ns. Phosphorescence, however, occurs from a spin-triplet state and, since it is spin-forbidden, the lifetime is much longer, of the order of μs to s. According to the Frank-Condon principle, the actual absorption process is so fast that the nuclei of the molecule do not have time to change their positions. Once the molecule is excited, however, the nuclei may find a new equilibrium position. The excitation usually leads to an excited vibrational state, but through the exchange of phonons the molecule loses its excess vibrational energy within ps and ends up in the lowest vibrational level. There it sits for about ns and then it emits a photon, decays by internal conversion, or changes to the triplet state. The fact that the molecule in the excited state reaches the vibrational ground state long before remission takes place has the following important consequences:

1. The fluorescence spectrum is independent of the exciting wavelengths; it is a characteristic of the molecule rather than a replica of the excitation.

2. The fluorescence spectrum is red-shifted relative to the absorption (= excitation) spectrum.

The lifetime of the excited state is not only long enough for the vibrational energy to reach thermal equilibrium, but it also allows the surrounding solvent molecules to adjust their positions so as to minimize the total energy of the system. Solvent relaxation has the effect of lowering the energy of the excited singlet state and causes therefore a red shift of the fluorescence spectrum. The larger the red shift is, the more polar the solvent, that is, it is largest in aqueous solution. In fact, this red shift is a very good indicator of the polar or nonpolar character of the environment of the fluorophore [10, 11].

We have already listed the types of chromophores that occur naturally in proteins. In fluorescence experiments, additional chromophores are introduced chemically (or through genetic engineering). Thus four types can be distinguished:

1. *Intrinsic*, such as the aromatic amino acids Phe, Tyr, and Trp;
2. *coenzymes* (prosthetic groups), such as NADH, FAD, and heme;
3. *extrinsic*, such as dye molecules specifically attached to biomolecules, or
4. *analogues*, such as the ε-adenosine derivatives, εAMP, εADP, εATP and εNAD^+ developed by Leonard [12].

Extrinsic probes were introduced in 1952 by Weber. The following example demonstrates the sensitivity of such a probe to its environment. ANS (Fig. 26.6) has a quantum yield of 4×10^{-3} and a peak wavelength $\lambda_{\text{max}} = 515\,\text{nm}$ in aqueous solution. It binds to apomyoglobin, that is, myoglobin from which the heme group has been removed. ANS apparently binds in place of the heme in the hydrophobic heme pocket. In this nonpolar environment the quantum yield increases to 0.98 and the wavelength of the fluorescence maximum shifts to $\lambda_{\text{max}} = 454\,\text{nm}$.
To characterize fluorescence, we note that the rate coefficient for decay of an excited electronic level can be written as (Fig. 26.4)

$$k_{tot} = k_{F\ell} + k_{IC} + k_{ISC} + k_q(Q). \tag{26.4}$$

Here $k_{F\ell}$ is the rate coefficient describing radiative decay to the ground state, k_{IC} characterizes internal conversion, k_{ISC} describes intersystem crossing. The process characterized by $k_q(Q)$ is not included in Fig. 26.4; it is *fluorescence quenching* [13]. A quenching molecule, often dioxygen, comes into contact with the chromophore and causes nonradiative deexcitation. The rate with which quenching occurs is proportional to the concentration (Q) of the quencher molecule. The lifetime of the excited state (S_1 in Fig. 26.4) is given by

$$\tau_F = 1/k_{\text{tot}}. \tag{26.5}$$

In general, only a small fraction of the decays will occur with the emission of a photon; this fraction is called the fluorescence quantum yield and is given by

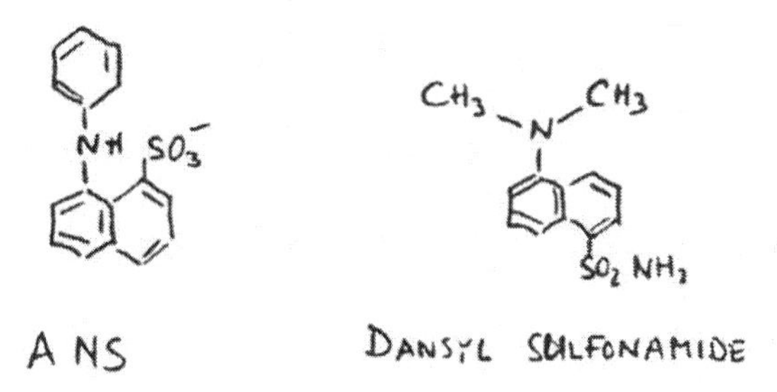

Fig. 26.6. Extrinsic probes.

$$\phi_F = k_{F\ell}/k_{tot} = \tau_F \, k_{F\ell}. \tag{26.6}$$

If $k_{F\ell}$ can be calculated and τ_F measured, ϕ_F can be determined. Typical values are given in Table 26.2.

Table 26.2. Fluorescence characteristics of protein and nucleic acid constituents and coenzymes (after Cantor and Schimmel [9]).

		Absorption		Fluorescence[a]			Sensitivity
Substance	Conditions	λ_{max} (nm)	$\epsilon_{max} \times 10^{-3}$	λ_{max} (nm)	ϕ_F	τ_F (nsec)	$\epsilon_{max}\phi_F \times 10^{-2}$
Trypotoplan	H_2O, pH 7	280	5.6	348	0.20	2.6	11.0
Tryosin	H_2O, pH 7	274	1.4	303	0.14	3.6	2.0
Phenylalanin	H_2O, pH 7	257	0.2	282	0.04	6.4	0.08
Y base	Yeast $tRNA^{Phe}$	320	1.3	460	0.07	6.3	0.91
Adenine	H_2O, pH 7	260	13.4	321	2.6×10^{-4}	<0.02	0.032
Guanine	H_2O, pH 7	275	8.1	329	3.0×10^{-4}	<0.02	0.024
Cytosine	H_2O, pH 7	267	6.1	313	0.8×10^{-4}	<0.02	0.005
Uracil	H_2O, pH 7	260	9.5	308	0.4×10^{-4}	<0.02	0.004
NADH	H_2O, pH 7	340	6.2	470	0.019	0.40	1.2

[a]Values shown for ϕ_F are the largest usually observed. In a given case actual values can be considerably lower.

26.6 Experimental Observation of Fluorescence

Fluorescence can be observed in three different ways: in the steady state, in the time domain, and in the frequency domain. We discuss some examples of time and frequency experiments.

The principle of the observation of fluorescence in the *time domain* is simple: The sample is hit with a short laser pulse, and the subsequent photon emission is determined as a function of time after the pulse [14]–[18] (Fig. 26.7).

Observation of fluorescence in the frequency domain, called *phase fluorometry*, dates back to at least 1926 [19]. The literature is extensive; we quote

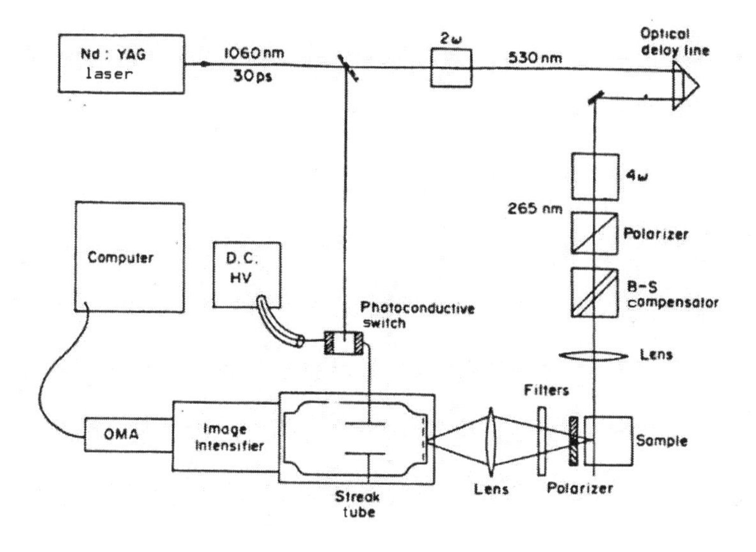

Fig. 26.7. Signal-averaging picosecond anisotropy detection system [17].

some reviews [20, 21]. The basic principle is simple [22]: The chromophore is excited by light with an intensity that is sinusoidally modulated in intensity with angular frequency ω. If the fluorophore is characterized by a single exponential decay with lifetime τ, the emitted light will also be modulated with the same frequency, but delayed in phase and demodulated with respect to the excitation. Excitation, $E(t)$, and fluorescence, $F(t)$, are shown in Fig. 26.8. The relations between phase shift ϕ and modulation ratio M and the lifetime τ are given by

$$\phi = \tan(\omega\tau) \, , \quad M = \left[1 + (\omega\tau)^2\right]^{-\frac{1}{2}} \, . \tag{26.7}$$

If the system decays with more than one lifetime, Eqs. (26.7) must be generalized [20, 21].
The demodulation is given by the ratio of the AC component to the average, DC. The modulation ratio is

$$M = (AC_{EM}/DC_{EM})/(AC_{EX}/DC_{EX}).$$

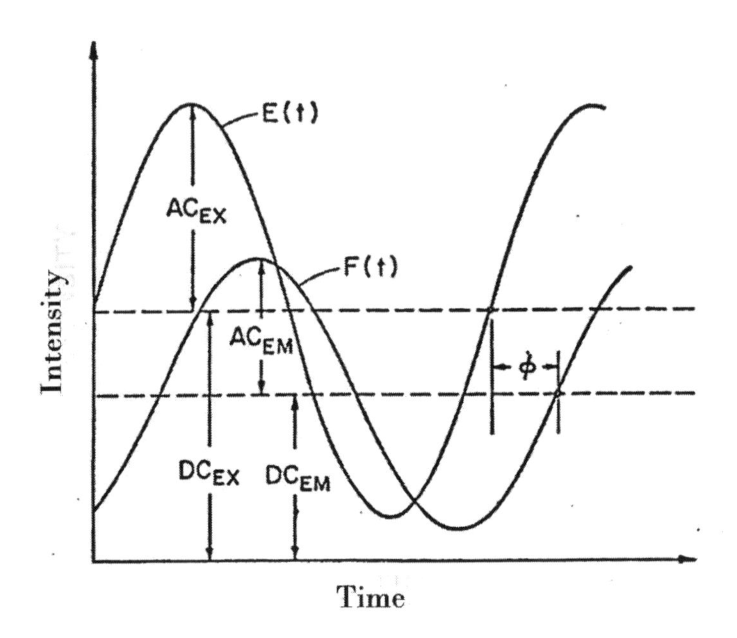

Fig. 26.8. Excitation $E(t)$ and fluorescence response $F(t)$ in the harmonic response method. Fluorescence is delayed by an angle ϕ and demodulated.

For additional literature, readers can refer to the book *Principles of Fluorescence Spectroscopy* by J. R. Lakowicz [23].

References

1. C. R. Cantor and P. R. Schimmel. *Biophysical Chemistry, Vol. II.* W. H. Freeman, San Francisco, 1980.
2. M. Weissbluth. *Hemoglobin.* Springer, New York, 1974.
3. A. S. Brill. *Transition Metals in Biochemistry.* Springer, Berlin, 1977.
4. W. A. Eaton and J. Hofrichter. Polarized absorption and linear dichroism spectroscopy of hemoglobin. In *Methods in Enzymology*, 76:175–261, 1981.
5. W. Hoppe et al. editors. *Biophysics.* Springer Verlag, Berlin, 1983.
6. M. W. Makinen and A. K. Churg. Structural and analytical aspects of the electronic spectra of hemeproteins. In A. B. P. Lever and H. B. Gray, editors, *Iron Porphyrins.* Addison-Wesley, Reading, MA, 1983, pp. 141–235.
7. W. A. Eaton, L. K. Hanson, P. J. Stephens, J. C. Sutherland, and J. B. R. Dunn. Optical spectra of oxy- and deoxyhemoglobin. *J. Amer. Chem. Soc.*, 100:4991–5003, 1978.
8. C. Tanford. Protein denaturation. *Adv. Protein Chem.*, 23:121–282, 1968.
9. C. R. Cantor and P. R. Schimmel. *Biophysical Chemistry, Vol. II.* W. H. Freeman, San Francisco, 1980. Section 8.2.
10. L. Brand and J. Gohlke. Fluorescence probes for structure. *Ann. Rev. Biochem*, 41:843–68, 1972.
11. R. B. MacGregor and G. Weber. Estimation of the polarity of the protein interior by optical spectroscopy. *Nature*, 319(6048):70–3, 1986.
12. J. A. Secrist III, J. R. Barrio, and N. J. Leonard. A fluorescent modification of adenosine triphosphate with activity in enzyme systems: 1,N^6-ethenoadenosine triphosphate. *Science*, 175(4022):646–7, 1972.
13. M. E. Eftink and C. A. Ghiron. Fluorescence quenching studies with proteins. *Anal. Biochem.*, 114:199–227, 1981.
14. R. Rigler and M. Ehrenberg. Fluorescence relaxation spectroscopy in the analysis of macromolecular structure and motion. *Q. Rev. Biophys.*, 9:1–19, 1976.
15. G. R. Fleming, J. M. Morris, and G. W. Robinson. Direct observation of rotational diffusion by picosecond spectroscopy. *Chem. Phys.*, 17:91–100, 1976.
16. I. Munro, I. Pecht, and L. Stryer. Subnanosecond motions of tryptophan residues in proteins. *Proc. Natl. Acad. Sci. USA*, 76:56–60, 1979.
17. G. S. Beddard, T. Doust, and J. Hudales. Structural features in ethanol-water mixtures revealed by picosecond fluorescence anisotropy. *Nature*, 294(5837):145–6, 1981.

18. T. M. Nordlund and D. A. Podolski. Streak camera measurement of tryptophan and rhodamine motions with picosecond time resolution. *Photochem. Photobiol.*, 38:665–9, 1983.
19. E. Gaviola. The time of decay of the fluorescence of dye solutions. *Ann. Phys. (Leipzig)*, 81:681–710, 1926.
20. E. Gratton, D. M. Jameson, and R. D. Hall. Multifrequency phase and modulation fluorometry. *Ann. Rev. Biophys. Bioeng.*, 13:105–24, 1984.
21. D. M. Jameson, E. Gratton, and R. D. Hall. Measurement and analysis of heterogeneous emission by multifrequency phase and modulation fluorometry. *Appl. Spectros. Rev.*, 20:55–106, 1984.
22. F. Duschinsky Der zeitliche Intensiätsverlauf von intermittierend angeregter Resonanzstrahlung. *Z. Phys.*, 81:7–22, 1933.
23. J. R. Lakowicz. *Principles of Fluorescence Spectroscopy.* Plenum Press, New York, 1983.

27

Vibrations

Assume that the energy landscape of a biomolecule is as shown in Fig. 27.1, where the energy of the biomolecule is plotted as a function of the conformational coordinate *cc*. We note again that the figure is misleading; the actual energy hypersurface is a function of a very large number of coordinates, and Fig. 27.1 only gives a one-dimensional cross section.

Two types of excitations can occur. The system can either stay within one conformational substate (CS) or jump from one CS to another CS. In the first case, the system can be excited within its own CS vibrational levels; in the second a conformational excitation occurs. Both are important for the function of biomolecules. Here we first discuss vibrations, begining with diatomic molecules.

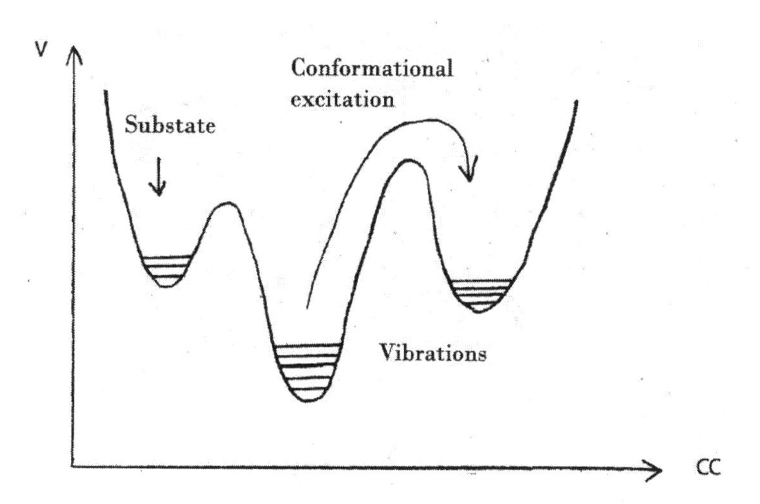

Fig. 27.1. Vibrational and conformational excitations.

H. Frauenfelder, *The Physics of Proteins*, Biological and Medical Physics, Biomedical Engineering, DOI 10.1007/978-1-4419-1044-8_27,

27.1 Vibrational States

For a diatomic molecule consisting of two atoms with masses M_1 and M_2, the Schrödinger equation can be written in their center of mass, M, as

$$\frac{\hbar}{2M}\Delta^2\psi + [E - V(R)]\psi = 0. \tag{27.1}$$

Here $M = M_1M_2/(M_1+M_2)$ is the reduced mass and R the distance between the two nuclei. To find the vibrational energy levels, the form of $V(R)$ must be known [1]–[3]. In the simplest case, we can take $V(R)$ to be parabolic,

$$V(R) = V_0 + \frac{1}{2}k_v(R - R_e)^2, \tag{27.2}$$

where k_v is a force constant and R_e the equilibrium distance. With this potential, Eq. (27.1) can be solved easily; the result is

$$E_v = \nu + \left(\frac{1}{2}\right)\hbar\omega_e, \tag{27.3}$$

where ω_e is the classical vibration frequency of the oscillator and $\nu = 0.1, 2, \ldots$ is the vibrational quantum number. In transitions, ν is governed by the selection rule

$$\Delta\nu = \pm 1. \tag{27.4}$$

A more realistic potential was introduced by Morse:

$$V(R) = D_e\{1 - \exp[-a(R - R_e)]\}^2, \tag{27.5}$$

where D_e is the dissociation energy of the diatomic molecule, and the coefficient a is connected to the force constant $\mathrm{k_v}$ by

$$D_e a^2 = \mathrm{k_v}/2.$$

The shape by $V(R)$ is shown in Fig. 27.2; the energy levels are given by [4, 5]

$$E_{vib} = \hbar\omega_e\left(\nu + \frac{1}{2}\right) - x_e\hbar\omega_e\left(\nu + \frac{1}{2}\right)^2, \tag{27.6}$$

with

$$\omega_e = a(2D_e/M)^{1/2} \tag{27.7}$$

$$x_e = (\hbar\omega_e/4D_e)\ , \tag{27.8}$$

where x_e is the internuclear distance.

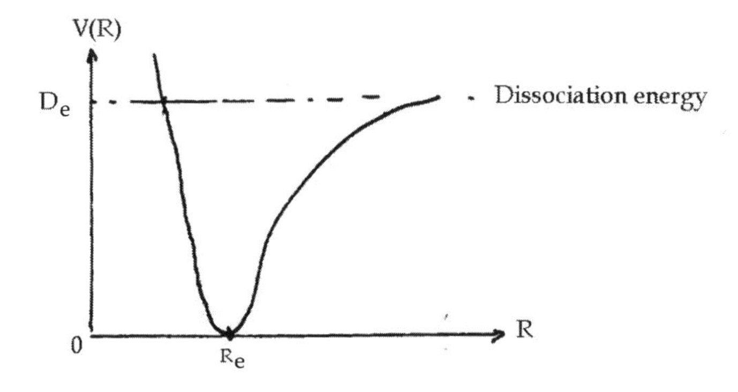

Fig. 27.2. More potential.

In a diatomic molecule, only vibrations along the axis are possible; in molecules containing more than two atoms, additional normal modes exist [6]. A few are shown in Fig. 27.3.

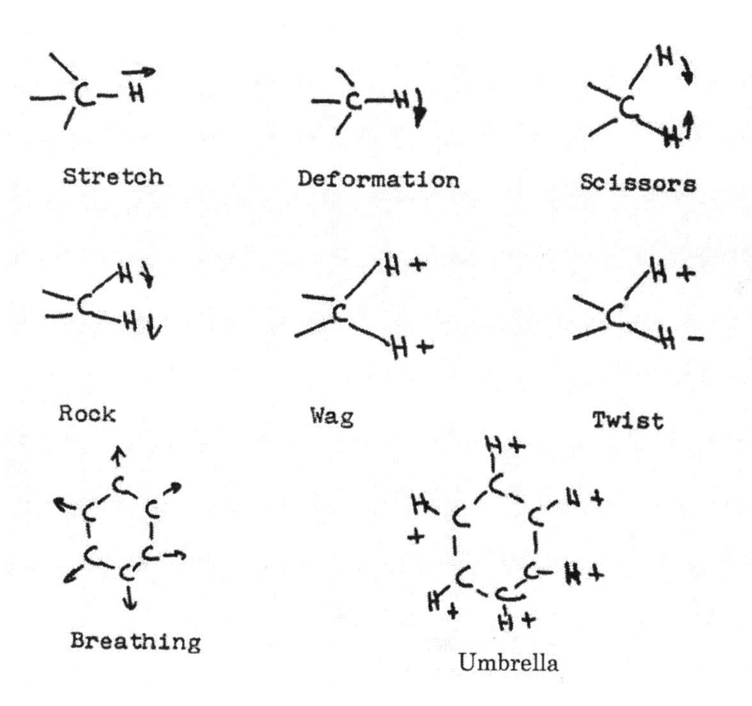

Fig. 27.3. Some vibrational modes: + and – refer to motions perpendicular to the plane.

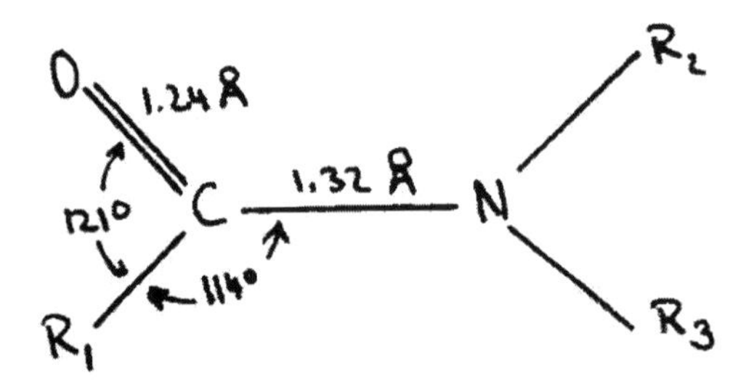

Fig. 27.4. A typical amide

While it is difficult to calculate the frequencies of complex vibrations *ad initio*, a great deal of empirical and semiempirical information is available. We will discuss some important cases in detail later but mention here only some general observations.

A few remarks about amides are in order here. A typical amide is a planar molecule of the form (Fig. 27.4). If R_3 is a hydrogen, two configurations occur, *trans* and *cis* (Fig. 27.5). Normally, the *trans* configuration is more stable than the *cis*. The number, position, and intensity of the N–H bands of many amides depend on the configuration.

The peptide groups in the protein backbone possess five in-plane (CONH plane) and three out-of-plane vibrations. A strong band occurs at about 3300 cm^{-1} and a somewhat weaker one at 3100 cm^{-1}. These are called amide A and amide B bands. The amide A is caused by the NH stretching vibration; the amide B is the first overtone of the amide II vibration. The amide I band occurs at about 1650 cm^{-1}; it is mainly caused by $C = 0$ stretching vibration. The amide II bond at 1540 cm^{-1} has C–N stretching and NH bending character. The positions of the various bands depend on the conformation of the backbone, as indicated in Table 27.1.

The positions of various IR bands are also shown in Fig. 27.6. The regions of strong water absorption are indicated. To obtain data in these regions,

Fig. 27.5. cis and trans amide bonds

Table 27.1. Influence of conformation on the position of the infrared bands.

Band	Assignment	Frequency $\bar{\nu}(\mathrm{cm}^{-1})$ Coil	Helix	Sheet
Amide A	NH stretch	3250	3290	3290-3260
	Half-width	150–200	55–90	55–90
Amide I	$C = 0$ stretch	1655	1650	1630
Amide II	C–N stretch NH bend	1520–1545	1545–1550	1520–1530

dried samples or samples in D_2O are used, but both methods can affect the function of the biomolecule.

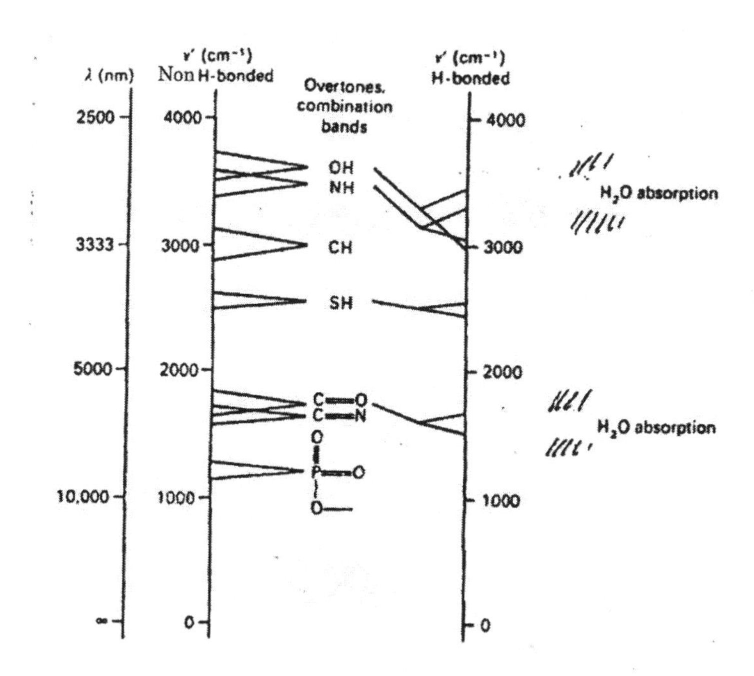

Fig. 27.6. Some typical infrared absorption bands found in biological molecules. Approximate shifts on H-bonding are shown, and the regions of water absorption are within the white area.

27.2 Fourier-Transform Infrared Spectroscopy

As shown in Fig. 27.6, the important vibrational modes lie in the infrared. In principle they can be studied by a "standard" IR spectrometer as shown in Fig. 27.7(a). For most applications, it is much more convenient to use Fourier-transform (FT) techniques [7]–[10]. Since this technique can also be applied to other spectroscopies, we discuss the method here. The basic arrangement of an FT spectrometer is shown in Fig. 27.7(b). In the standard approach, a white light source is used, but a monochromator selects a particular frequency. The transmission is then determined as a function of frequency.

In the FT approach, the entire spectrum of a white or polychromatic light source passes through the sample. The frequency spectrum $S_o(\nu)$ without sample and $S(\nu)$ with sample are determined with a Michelson interferometer. The spectrum of the sample is then extracted. To discuss the idea in more detail, consider the Michelson interferometer and the setup in Fig. 27.8.
The polychromatic light, after passing through the sample, hits beam splitter C. Half of the light is reflected to mirror A and from there returns and passes through the beam splitter. The light that initially passes through the beam splitter is reflected at the movable mirror B and again at the beam splitter. The two beams recombine, and their intensity is measured as a function of path difference, $\delta = 2L$. L is the distance by which the mirror B is moved from the initial position L_o. At $L = 0$, all frequencies will interfere constructively, and $I(0)$ will be a maximum. To find $I(\delta)$ for $\delta \neq 0$, consider a wave with frequency ω, wave vector k, and wave number $\nu = 1/\lambda = k/2\pi$, $F(x, \nu)$, the amplitude of this wave is given by

$$F(x, v) = e^{i(kx-\omega t)} = e^{i(2\pi\nu x-\omega t)}.$$

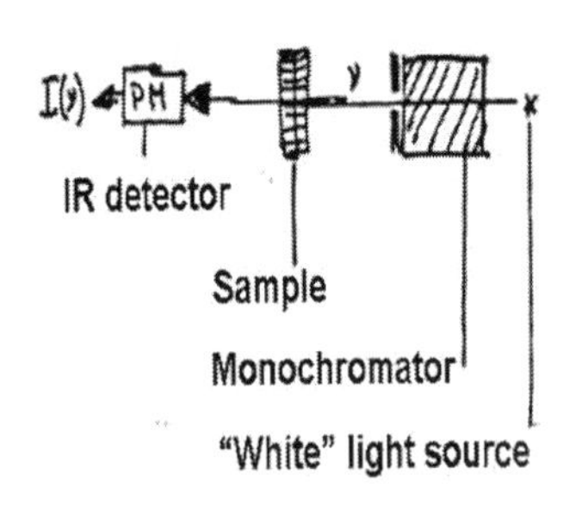

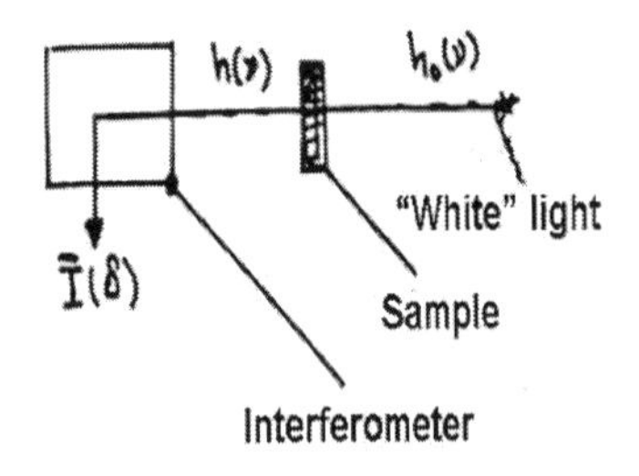

Fig. 27.7. Standard and Fourier-Transform spectrometers.

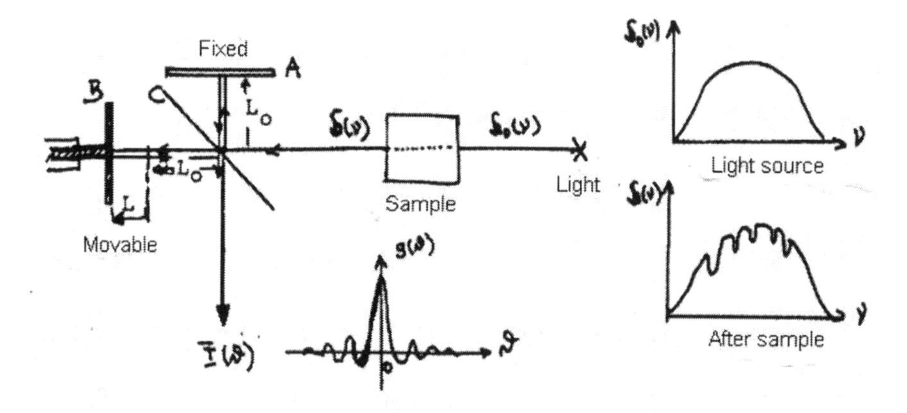

Fig. 27.8. Michelson interferometer and spectra before and after sample. C beam splitter, A and B mirrors.

Half of the beam will arrive at the detector after a path length $x = x_o + \delta$. The amplitude of the wave with wave number ν arriving at the detector is given by

$$F(\delta, \nu) = \frac{1}{2}[e^{i(\pi \nu x_o - \omega t)} + e^{i(2\pi\nu(x_o+\delta)-\omega t}]$$
$$= \frac{1}{2}e^{i(2\pi\nu x_o - \omega t)}[1 + e^{2\pi i \nu\delta}].$$

The intensity of this wave is

$$I(\delta, \nu) = F^*(\delta, \nu)F(\delta, \nu) = \frac{1}{2}\left[1 + \cos(2\pi\nu\delta)\right].$$

f the contribution of the wave with wave number ν has the weight $S(\nu)$, the total intensity measured at position δ is given by

$$I(\delta) = \int_0^\infty I(\delta, \nu)S(\nu)d\nu = \frac{1}{2}\int_0^\infty S(\nu)d\nu + \frac{1}{2}\int_0^\infty \cos(2\pi\nu\delta)S(\nu)d\nu.$$

With $(1/2)I(0) = \int_0^\infty S(\nu)d\nu$, we rewrite the expression as

$$I(\delta) - \frac{1}{2}I(0) = \frac{1}{2}\int_0^\infty \cos(2\pi\nu\delta)S(\nu)d\nu. \tag{27.9}$$

The expression on the right-hand side is a Fourier cosine integral, with the inverse

$$S(\nu) = \text{const} \int_0^\infty [2I(\delta) - I(0)] \cos(2\pi\nu\delta) d\delta. \quad (27.10)$$

The integral extends from 0 to ∞. It is impossible to design a system with these limits; travel is limited to distances $2L_{\text{max}}$. This restriction limits the obtainable resolution; two lines with separation $\Delta\nu$ can be resolved if

$$\Delta\nu \gtrsim 1/(2L_{\text{max}}). \quad (27.11)$$

Equation (27.10) gives the theoretical resolution of the FT system. To achieve a resolution of 1 cm^{-1}, $L_{\text{max}} = 0.5\,\text{cm}$.

The measurement with a FT spectrometer proceeds as follows: The interferometer is illuminated with a polychromatic source of IR radiation that passes through the sample. The mirror B is moved in small steps from 0 to $+L_{\text{max}}$ and the interferometer signal, observed with an IR detector, is sampled at the desired values of L, $S(\nu)$ and then extracted by computer, often one line. $S(\nu)$ for the sample alone is found by subtracting the source spectrum.

FT spectroscopy has a number of advantages. We mention two, Fellgett's and Jacquinot's advantages. Fellgett's or multiplex advantage comes from the fact that a standard instrument measures one wavelength at a time; an FT system takes all at once. Jacquinot's or the throughput advantage favors the FT system because no input energy is lost between source and detector to slits and dispersive elements (gratings).

The availability of powerful commercial FTIR systems with online data analysis has made studies of proteins much easier. Nevertheless, some difficulties remain. One of the most important arises from the richness of data. A protein possesses so many IR bands that it is often difficult or nearly impossible to assign a particular band to a specific bond within the protein. Two approaches then are useful. In one, atoms in strategic position are replaced by isotopic substitution. Since the nuclear mass affects the vibrational frequencies, as shown by Eq. (27.7), the substitution fingerprints lines and unambiguous assignments become possible. The second technique is FTIR difference spectroscopy. The FTIR spectra are taken on the same protein in two different states, for instance, before and after a bond has been broken by a laser flash. Lines or bands that have not changed then disappear, and only the groups involved in the reaction under investigation show up.

27.3 Raman Scattering—General

Raman scattering was predicted in 1923 by Smekal, found in 1928 by Raman and Krishnan, and the fundamental theory produced by Placzek in 1934 [2, 11, 12]. The importance for biomolecular studies lies in the fact that IR and Raman spectroscopy are complementary. Equations (27.9)–(27.11) show that a molecule without a permanent electric dipole moment does not absorb radiation and hence cannot be studied by IR. As we will see, this restriction does not hold for Raman spectroscopy, and thus important molecules like O_2 can be investigated.

Consider light scattering as sketched in Fig. 27.9. The scattered light consists of a line at the incident frequency and some weaker lines with higher and lower frequencies as indicated in Fig. 27.10. As an order of magnitude value, the Rayleigh scattering occurs with about 10^{-3} I_o, Raman scattering with about $10^{-6} I_o$. The physical explanation of the Raman lines (or bands) is straightforward: In Stokes lines, some energy is given to the scattering system (production of phonons); in anti-Stokes scattering, energy is transferred from phonons to the outgoing photon. The three important processes as sketched in Fig. 27.11.

An experimental setup is shown in Fig. 27.12.

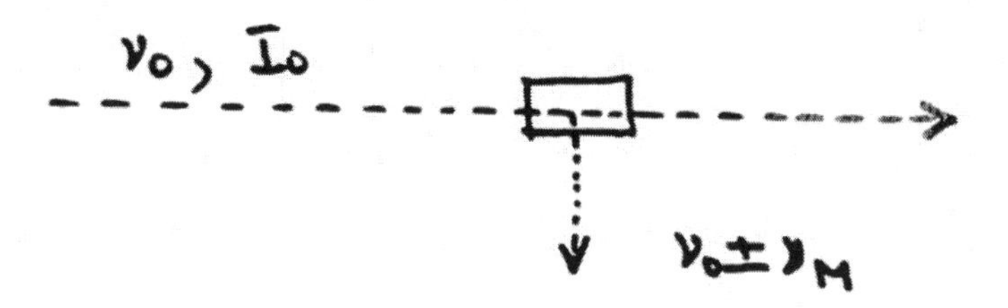

Fig. 27.9. Scattering of light from a molecular sample.

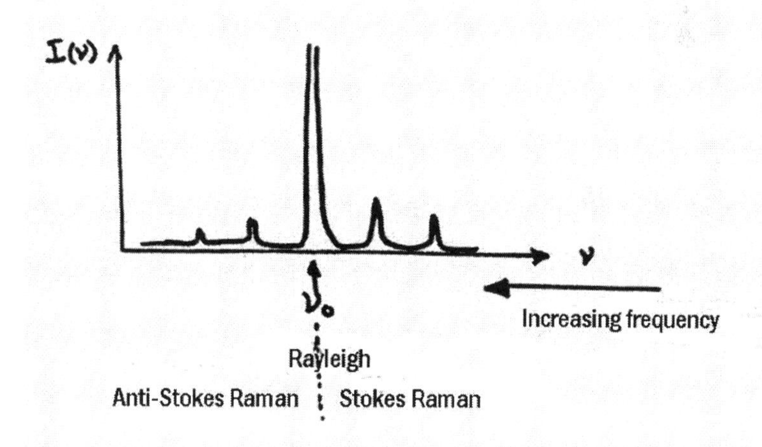

Fig. 27.10. Spectrum of Rayleigh and Raman scattering.

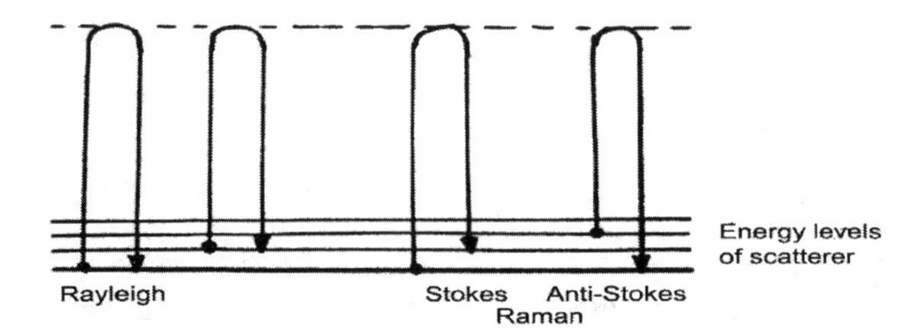

Fig. 27.11. Change of energy levels from Rayleigh scattering.

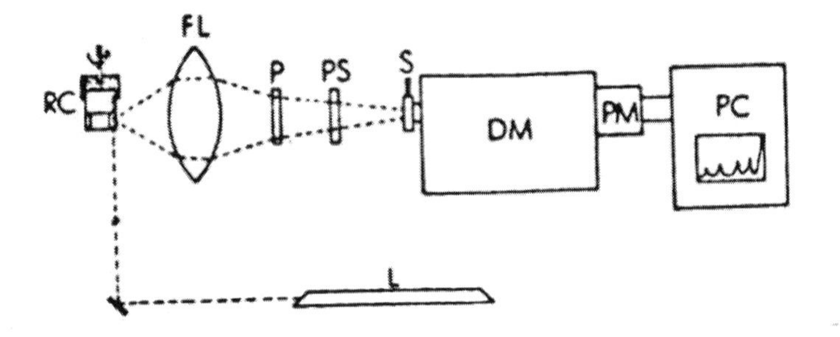

Fig. 27.12. Schematic diagram of a Raman spectrometer. Light from the laser L is passed through a thin layer of solution coating the walls of the rotating cell RC. Scattered light is collected by a lens FL and passes through a polarizer P to yield $I_{\|}$ or $I_{\perp}$. The polarization scrambler PS provides randomly polarized light for detection by the photomultiplier PM. A double monochromator DM and photon counter PC are used to detect the Raman lines.

The complete quantum-mechanical theory of the Raman effect is lengthy and can be found in [2] and some of the papers given later. Here we sketch the classical approach to bring out some of the essential physical ideas.

As incident electromagnetic wave, $E = E_o \exp(i\omega_o t)$, induces an oscillating dipole μ in a molecule of polarizability α,

$$\mu = \alpha E = \alpha E_o \exp(i\omega_o t).$$

The oscillating dipole radiates with frequency ω_o and thus gives rise to Rayleigh scattering. If the molecule vibrates with a frequency ω_{vib}, the electronic and nuclear coordinates will also oscillate and the polarizability will vary with ω_{vib},

$$\alpha = \alpha_o + \frac{\partial \alpha}{\partial Q} Q = \alpha_o + \frac{\partial \alpha}{\partial Q} Q_o e^{i\omega_{\text{vib}} t} .$$

Here the coordinate $Q = Q_o \exp(i\omega_{\text{vib}} t)$ describes the internuclear displacement. Inserting α into μ shows that the scattered radiation now also includes Raman-shifted frequencies with $\omega_o \pm \omega_{\text{vib}}$. The intensity of these lines depends on $\partial\alpha/\partial Q$.

27.4 Raman Scattering—An Example

Figure 27.13 gives the Raman spectrum of a typical protein, human carbonic anhydrase-B. This enzyme catalyses the reaction

$$CO_2 + H_2O \rightleftharpoons H_2CO_3 .$$

This reaction is important in the transfer of CO_2 from the tissues into the blood.

Vibrational modes of the backbone and the constituent amino acids are apparent. Information on conformations, conformational changes, hydrogen bonding, and the environment of individual residues can be extracted from such spectra.

Two modes associated with the amide bond are shown, amide I (peptide C=O stretching) and amide II (C–N stretching and N–H bending). The position of these bands provides information about the secondary structure.

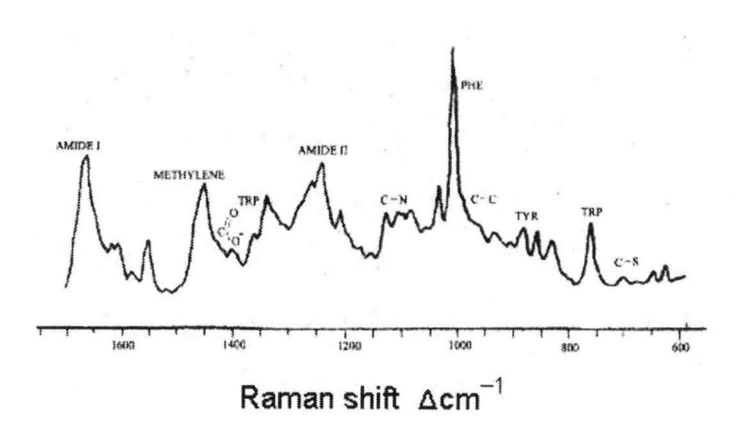

Fig. 27.13. Raman spectrum of human carbonic anhydrase.

27.5 Resonance Raman Scattering

In the Raman scattering discussed so far, the frequency of the incident light is not too important; the essential information is contained in $\Delta\nu = \nu_o \pm \nu_{\text{vib}}$. In resonance Raman (RR) [13]–[17], the incident energy coincides with the electronic activation energy of the target molecule. The dashed level in Fig. 27.11 is no longer at an arbitrary energy. The situation thus is as sketched in Fig. 27.14.

Some essential differences between R and RR are:

1. In R, the Rayleigh line is strong and the vibrational lines are weak. In RR, the overtones may be as strong as the Rayleigh line.

2. In R, the scattering intensity varies slowly and smoothly (ν^4) with the excitation frequency. In RR, the intensity varies strongly with excitation frequency.

3. In R, the Stokes–to–anti-Stokes ratio follows from the Boltzmann factor. In RR, the ratio is not given by this factor, and anti-Stokes lines may be more intense than Stokes lines.

4. In R, the scattering time is "fast"; in RR, it may be slow.

Heme proteins provide a very good testing ground for RR [14, 15]. The optical spectrum of iron porphyrin is given in Fig. (21.30). The excitation frequency can be chosen to excite either the Q or the B (Soret) band. An example is shown in Fig. 27.15.

The Raman lines provide considerable information about the physical properties of the heme group. The positions of some Raman lines correlate very well with structural parameters of the heme, namely with oxidation state, spin state, metal displacement distance, and core expansion.

RR lines can be assigned to various modes of the protein-heme system: in-plane and out-of-plane heme vibrations, vibrations of the Fe-axial bonds, and charge transfer modes [17].

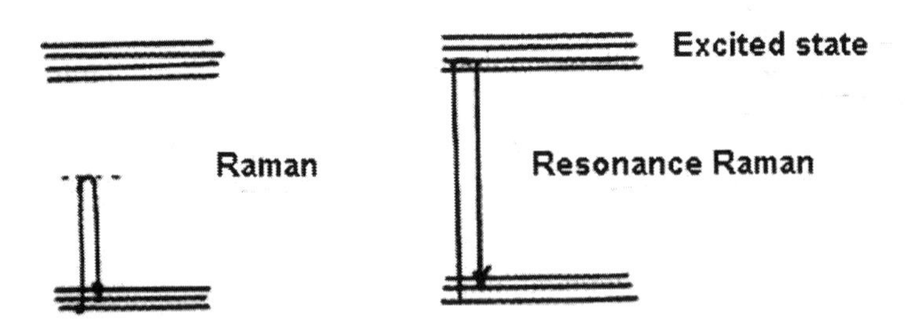

Fig. 27.14. The transitions of Raman and resonance Raman scattering.

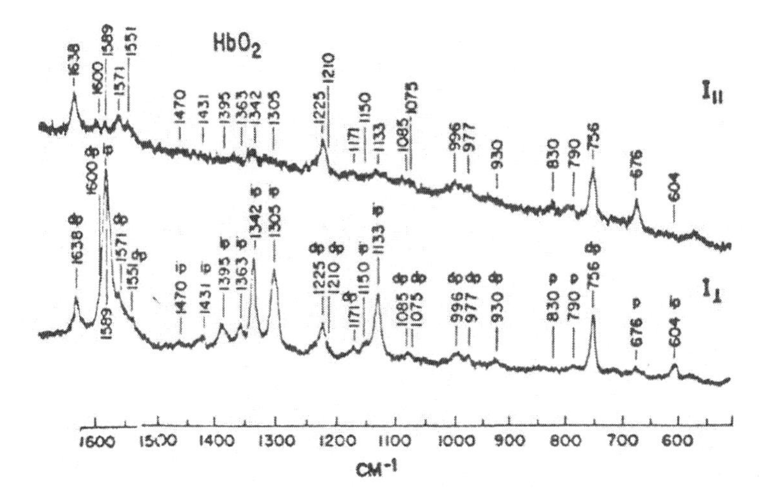

Fig. 27.15. RR spectrum of oxyhemoglobin with exciting line at 568.2 nm (from [16]). The two spectra refer to two different polarizations.

The intensity of the RR lines makes it possible to study time-dependent processes. Transient Raman investigations have become very important in the exploration of the structure-function relation in heme proteins and in studies of the dynamics [18, 19].

References

1. G. Herzberg. *Spectra of Diatomic Molecules.* Van Nostrand, New York, 1950.
2. G. Herzberg. *Infrared and Raman Spectra.* Van Nostrand, New York, 1945.
3. E. B. Wilson, J. C. Decius, and P. C. Cross. *Molecular Vibrations.* Dover, New York, 1955, 1980.
4. M. Tinkham. *Group Theory and Quantum Mechanics.* McGraw-Hill, New York, 1964. pp. 234–8.
5. L. I. Schiff. *Quantum Mechanics.* McGraw-Hill, New York, 1968, pp. 451–3.
6. L. J. Bellamy. *The Infra-red Spectra of Complex Molecules*, 3rd edition. Wiley, New York, 1975.
7. R. J. Bell. *Introductory Fourier Transform Spectroscopy.* Academic Press, New York, 1972.
8. J. R. Ferraro and L. J. Basile, editors. *Fourier Transform Infrared Spectroscopy.* Academic Press, New York, 1982.
9. P. R. Griffiths. Fourier-transform infrared spectrometry. *Science*, 222(4621): 297–302, 1983.
10. J. O. Alben and F. Fiamingo. Fourier transform infrared spectroscopy. In D. L. Rousseau, editor, *Optical Techniques in Biological Research.* Academic Press, New York, 1985.
11. For introductions, see M. Karplus and R. N. Porter. *Atoms and Molecules*, W. A. Benjamin, New York, 1970; or J. Steinfeld. *Molecules and Radiation*, 2nd edition. MIT Press, Cambridge, 1985.
12. D. A. Long. *Raman Spectroscopy.* McGraw-Hill, New York, 1977.
13. A. Weber, editor. *Raman Spectroscopy of Gases and Liquids* (Topics in Current Physics, Vol. 11). Springer, Berlin, 1979.
14. R. H. Felton and N. T. Yu. Resonance Raman scattering from mtealloporphyrins and hemoproteins. In D. Dolphin, editor, *The Porphyrins, Vol. III.* Academic Press, New York, 1978. Chapter 8.
15. T. G. Spiro. The resonance Raman spectroscopy of metalloporphyrins and hemoproteins. In A. B. P. Lever and H. B. Gray, editors, *Iron Porphyrins, Vol. II.* Addison-Wesley, Reading, MA, 1983. Chapter 3..
16. T. G. Spiro and T. C. Strekas. Resonance Raman spectra of hemoglobin and cytochrome *c*: Inverse polarization and vibronic scattering. *Proc. Natl. Acad. Sci. USA*, 69:2622–6, 1972.

17. D. Rousseau, editor. *Optical Techniques in Biological Research.* Academic Press, New York, 1985.
18. D. L. Rousseau and M. R. Ondrias. Resonance Raman scattering studies of the quaternary structure transition in hemoglobin. *Ann. Rev. Biophys. Bioeng.*, 12:357–80, 1983.
19. J. M. Friedman. Structure, dynamics and reactivity in hemoglobin. *Science*, 228(4705):1273–80, 1985.

28

The Nucleus as a Probe (C. E. Schulz[1])

The nucleus is a nearly ideal tool to study fields and processes within biomolecules. First, a nucleus is heavy and hence localized in a small region of space; it occupies a well-defined position. Second, the nucleus has electromagnetic moments, and these interact with the fields, magnetic and electric, produced by its surroundings. The nucleus can thus be used to probe these fields. There are two important features: Static fields lead to a splitting and shift of the nuclear levels. From these, conclusions about the nature of the surrounding fields can be made. Time-dependent fields lead to a relaxation, and from such measurements, dynamic behavior can be elucidated.

The two important methods that we discuss here are nuclear magnetic resonance (NMR) and Mössbauer spectroscopy. In NMR, the system is placed in a magnetic field and the splitting and shifts are observed with *rf* fields. In a Mössbauer effect, gamma rays with essentially the natural line width are observed. A very accurate measurement of the energies of the emitted lines provides information on the line splittings and shifts; the intensity of the lines yields, just like the Debye-Waller factor, information concerning the dynamics.

28.1 Hyperfine Interactions

Three properties of nuclei are involved in the hyperfine interaction. First, a nucleus is not a point particle but has a finite extension. For the discussion here it is sufficient to describe the size by the mean square radius,

$$\langle r^2 \rangle = \int Z\rho(r) r^2 dv, \tag{28.1}$$

where $Z\rho(r)$ is the charge density at the distance r from the nuclear center with an atomic number Z. Second, many nuclei possess nonzero spins and magnetic moments. We define the factor g through the relation

[1] Department of Physics, Knox College, Galesburg, IL 61401, USA.

H. Frauenfelder, *The Physics of Proteins*, Biological and Medical Physics, Biomedical Engineering, DOI 10.1007/978-1-4419-1044-8_28,

$$\boldsymbol{\mu} = g\, \mu_N \frac{\mathbf{I}}{\hbar} \tag{28.2}$$

with

$$\mu_N = 3.15 \times 10^{-18}\, \text{MeV/G}. \tag{28.3}$$

Here $\mathbf{I}$ is the nuclear spin, $\boldsymbol{\mu}$ the magnetic moment, g a characteristic constant called the g factor, and μ_N the nuclear magneton. Equation (28.2) expresses the fact that the magnetic moment is parallel to the spin. Third, nuclei with spin larger than 1/2 may possess nuclear electric quadrupole moments, defined by

$$Q = \int dv Z\rho(\mathbf{r})(3z^2 - r^2). \tag{28.4}$$

Q is zero for a spherical nucleus, and it measures the deviation from sphericity. A simple example of a charge distribution having a quadrupole moment is shown in Fig. 28.1. Relevant properties of a number of nucleides are listed in Table 28.1.

To describe the influence of the surroundings, we also require three quantities; the magnetic field $\mathbf{B}$, the electric charge density $q\rho_e(0)$ at the nucleus produced by the atomic and molecular electrons, and the electric field gradient tensor V_{ij}. The charge density is given by

$$q\rho_e(0) = -e|\psi(0)|^2 \tag{28.5}$$

where $\psi(0)$ is the electron probability density at the nucleus and $e > 0$. The electric field gradient is defined by

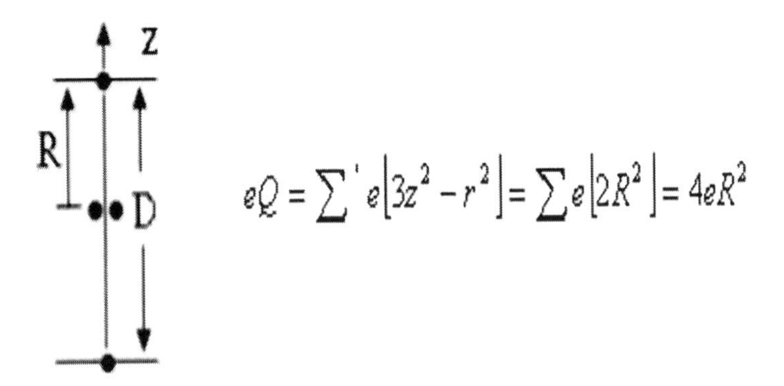

Fig. 28.1. Simple example for a charge distribution with nonvanishing electric quadrupole moment; e is the charge of an electron.

Table 28.1. Nuclear Properties of Selected Isotopes.

Isotope	Nuclear Spin I	Natural Abundance (%)	Relative Sensitivity for Equal Number of Nuclei at Constant Field	NMR Frequency (MHz) at 10 kG	Magnetic Moment μ ($\times eh/\pi Mc$, the nuclear magneton)	Electric Quadrupole Moment Q (in $e \times 10^{26} cm^2$)
^{1}H	1/2	99.98	1.000	42.57	2.7927	—
^{2}D	1	0.015	9.65×10^{-3}	6.53	0.8473	0.00277
^{11}B	3/2	81.17	1.65×10^{-1}	13.66	1.6880	0.041
^{13}C	1/2	1.11	1.59×10^{-2}	10.70	0.7022	—
^{14}N	1	99.63	1.01×10^{-3}	3.08	0.4036	0.01
^{15}N	1/2	0.37	1.04×10^{-3}	4.31	–0.2830	—
^{17}O	5/2	0.04	2.91×10^{-2}	5.77	–1.8930	–0.026
^{19}F	1/2	100	8.33×10^{-1}	40.05	2.6273	—
^{23}Na	3/2	100	9.25×10^{-2}	11.26	2.2161	0.12
^{25}Mg	3/2	10.05	2.68×10^{-3}	2.61	–0.8547	0.22
^{29}Si	1/2	4.70	7.85×10^{-2}	8.46	–0.5548	—
^{31}P	1/2	100	6.63×10^{-2}	17.23	1.1305	—
^{33}S	3/2	0.74	2.26×10^{-3}	3.27	0.6427	–0.05
^{35}Cl	3/2	75.4	4.70×10^{-3}	4.17	0.8209	–0.08
^{37}Cl	3/2	24.6	2.71×10^{-3}	3.47	0.6833	–0.063
^{39}K	3/2	93.08	5.08×10^{-4}	1.99	0.3909	0.055
^{43}Ca	7/2	0.13	6.40×10^{-2}	2.86	–1.3153	–0.045
^{79}Br	3/2	50.57	7.86×10^{-2}	10.67	2.0990	0.33
^{81}Br	3/2	49.43	9.85×10^{-2}	11.50	2.2626	0.28
^{111}Cd	1/2	12.86	9.54×10^{-3}	9.03	–0.5922	—
^{123}Cd	1/2	12.34	1.09×10^{-2}	9.44	–0.6195	—
^{127}I	5/2	100	9.34×10^{-2}	8.52	2.7939	–0.79

$$V_{ij} = \frac{\partial^2 V}{\partial x_i \partial x_j} \tag{28.6}$$

where V is the scalar potential. If the coordinate system is chosen so that mixed terms of the form $(\partial^2 V/\partial x \partial y)$ vanish, only the three components V_{xx}, V_{yy}, and V_{zz} appear. To get a feeling for field gradients, assume a potential $V(x) = \exp(-\alpha x)$. The field gradient at $x = 0$ then is given by $V_{xx}(0) = \alpha^2$.

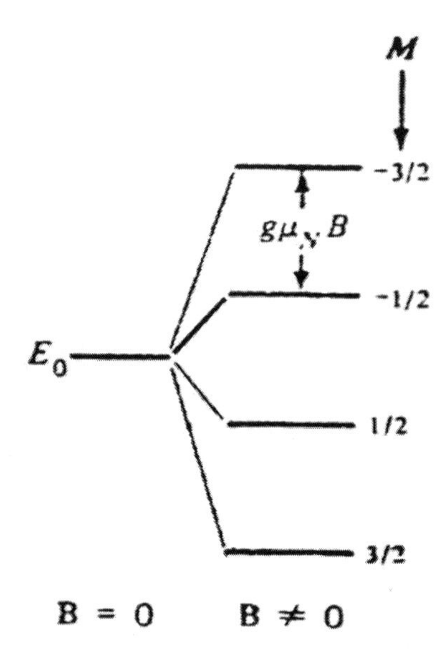

Fig. 28.2. Zeeman splitting of the energy levels of a state with spin $I = 3/2$ and g-factor g in an external magnetic field B. B points along the z-axis and $g > 0$.

The three important interactions behave as follows:
Magnetic dipole interaction. The interaction energy is given by

$$H_m = -\boldsymbol{\mu} \cdot \mathbf{B}. \tag{28.7}$$

A level with spin I splits into $(2I + 1)$ magnetic substates as shown in Fig. 28.2. The energy difference between two adjacent substates is given by

$$E = g\mu_n B. \tag{28.8}$$

Electric monopole. Consider a nucleus with mean square radius $\langle r^2 \rangle$ and charge Ze. In the absence of an external electric field, the energy of the nucleus will be at a level that we set to zero by definition. In the presence of the charge density $-e|\psi(0)|^2$ of the surrounding electrons, and the electrostatic potential created by that charge, the nucleus' energy is lowered by the amount

$$H_M = \frac{2\pi}{3} Ze^2 \langle r^2 \rangle |\psi(0)|^2 \tag{28.9}$$

called the electric monopole energy. H_M produces a shift of the nuclear energy levels when compared to that of a free nucleus (Fig. 28.3). The shift is proportional to the nuclear mean-square radius and the electron probability density at the nucleus.

The electric quadrupole term describes the interaction between an electric field gradient and the nuclear quadrupole moment. Consider a nucleus as

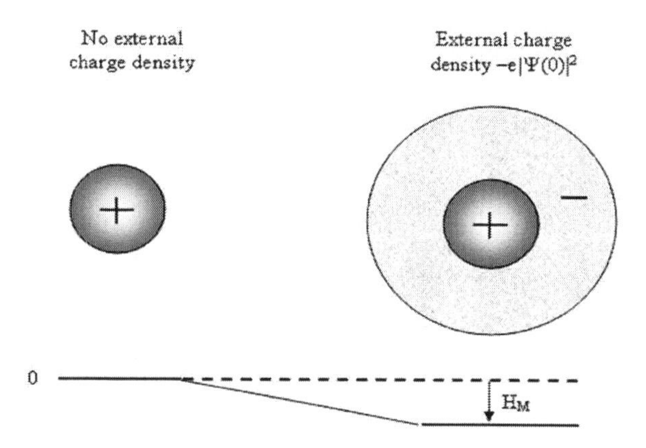

Fig. 28.3. Shift of the nuclear energy levels by the electric monopole interaction.

sketched in Fig. 28.4, with spin J and orientation described by $J_z = m$. For an axially symmetric field gradient, $V_{xx} = V_{yy}$, the quadrupole energy becomes

$$H_Q = \frac{1}{4} eV_{zz} Q \left[\frac{3m^2 - J(J+1)}{J(2J-1)} \right] . \tag{28.10}$$

The quadrupole interaction gives rise to a splitting of the nuclear energy levels with spins $J \geq 1$. The splitting is proportional to the nuclear quadrupole moment, Q, and the electric field gradient, V_{zz}. States m and $-m$ have the

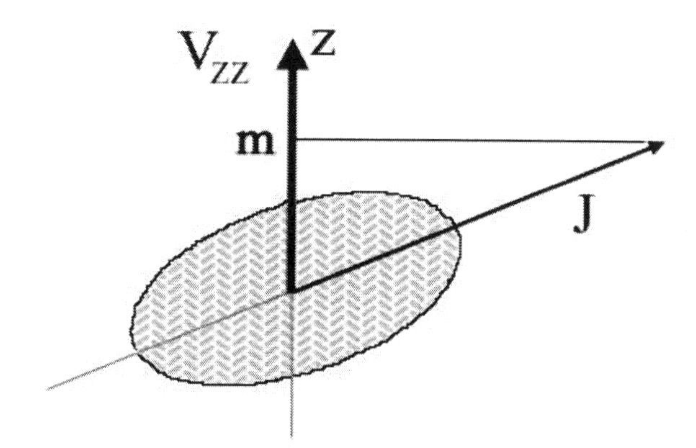

Fig. 28.4. The energy of a nucleus with quadrupole moment Q in an electric field gradient V_{zz} depends on the orientation of the spin with respect to the direction of the field gradient.

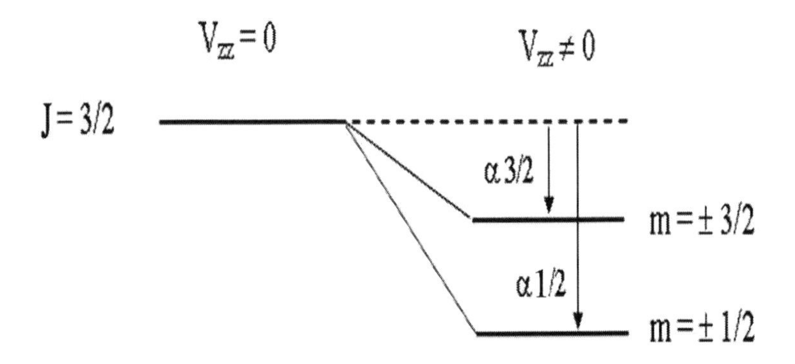

Fig. 28.5. Splitting of the nuclear energy levels caused by the quadrupole interaction.

same energy, but levels with different m^2 split as sketched for the case $J = 3/2$ in Fig. 28.5.

As a simple example of the quadrupole energy, consider the system shown in Fig. 28.1 in an inhomogeneous electric field as indicated in Fig. 28.6. We normalize the energy so that charges at $z = 0$ do not contribute to the energy; only charges at $z \neq 0$ count. The energy for the spin at right angle to the z-axis then is $H_Q = 0$; for the parallel case $H_Q = 2e\{1/2\ R^2(\partial^2 V/\partial z^2)\} = 1/4\ V_{zz}Q$, in accord with Eq. (28.10). Figure 28.6 demonstrates that H_Q depends on the orientation of the molecule in the field and on the sign of the quadrupole moment, but not on the sign of m.

In nuclear physics we are interested in the nuclear properties $\langle r^2 \rangle$, I, $\boldsymbol{\mu}$, and Q. In the application of nuclear tools to biomolecules we assume that

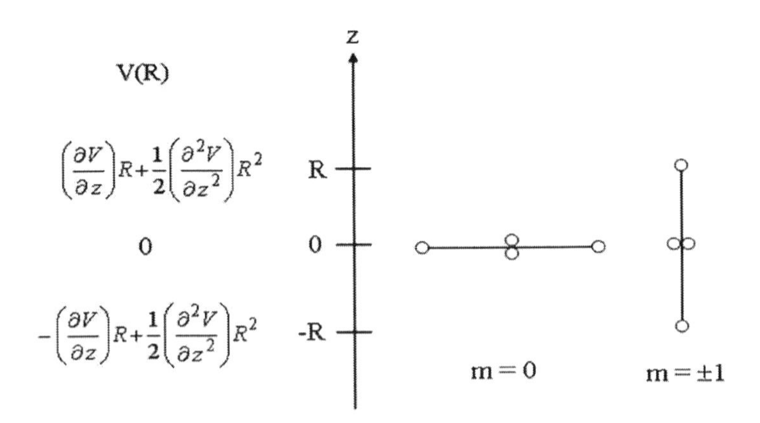

Fig. 28.6. The energy of a simple quadrupole in an electric field gradient.

these properties are known and try to find the sources of the field. The values of these properties ($\mathbf{B}$, $\psi(0)$, V_{xx}, V_{yy}, V_{zz}) provide us with information about biomolecular characteristics. Before discussing the sources, we sketch two important techniques, Mössbauer effect and nuclear magnetic resonance.

28.2 The Mössbauer Effect

Further details are available in a number of excellent reviews [1]–[4].

In Mössbauer spectroscopy, also known as nuclear gamma-ray resonance absorption spectroscopy, one observes the spectrum of energies at which gamma rays are absorbed by an appropriate nucleus. A source emits gamma rays of a specific energy, and one observes absorption of those gamma rays by a sample being studied. Figure 28.7 illustrates the energy level diagram for Co^{57}, the nuclide that acts as a source for the 14.4-keV Mössbauer transition in Fe^{57}.

So if one Fe^{57} nucleus (the source) is emitting a gamma ray (and the nucleus could be held in place so it would recoil only slightly due to the gamma ray emission), and a second nucleus in an identical environment were similarly held in place, the amount of resonance absorption we would see would be proportional to the extent the emission and absorption lines overlap, as illustrated by the gray area in Fig. 28.8. This is analogous to the spectral criteria for fluorescence resonant energy transfer (FRET) for a donor-acceptor pair. Note that the physics for these two spectroscopies are different. FRET involves a dipole-dipole interaction and thus is a strong function of distance, r^{-6}, where Mössbauer effect is an absorption of a real photon, so the only spatial dependence is an r^{-2} variation of the count rate due to solid angle effects.

However, it's not really convenient to reach in and hold down the nuclei (our fingers being rather too big). So the nuclei will be expected to recoil upon

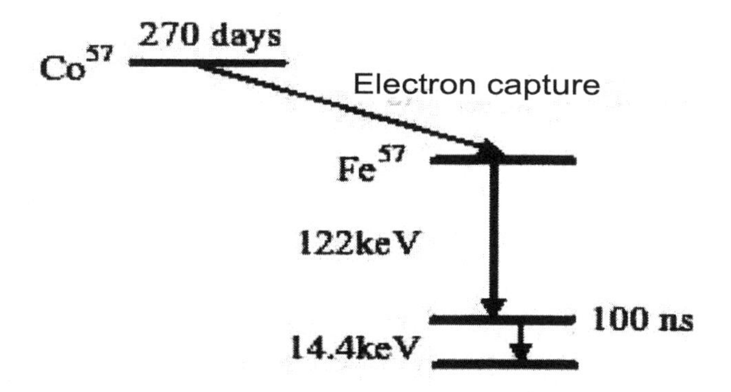

Fig. 28.7. Energy level diagram for Co^{57} decay.

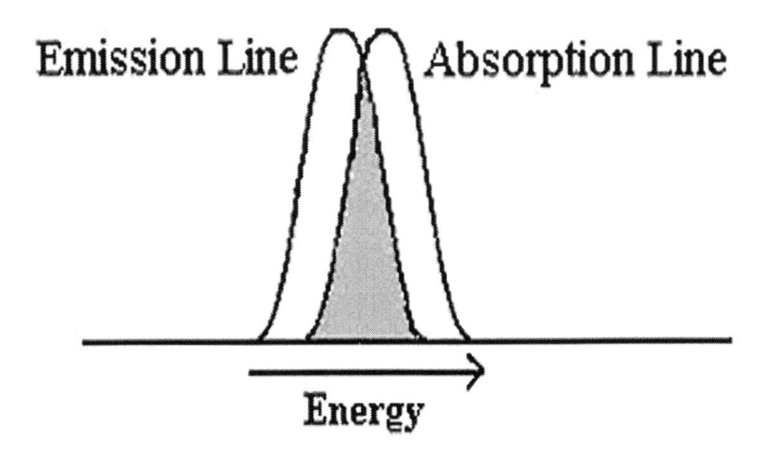

Fig. 28.8. Illustration of the requirement for overlapping emission and absorption lines for resonant absorption of gamma rays.

gamma ray emission to conserve momentum. This recoil on emission will steal an amount of energy R from the gamma ray. Similarly, the absorbing nucleus momentum must recoil, so to conserve energy the incoming gamma ray must have a similar excess energy R. Given the line width Γ of the Fe^{57} transition ($\sim 6 \times 10^{-9}$eV) and R expected for the 14.4-keV emission and absorption ($\sim 2 \times 10^{-3}$ eV), the lines are far from overlapping, and one expects resonant absorption to be impossible, as in Fig. 28.9.

So the first issue to be confronted is why Mössbauer spectroscopy works at all. In emitting or absorbing a gamma ray, momentum must be conserved.

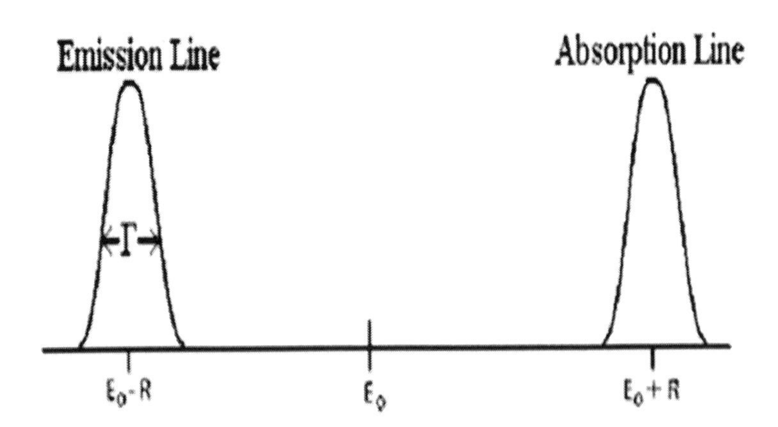

Fig. 28.9. In the presence of recoil on emission and absorption, there will be little or no overlap.

As noted earlier, a free atom, in emitting or absorbing a nuclear gamma ray, must recoil. The kinetic energy of this recoil must come from somewhere, thus the energy of an emitted gamma ray will be less than the energy difference between the nuclear ground and excited states by the amount of recoil kinetic energy. For a nucleus of mass m, this recoil energy will be approximately

$$R = \frac{E_o^2}{2mc^2}, \tag{28.11}$$

where E_o is the difference in energy between the nuclear excited and ground states.

The key to resolving this dilemma is that the nucleus of interest in a biological molecule (or in any sample one might like to study) is not floating free in space; rather the atom containing the nucleus of interest is bound in place at some site in the molecule, and the molecule itself may be in a frozen solution. While momentum must still be conserved, it is possible for the recoil momentum to be taken up by the entire sample, for which the mass m in the denominator of Eq. (28.11) is enormously larger than that of a single atomic nucleus. If this happens, R can become negligibly small, the absorption and emission lines can overlap, and resonant absorption of the gamma rays by the sample can occur. The physics of this "recoilless" process is of special interest.

The molecule and its internal state of vibration can be described in a method familiar from condensed-matter physics: It has a set of quantized lattice vibrations (phonons) having specific frequencies. The state of the lattice $|L>$ is described in terms of the number of quanta of excitation of each of its phonon modes and is obviously a function of temperature. The recoil of a nucleus upon emission of a gamma ray can be described as causing the creation of an additional group of phonons in order to conserve momentum. Thus the lattice makes a transition from some initial state $|L_i>$ to a final state $|L_f>$ after the recoil. Recoilless emission results from the special case in which the lattice state is unchanged by the emission process: No phonons are created so the lattice as a whole must recoil to conserve momentum. The probability f for recoilless emission of a gamma ray is [1]:

$$f = |\langle L_i | e^{i\mathbf{k}\cdot\mathbf{X}} | L_i \rangle|^2. \tag{28.12}$$

The evaluation of the recoilless fraction f, called the Lamb-Mössbauer factor, requires a specific model for the vibrational modes of the lattice. For example, in an Einstein solid, for which there is a single vibrational mode with the oscillator having an effective mass M and angular frequency ω_E, one gets

$$f = \exp(-R/\hbar\omega_E), \tag{28.13}$$

while for a Debye solid, the result is

$$f = \exp(-2w) \tag{28.14}$$

with w given by

$$w = 3\frac{R}{k\theta_D}\left[\frac{1}{4} + \left(\frac{T}{\theta_D}\right)^2 \int_0^{\theta_D/T} \frac{x dx}{e^x - 1}\right] \tag{28.15}$$

where R is the recoil energy, k is Boltzmann's constant, and θ_D is the Debye temperature for the lattice: $k\theta_D$ is the energy of the highest-frequency vibrational mode of the lattice. For both of these expressions it can be shown that the argument of the exponential function in f is essentially proportional to $\langle x^2 \rangle$, the mean-squared displacement of the nucleus from its equilibrium position.

It should be noted that the Lamb-Mössbauer factor is similar in nature to the Debye-Waller factor from X-ray diffraction. The Debye-Waller factor is important for describing the temperature dependence of the intensities of diffraction spots:

$$\mathrm{I} = \mathrm{I}_o \exp(-2\mathrm{W}), \tag{28.16}$$

with

$$\mathrm{W} = 2\langle \mu_z^2 \rangle \sin^2\left[\varphi\left(\frac{2\pi}{\lambda}\right)^2\right] \tag{28.17}$$

where φ is the Bragg angle, λ is the X-ray wavelength, and $\langle \mu_z^2 \rangle$ is the mean-square deviation of the component of displacement of the atoms along a direction z perpendicular to the reflecting planes. This Debye-Waller factor also applies to describing *nonresonant* (Rayleigh) scattering of nuclear gamma rays from a resonant source.

X-ray diffraction and Mössbauer scattering are related in depending on the mean-square displacements of the scattering atoms. The characteristic time scales for these two experiments are very different: X-ray diffraction has a characteristic time scale much shorter than lattice vibrations, while Mössbauer scattering's time scale is similar to or longer than the period of typical lattice vibrations. Nonetheless, their similarity in the dependence on the mean-square displacement of the nucleus is an interesting one. Further details on the calculation of f, the fraction of recoilless gamma ray emission, are given in Frauenfelder's review [1]. Similar considerations apply to the calculation of f', the corresponding fraction for recoilless gamma ray absorption.

Now we move on to experimental details. How does one take a Mössbauer spectrum? The gamma rays are emitted by the source with an energy of 14.4 keV and an incredibly narrow line width, but we're interested in the absorption of those gamma rays by our biological sample *as a function of*

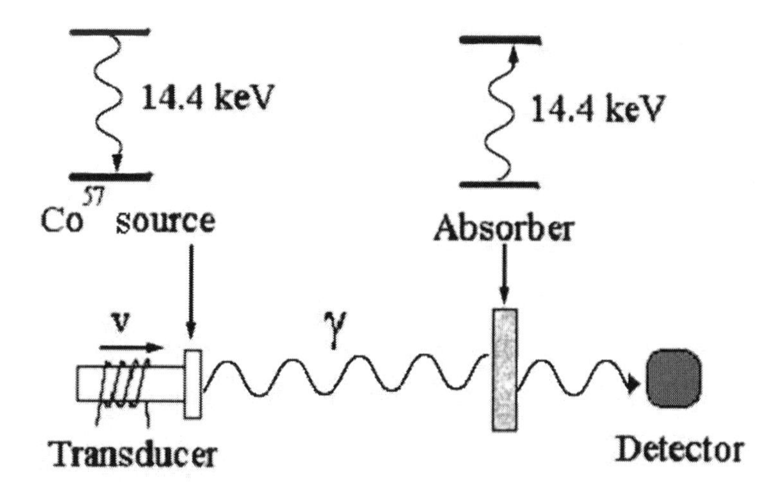

Fig. 28.10. Experimental setup for a Mössbauer transmission experiment.

the gamma ray energy. So how does one vary the gamma rays' energy in a controlled way? Rudolf Mössbauer's idea was to give the source a velocity, which would Doppler-shift the gamma rays to higher energy (source moving toward the absorber) or lower energy (source moving away from the absorber). An apparatus to do this is shown schematically in Fig. 28.10.

The transducer gives the source a velocity, thus Doppler-shifting the gamma rays to higher or lower energies. The amount of energy by which the gamma rays are Doppler-shifted is

$$\Delta E = \frac{v}{c} E_o. \tag{28.18}$$

For Fe^{57} Mössbauer spectroscopy, to sweep out the entire range of velocities needed to see all of the sample's nuclear transitions normally requires at most a range of velocities $\Delta v = \pm 10$ mm/sec, corresponding to a fractional change in energy of the gamma rays of only $\Delta E/E_o < 3 \times 10^{-11}$.

The number of gamma rays counted by the detector is then measured as a function of the source velocity. A typical Mössbauer spectrum for ^{57}Fe is illustrated in Fig. 28.11. The vertical axis gives the percentage of the gamma ray beam absorbed by the sample, and the horizontal axis gives the velocity of the source. To sweep out the velocity range of interest, the transducer is driven in such that its velocity varies linearly with time (i.e., has constant acceleration).

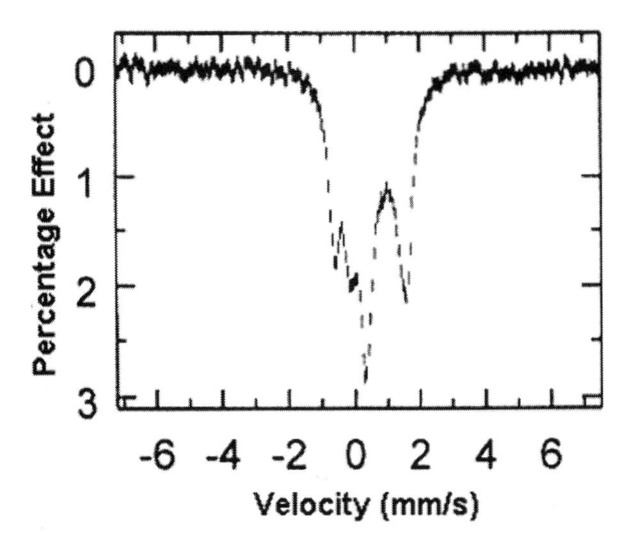

Fig. 28.11. Typical ^{57}Fe for Mössbauer spectrum.

Not all elements have a nuclide for which Mössbauer spectroscopy is possible. The key characteristics of a good Mössbauer nuclide are:

- A relatively low-energy excited state. The higher the energy of the emitted gamma rays, the lower the probabilities for recoilless emission and absorption.
- A relatively long-lived excited state. By the uncertainty principle, the shorter the lifetime of the excited state, the greater will be the uncertainty in the energy of the gamma rays emitted and absorbed. A short lifetime corresponds to a broad Lorentzian line shape, which makes resolution of closely spaced absorption lines more difficult.
- A parent nuclide with a relatively long half-life. This helps by lengthening the amount of time before one has to purchase a new source.
- A relatively high natural abundance for the nuclide. This increases the sensitivity since one can study less-concentrated samples.

To see that Fe^{57} satisfies these criteria, we note:

- It has a low-energy excited state, corresponding to a gamma ray energy of only 14.4 keV. One can easily get f factors in the range of 70 to 80 percent.
- The lifetime of the excited state is relatively long, at 98 ns. Even given the fact that both the emission from the source and absorption by the sample contribute to the experimentally observed width, one can get experimental line widths of 0.23 mm/sec for this nuclide.
- A parent nuclide of Fe^{57} is Co^{57}, which has a half-life of 272 days. It decays by electron capture to an excited state of Fe^{57}. Depending on the needed count rate, a source may last a couple of years.
- Note also that although Fe^{57} is a stable nuclide, it has a natural abundance of only 2.2 percent. Thus, for some experiments it may be desirable to enrich the sample being studied in Fe^{57}. This fourth criterion is only marginally satisfied.

Iron is not the only element for which Mössbauer can be usefully done in biological systems. One can also use Sn^{119}, Ni^{61}, I^{127}, I^{129}, and (in principle, although with great difficulty) Zn^{67}. We now turn to describing what information the Mössbauer spectrum tells you about the sample.

28.3 Isomer Shift

This isomer shift, also known as chemical shift, of a Mössbauer spectrum measures the velocity of its center of gravity. For an absorber with a single absorption line, the isomer-shift parameter δ is illustrated in Fig. 28.12. It is common practice for Fe^{57} Mössbauer to take the center of gravity of the spectrum of a metallic iron foil absorber as the zero of the velocity scale in displaying Mössbauer spectra and in calculating isomer shifts.

The isomer shift originates from the electric monopole interaction between the nucleus and the electron charge density at the nucleus. As the nucleus interacts with the electron charge density at the site of the nucleus, a change in the charge density will change slightly the energies of the nuclear states. A detailed theory of the isomer shift was given by Solomon [5], and its application to Fe^{57} was published by Walker et al. [6]. Solomon's formula for the isomer shift is

$$\delta = \frac{2}{3}\pi Z e^2 \left[\langle R_{ex}^2 \rangle - \langle R_{gr}^2 \rangle\right] \left[|\psi(0)_a|^2 - |\psi(0)_e|^2\right], \tag{28.19}$$

where the $\langle R^2 \rangle$ values are the expectation values of the mean-squared radius of the nuclear charge for the ground and excited states, and the $|\psi(0)|^2$ functions give the total s-electron density at the nucleus for the absorbing and emitting nucleus. The effective radius of the ground state is in fact larger than that of the excited state for Fe^{57}.

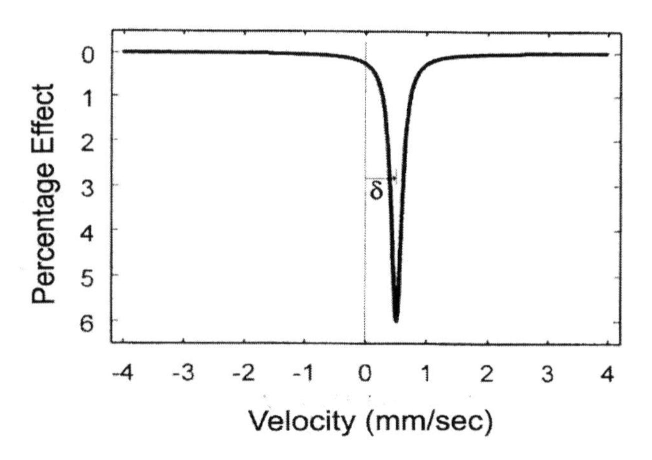

Fig. 28.12. Unsplit Mössbauer spectrum illustrating isomer shift measurement.

The charge density at the nucleus will depend on the number of valence electrons: Adding an extra 3d electron (say going from Fe^{3+} to Fe^{2+}) will shield some of the s-electron cloud from the nucleus, causing it to expand slightly. This reduces the s-electrons density seen by the nucleus, resulting in a more positive isomer shift. Addition of electron density to 4s oribitals through covalent bonding to ligands will similarly affect the isomer shift. In addition, the configuration of electrons among the 3d-orbitals will change the value of the isomer shift, with low-spin ferric ($S = 1/2$) shift being slightly lower than high-spin ferric ($S = 5/2$), and low-spin ferrous ($S = 0$) shift lower than high-spin ferrous ($S = 2$).

Table 28.2 gives the isomer-shift values for a variety of iron-containing compounds, and in general the pattern holds: A greater number of d-electrons corresponds to a greater isomer shift. Obviously, the extent to which 4s orbitals are populated by electrons from neighboring atoms covalently bonded will also affect the isomer shift. In short, the isomer shift determined from an ion's Mössbauer spectrum is a good indicator of its ionization state.

A second effect will also influence the center of gravity of a molecule's Mössbauer spectrum: the second-order Doppler shift. As the temperature of a sample increases, its mean-square velocity will also increase due to its greater vibrational motion. This will cause the isomer shift to move to the left on the velocity scale, toward more negative velocities. There is a simple relativistic explanation of this effect: As the mean-square velocity of the nucleus increases, its "clock" is seen to run slowly in the lab frame, so the nucleus absorbs gamma rays of lower frequency/energy. Its absorption spectrum will be seen to shift to the left. This second-order Doppler shift is generally a small effect. For Fe^{57} in metalloproteins, δ will decrease by about 0.14 mm/sec as the temperature

Table 28.2. Representative isomer shifts and quadropole splittings for iron-containing proteins at 4.2 K.

Metalloprotein	Iron State	Electron Spin	Isomer Shift (mm/s)	Quadrupole Splitting (mm/s)	η
Myoglobin CO	2+	0	0.27	0.35	<0.4
Myoglobin O_2	2+	0	0.27	–2.31	0
Cytochrome c	2+	0	0.45	1.17	~0.5
Cyt. P450 O_2	2+	0	0.31	–2.15	~0.5
Myoglobin	2+	0	0.92	–2.22	
Reduced horseradish peroxidase (HRP)	2+	2	0.89	2.68	
Cyt. P450	2+	2	0.83	2.42	0.8
Chloroperoxidase	2+	2	0.86	2.50	1
Reduced rubredoxin	2+	2	0.70	–3.25	0.65
Myoglobin CN	3+	1/2	0.16	–1.46	0.3
Myoglobin N_3	3+	1/2	0.24	–2.25	0.06
Cyt. $P450_{Cam}$	3+	1/2	0.38	–2.85	0.4
Metmyoglobin	3+	5/2	0.42	1.24	
Native HRP	3+	5/2	0.40	1.70	0
Cyt $P450_{Cam}$	3+	5/2	0.44	0.79	0.59
Catalase pH 6.3	3+	5/2	0.27	1.14	0.15
Rubredoxin	3+	5/2	0.32	–0.5	0.2
Myoglobin(H_2O_2)	4+	1	0.09	1.43	0
HRP compound II	4+	1	0.03	1.61	0

is raised from 4.2 K to room temperature. But it must be accounted for if one is comparing the isomer shifts measured at different temperatures.

28.4 Zeeman Effects

Mössbauer nuclides commonly have nuclear magnetic moments. Fe^{57} has a ground state of nuclear spin $I = 1/2$ with a relatively weak moment (making NMR difficult), while the 14.4-keV excited state has $I = 3/2$. The nuclear moment will interact with magnetic fields, which can arise from several sources:

- *External applied magnetic fields.* This interaction has the form $-g_N \mu_N \mathbf{H} \cdot \mathbf{I}$, where μ_N is the nuclear magneton, 3.15×10^{-12}eV/Gauss, and g_N is the nuclear g-factor, which for the nuclear ground state in Fe^{57} is $g_{\text{gnd}} = 0.18$, and for the excited state $g_{\text{exc}} = -0.10$.

- *Magnetic fields created by electron spins on the same atom.* This interaction has two parts:

  1. *The dipolar interaction.* Due to its spin, the electron has a magnetic moment $\boldsymbol{\mu}_e = g\mu_B\mathbf{S}$, with $g = -2.0023$. The dipole-dipole interaction between the nucleus and each of the atom's electrons is $2\mu_B g_N \mu_N\{3(\hat{r}\cdot\mathbf{I})(\hat{r}\cdot\mathbf{s}) - \mathbf{I}\cdot\mathbf{s}\}/r^3$.

  2. *The Fermi, or contact interaction.* The exchange interaction of the unpaired 3d-electrons with the s-electrons causes a polarization of the (core) s-electrons. Their radial wave functions for spin up and spin down therefore differ and, in particular, the spin-up and spin-down density at the origin change. As a result, there is an interaction of the form $-8\pi/3\ \rho(0)2\mu_B\mathbf{S}\cdot g_N\mu_N\mathbf{I}$ between the nucleus and the d-electron's spin.

- *Magnetic fields created by electron orbital moments on the same atom.* An electron in an orbital with an angular momentum $\boldsymbol{\ell}$ will have a magnetic moment proportional to $\boldsymbol{\ell}$, and thus the nucleus will experience a magnetic interaction $2\mu_B g_N \mu_N \sum_i(\boldsymbol{\ell}_i\cdot\mathbf{I})/r^3$.
- *Magnetic fields due to electrons on another atom or atoms.* This sort of magnetic interaction can be important for molecules having two or more atoms with unpaired electrons that are exchange coupled through covalent bonds. Numerous examples exist, particularly among the iron-sulfur proteins, of iron-containing proteins with several irons that interact magnetically with each other. Systems with long-range magnetic order also fall into this category.

The first three of these interactions are normally combined into a single Zeeman interaction Hamiltonian using the equivalent total spin operator S: $H_Z = \mathbf{S}\cdot\tilde{A}\cdot\mathbf{I}$, where A is the magnetic hyperfine tensor.

Figure 28.13 illustrates the energy level diagram for Fe^{57} in a large magnetic field with the six allowed gamma ray transitions between states in the ground doublet and the excited quartet. (Eight transitions would seem to be expected, but the selection rule $\Delta I_z = \pm1, 0$ forbids two transitions.) A six-line spectrum of this sort is characteristic of the iron nucleus interaction being dominated by the Zeeman interaction. The relative intensities of the six lines will depend on the direction of the magnetic field relative to the direction of propagation of the gamma ray beam. If the field is parallel to the beam, lines 2 and 5 will be suppressed. If the field is perpendicular to the beam, they will be enhanced.

A six-line pattern as in Fig. 28.13 is typical of magnetically ordered systems, as in ferromagnetic or antiferromagnetic substances, for which the elctron spin magnetization is strongly polarized.

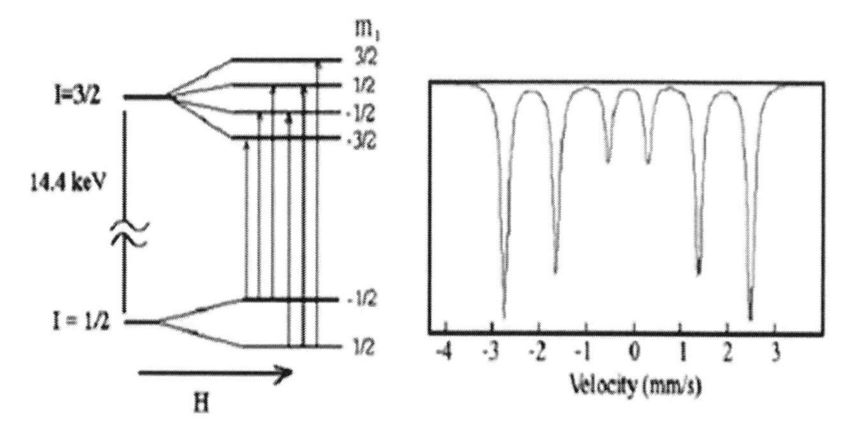

Fig. 28.13. Illustration of Fe^{57} Mössbauer spectrum in the presence of a large magnetic field.

28.5 Quadrupole Interaction

In the absence of an applied field, but in the presence of an electric field gradient (EFG) caused by the electrons on the iron atom itself and on neighboring atoms, the Fe^{57} nuclei will have a quadrupole splitting, ΔE_Q, in the $I = 3/2$ nuclear excited state between the $m_I = \pm 3/2$ doublet and the $m_I = \pm 1/2$ doublet, as shown in the energy levels of Fig. 28.14(b), which will result in a two-line spectrum like that in Fig. 28.15.

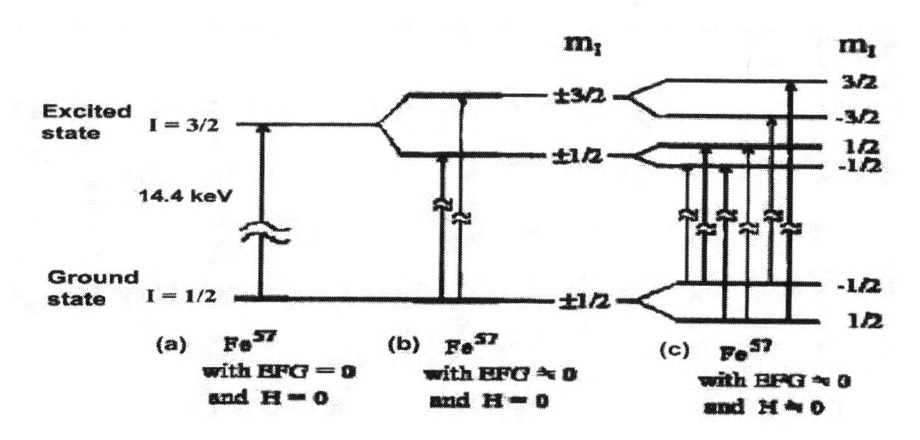

Fig. 28.14. Energy levels of Fe^{57} and allowed transitions with EFG and magnetic effects.

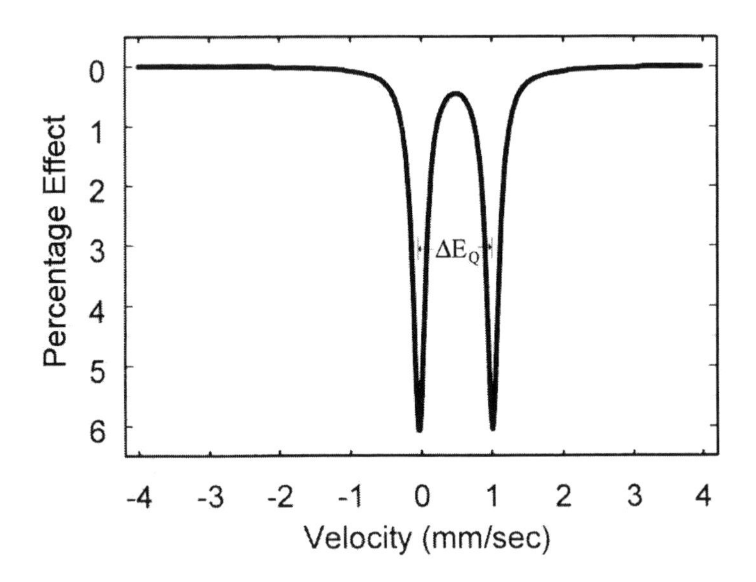

Fig. 28.15. Mössbauer spectrum of Fe^{57} with a quadrupole splitting.

The quadrupole interaction of nuclei was discussed earlier in this chapter. Figure 28.5 illustrates the sort of splitting that will occur in the nuclear excited state if the nucleus experiences a positive electric field gradient ($V_{zz} > 0$). The nuclear ground state will be unaffected, as it has no quadrupole moment. The quadrupole interaction, which for an axial EFG is given by Eq. (28.10), generalizes in the lower-symmetry rhombic case to

$$H_Q = \frac{e^2 V_{zz} Q[3I_z^2 - I(I+1) + \eta(I_x^2 - I_y^2)]}{4I(2I-1)}, \tag{28.20}$$

where η is the rhombicity parameter, given by $\eta = (V_{xx} - V_{yy})/V_{zz}$.

A two-line Fe^{57} Mössbauer spectrum represents a case in which the quadrupole interaction is dominant, with negligible magnetic splitting. The splitting between the two lines is proportional to V_{zz}, the largest component of the EFG. The standard convention is to choose a coordinate system for the traceless EFG such that $|V_{zz}| \geq |V_{yy}| \geq |V_{xx}|$. If follows that in such a standard coordinate system $1 \geq \eta \geq 0$. The quadrupole splitting ΔE_Q, the energy splitting between the two quadrupole lines, is given by $\Delta E_Q = 1/2eQV_{zz}\sqrt{1+\eta^2/3}$ The sign of the quadrupole splitting (the same as the sign of V_{zz}) can't be determined by a Mössbauer spectrum in the absence of a magnetic interaction.

Figures 28.14(c) and 28.16 show the effects of turning on a moderately sized magnetic field for the case of axial symmetry ($\eta = 0$). The three-line pattern on the left quadrupole line clearly indicates that the $m = \pm 1/2$ excited

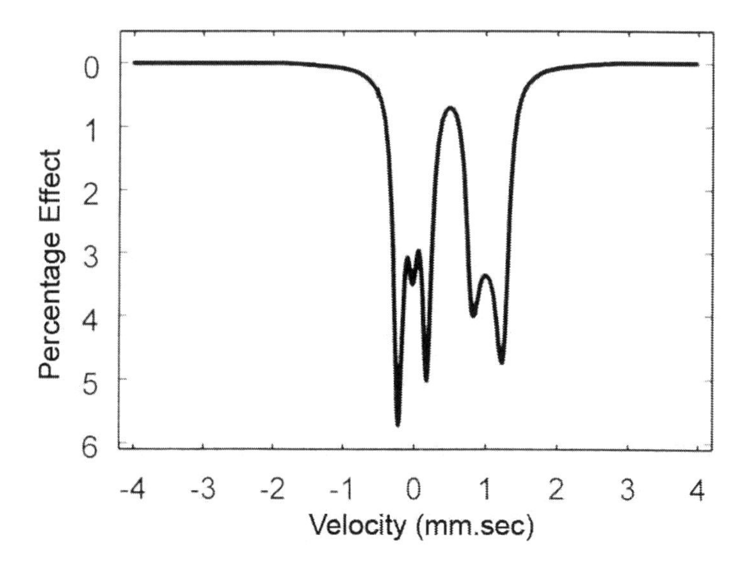

Fig. 28.16. Mössbauer spectrum with large axial quadrupole splitting ($\eta = 0$) and small Zeeman splitting.

states are lower in energy than the $\pm 3/2$ states, so the quadrupole splitting is positive. A negative quadrupole splitting would have the triplet on the right. As the symmetry is lowered toward rhombic, m_I is no longer a good quantum number for the $I = 3/2$ multiplet, the states of which get mixed by the η term. In the completely rhombic limit $\eta = 1$, the spectrum appears as two symmetric triplets, and the sign of the EFG is ambiguous, as the EFG has one component that is zero while the other two are equal in magnitude but opposite in sign.

Contributions to the EFG seen by the nucleus come from the electrons on its own atom or from unbalanced charge on neighboring atoms. A half-filled or completely filled electron shell will have a spherically symmetric charge distribution and therefore create no EFG at the nucleus. But a shell that's partially filled, or has unequal electron populations in its orbitals due to covalency effects, will result in a contribution to the EFG. For iron 3d-orbital electrons in the d_{z^2}, d_{xy}, and d_{yz} orbitals will make negative contributions to V_{zz} while electrons in d_{xy} and $d_{x^2-y^2}$ will make positive contributions. Thus the sign of the EFG is a good indicator of the d-electron configuration of the iron.

The quadrupole splitting often shows temperature dependence. A temperature-dependent quadrupole splitting is, according to a theory by Ingalls [7], the result of a low-lying excited state (or states) that can be thermally populated. As the temperature rises, the atom's electrons make rapid transitions into and out of that excited state, so the nucleus sees a thermal average of the ground and excited configurations' EFGs.

Examples of quadrupole splitting results for iron-containing proteins are given in Table 28.2.

For the most part, quadrupole splittings have been used as a qualitative measure of the relative strengths of the EFG in comparing similar proteins and model compounds. However, some heroic attempts have been made to calculate predicted quadrupole splittings through the use of density functional theory. As an example, see the paper by Zhang et al. [8].

28.6 Electronic Parameters

So far, the discussion has focused on the nucleus and the parameters that determine its energy levels. To fully understand the Mössbauer data of a metalloprotein, however, we must also have parameters that describe the electrons that interact with the Mössbauer nucleus. The spin Hamiltonian model is a straightforward way to characterize the energy levels of the elctron spins that are interacting with the nucleus of interest. A thorough treatment of the spin Hamiltonian approximation can be found in Abragam and Bleaney [9]. Here we present a brief summary. The energy of interaction of the electrons with the nucleus is generally quite weak compared to other interactions the electrons feel. So it is normally a very good approximation to ignore the nucleus when characterizing the electronic states of the atom. For a magnetically isolated atom (one that is isolated from magnetic fields due to other atoms) there are three principal interactions that are important: a Zeeman interaction with an external applied field, the electric field created by neighboring atoms, and spin-orbit coupling.

The Zeeman interaction, written in terms of the individual electron's spins and angular momenta, is $\mu_B \sum_i (2\mathbf{s_i} + k\boldsymbol{\ell}_\mathbf{i}) \cdot \mathbf{H}$, where k is an "orbital reduction factor" that in an ad hoc fashion takes into account covalency effects on the electrons' orbital angular momenta. If all of the unpaired electron spins $\mathbf{s}$ in the atom couple to a total $\mathbf{S}$, the Zeeman interaction can be written more simply as $\mu_B \mathbf{S} \cdot \tilde{g} \cdot \mathbf{H}$, where the g-tensor includes the effects of the orbital angular momenta into the equivalent operator representation of total spin.

Spin-orbit coupling is a magnetic interaction between the dipole moment due to an electron's spin and the moment due to its orbital angular momentum. It can be written in terms of the individual electrons' spin and angular momentum operators as $\sum_i \zeta_i (\boldsymbol{\ell}_\mathbf{i} \cdot \mathbf{s_i})$, where the ζ's are given by the expectation values $\langle \frac{\hbar^2}{2m^2c^2} \frac{1}{r} \frac{\partial V}{\partial r} \rangle$ for the spatial wave functions occupied by each electron. In terms of the spin and orbital angular momentum operators, the spin-orbit coupling can be written as $\lambda \mathbf{L} \cdot \mathbf{S}$, with $\lambda = \pm\zeta/2S$. The + sign applies to shells that are less than half-filled, and the – sign applies to shells more than half-filled. The magnitiude of λ is expected to be about 100 cm^{-1} for Fe^{2+} ions (1 cm^{-1} = 1.24×10^{-4}eV).

The 3d wave functions holding the unpaired electrons will in general be nondegenerate. Their degeneracy is lifted by electrostatic interactions with the electrons on neighboring atoms, or ligands. (The potential due to these neighbors is called a ligand field, or crystal field.) Crystal field splittings are typically hundreds to thousands of cm^{-1}. So normally the ground multielectron wave function is an orbital singlet with a (2S+1)-fold spin degeneracy. The principal of effect spin-orbit coupling is to split the spin degeneracy by mixing in excited orbital states. If the crystal field splitting is large relative to the spin-orbit coupling (and it normally is!), perturbation theory can be applied to include the effects of spin-orbit interaction into a spin Hamiltonian written only in terms of the equivalent total spin operator S:

$$H_s = D[S_z^2 - \frac{1}{3}S(S+1)] + E(S_x^2 - S_y^2). \tag{28.21}$$

The parameters D and E are called the axial and rhombic zero field splitting parameters, as they describe a splitting in energy of the ground spin multiplet even in the absence of an applied magnetic field. The zero field splitting is also known as *fine structure*. In principle, the spin Hamiltonian may also include axial, rhombic, and cubic terms of fourth order and higher in the S operator. But for most purposes D and E have been found to suffice as parameters to fit experimental Mössbauer (and EPR and magnetic susceptibility) data.

References

1. H. Frauenfelder. *The Mössbauer Effect.* W. A. Benjamin, New York, 1962.
2. G. Lang. Mössbauer spectroscopy of haem proteins. *Q. Rev. Biophys.*, 3:1–60, 1970.
3. E. Münck. Mössbauer spectra of hemoproteins. In D. Dolphin, editor, *The Porphyrins. Vol. IV.* Academic Press, New York, 1979. pp. 379–423.
4. P. G. Debrunner. Mössbauer spectroscopy of iron porphyrins. In A. B. P. Lever and H. B. Gray, editors, *Iron Porphyrins. Part III.* VCH Publishers, New York, 1989. pp. 139–234.
5. I. Solomon. Effet mössbauer dans la pyrite et la marcassite. *Compt. rend. herd acad sci*, 250:3828–30, 1960.
6. L. R. Walker, G. K. Wertheim, and V. Jaccarino. Interpretation of the Fe^{57} isomer shift. *Phys. Rev. Lett.*, 6:98–101, 1961.
7. R. Ingalls. Electric-field gradient tensor in ferrous compound. *Phys. Rev.*, 133:A787–95, 1964.
8. Y. Zhang, J. Mao, N. Godbout, and E. Oldfield. Mössbauer quadrupole splittings and electronic structure in heme proteins and model systems: A density functional theory investigation. *J. Amer. Chem. Soc.*, 124:13921–30, 2002.
9. A. Abragam and B. Bleaney. *Electron Paramagnetic Resonance of Transition Ions.* Clarendon Press, Oxford, 1970.

29

Nuclear Magnetic Resonance and Molecular Structure Dynamics (R. H. Austin[1])

In Part III we discussed in some detail the evidence for and importance of protein structural dynamics in light of a free-energy landscape. However, there is a major problem with the powerhouse technique of X-ray crystallography that we discussed in Chapter 25: The proteins are locked into a crystal structure. If many proteins really need to make large conformational changes for their biological function, then it is worrisome that the structures we obtain from X-ray crystallography are static and possibly not fully functional structures. While it is possible to obtain a fair amount of dynamic information about proteins from X-ray crystallography using the Debye-Waller factors, it still is by no means the whole picture. The analogy might be to a person tied down in a chair, with a gag in the mouth. The person can struggle to get out; by looking at the little wiggles of the body as the person struggles to get free (these are the Debye-Waller factors), you might get some idea of how that person moves when free, but you will have no idea if the person is a world-class sprinter or a world-class mountain climber, quite different motions! Until fairly recently, it was very difficult to obtain 3-D structures of biomolecules in their native habitat (that is, in a solvent) at high (0.1 nm) resolution. Now, we are beginning to do this, and much more: We are beginning to chart their motions. In this chapter we try to give the reader a brief introduction to this exciting development.

29.1 Coherent and Incoherent Processes in NMR

An introduction to NMR for physicists can be found in the classic text by Slichter [1], but it is not aimed at a biological physics audience, and much of the powerful ideas in NMR structural determination are not discussed. In fact, the NMR literature is extremely difficult to read because it is at heart quantum mechanical and highly developed. Some intuition can be gotten from classical

[1] Department of Physics, Princeton University, Princeton, NJ 08544, USA.

H. Frauenfelder, *The Physics of Proteins*, Biological and Medical Physics, Biomedical Engineering, DOI 10.1007/978-1-4419-1044-8_29,

analogies, but not much. One of the more accessible books that discusses in depth biological aspects of NMR is *NMR in Biological Systems*, by Chary and Govil [2].

The fundamental idea behind NMR is very simple when viewed classically and is shown in Fig. 29.1. A nucleus with a magnetic moment $\boldsymbol{\mu}$ and intrinsic angular momentum (or spin) $\mathbf{S}$ is placed in a static magnetic field $\mathbf{B}_o$ at some angle θ to the static field direction. If we ignore the substantial effects of temperature, the magnetic moment of the nucleus will align itself along the direction of the static field, but as it does so (losing potential energy $U = \boldsymbol{\mu} \cdot \mathbf{B}_o$ in the process), the spin not only aligns itself in the field direction (a process we call longitudinal relaxation with characteristic time T_1 later in this chapter) but will precess in a plane perpendicular to the field $\mathbf{B}_o$ due to the torque that the magnetic field exerts on the magnetic moment (the Bloch equation)

$$\frac{d\mathbf{S}}{dt} = \boldsymbol{\mu} \times \mathbf{B_o}. \tag{29.1}$$

The relationship between the angular momentum (spin) of the nucleus $\mathbf{S}$ and its magnetic moment $\boldsymbol{\mu}_\mathbf{B}$, where μ_B is the Bohr magnetic moment, is given by the gyromagnetic ratio γ:

$$\boldsymbol{\mu}_\mathbf{B} = \gamma \mathbf{S}. \tag{29.2}$$

Classically γ is equal to $e/2m$. Quantum mechanically this is changed by the g factor, which, for a proton, is $g_p = 5.591895$. Thus, the Larmor precessional frequency for a proton ω_p is given by:

$$\omega_p = \frac{g_p \mu_B B_o}{\hbar} \sim 267 \times 10^6 \text{rad} \left(\sec \cdot \text{T}\right)^{-1}, \tag{29.3}$$

where T is Telsa, the unit for magnetic fields.

In conventional NMR, a small RF coil at right angles to the main field $\mathbf{B}_o$ drives the spin resonantly at the Larmor precessional frequency ω_p. At resonance the precessing spins absorb the RF and an absorbance maximum is seen. The effective gyromagnetic moment of the proton is influenced by its surroundings in many ways that are structurally and chemically significant:

1. The frequency of the Larmor precession is shifted by an amount δ called the chemical shift due to shielding of the applied magnetic field by the diamagnetic susceptibility of the surrounding chemical groups, especially aromatic groups. These shifts are small, on the order of 1–10 parts per million, but because the line widths of NMR transitions are on the order of 1 Hz at high field strengths, it is in principle possible to decipher the sequence of amino acids in short oligopeptides based on the chemical shifts alone.

2. Through-bond spin-spin interactions, denoted by the coupling strength J, propagated by through-bond scalar interactions split the transitions and offer information about the local amino acid sequence.

3. The nuclear Overhauser enhancement (NOE) is a powerful tool we will discuss briefly. In NOE, narrow-band irradiation resonantly excites one subset

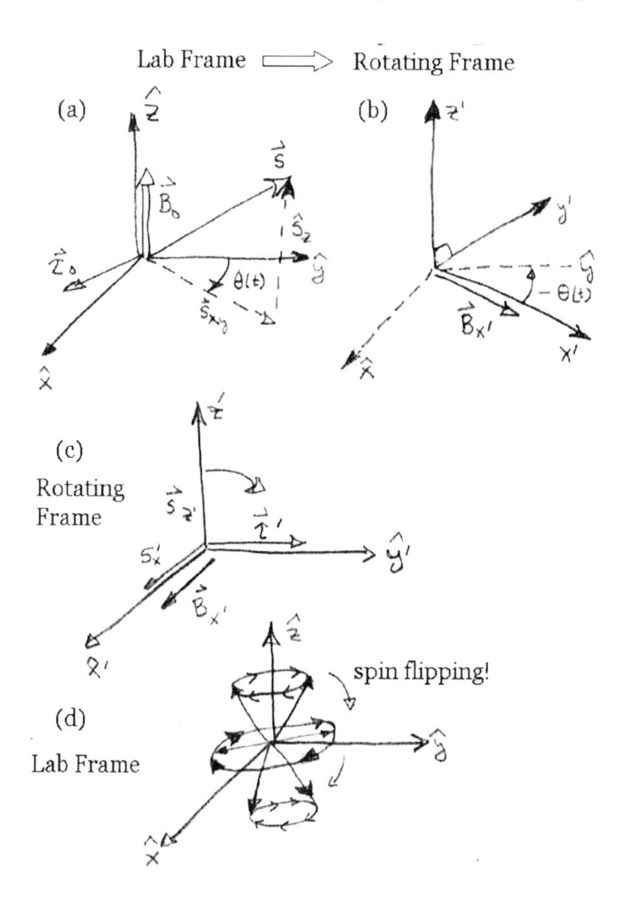

Fig. 29.1. Classical view of NMR. In (a) we show how the static magnetic field $\mathbf{B}_o$ gives rise to the Larmor precessional frequency through the torque $\boldsymbol{\tau}$. In (b) we boost to the primed frame of reference rotating around the z-axis at the Larmor precessional frequency. In (c) we show in the frame of reference rotating at the Larmor precessional frequency how the RF field $\mathbf{B}_1$ applies an additional torque. In (d) we go back to the lab frame and see the opening and closing of the spin envelope due to the field $\mathbf{B}_1$.

of the nuclei and changes the local fields; the influence of that local change can be sensed by the perturbation on surrounding nuclei.

4. The rate at which the induced magnetization changes due to dephasing of the spins (with the characterization relaxation time T_2) and relaxation along the static field direction (with the characterization relaxation time T_1) is influenced by the local surroundings of the spin, and hence is sensitive to the conformation of the protein. These relaxation processes are best understood as coherent quantum mechanical processes.

The NOE effects and the coherent spin relaxation processes are so fundamental to modern NMR structural imaging techniques that we now explain in a little detail what the basis behind these powerful ideas is.

29.2 Two-Level States, Density Matrices, and Rabi Oscillations

It is safe to say that all the recent progress in NMR imaging has come from doing NMR in the time domain through a series of pulses and reverting back to the frequency domain by the Fourier transform. If we present the basic principle of NMR in a general quantum mechanical way, it is fairly easy to see how the technique of multiple pulses can be applied to any two-level system, not just an abstract spin of 1/2 in an applied magnetic field. The quantum mechanics of the general-case coherently driven two-level systems can be found in [3]. Chapter 28 gives an account of the nuclear magnetic moment; for this simple example we consider just a spin-1/2 nucleus such as ^1H, so that in the presence of an externally applied magnetic field B_o the magnetic levels of the ^1H nucleus are split into two eigenstates $u_1(\mathbf{R})$ and $u_2(\mathbf{R})$, where the eigenstates are a function of their positions in space $\mathbf{R}$ with an energy difference ΔU between the two states. A critical concept is the generalized density matrix of the states as a function of time as perturbing time-dependent fields are applied to the system. Any general time- and space-dependent wave function $\psi_s(\mathbf{r}, t)$ of the two-level system can be expressed as a sum of the two eigenstates: In a state s, the particle will be defined by the wave function

$$\psi_s(\mathbf{r}, t) = \sum_n C_n^s(t) u_n(\mathbf{r}) \tag{29.4}$$

where the amplitudes $C_n^s(t)$ represent the time-varying occupation of the states under the influence of an external perturbation, such as an RF field in the case of NMR, which drives the occupation of the two states coherently. Perturbation connecting the ground and excited states can be written as:

$$H_{mn} = \int u_m^*(\mathbf{r}) \hat{H} u_n(\mathbf{r}) d^3r. \tag{29.5}$$

We define the matrix elements of a density matrix ρ_{nm} in terms of probability amplitudes $C_n^s(t)$ within a given state s and the probability the system is in fact in state s, $p(s)$. The overbar denotes an ensemble average

$$\rho_{nm} = \sum_s p(s) C_m^{s*} C_n^s = \overline{C_m^* C_n}. \tag{29.6}$$

To build a complete density matrix, ρ_{nm} must be calculated for every pair nm of eigenstates of the system. An n-level system will therefore be described by an n-by-n matrix. The diagonal elements of which ρ_{nn} give the probability that the system is in a state n while the off-diagonal elements ρ_{nm} for $n \neq m$ denote a coherence between eigenstates n and m. That is, the system is in a coherent superposition of two states n and m.

We won't bother to show the details here (they can be found in [3]), but we will state some important results and terminology. If an external perturbation is applied to our simple two-level system the terms of the density matrix will evolve in time. The time dependence of the matrix elements is given by:

$$\dot{\rho}_{nm} = \frac{i}{\hbar} \sum_{\nu} (\rho_{n\nu} H_{\nu.m} - H_{n\nu} \rho_{\nu m}) . \tag{29.7}$$

The on-diagonal matrix elements ρ_{11} and ρ_{22} have the simple interpretation of being simply the occupation of the ground and excited states, while the off-diagonal matrix elements represent more complex "entangled states" where the system has occupation in both the ground and excited states. One way to view the density matrix is in terms of the phase of the state made of the two eigenstates. If $\rho_{11} = 1$ and $\rho_{22} = 0$, then we say the state has a phase $\phi = 0$. If $\rho_{11} = 0$ and $\rho_{22} = 1$, we say the state has $\phi = \pi$, or 180°. If $\rho_{12} = 1$, we say that the state has $\phi = \pi/2$ and is a superposition of states u_1 and u_2. This system is said to have a phase of 90°.

Rabi oscillation [4] is the name given to what happens to the density matrix when the two-level system is driven by a coherent photon source of frequency ω_o that is resonant with the splitting, that is:

$$\hbar\omega_o = \Delta U. \tag{29.8}$$

What happens with coherent excitation and no dephasing is rather surprising. If we start with everything in the ground state ($\rho_{11} = 1$), then the occupation of the excited state (ρ_{22}) oscillates between 1 (complete population inversion!) and 0 (everything in the ground state with a frequency ω_R), the Rabi frequency [4]. In the case if NMR this frequency would be the Larmor precessional frequency ω_p, but the physics is more general than that. For example, in the case of optical transitions as discussed in Chapter 26, it would be the frequency of photons needed to create an excited electronic state. In the case of vibrational transitions in the infrared as discussed in Chapter 27, it would be the frequency of a photon needed to make a vibrational excited state. If the exciting photon field has magnetic field value B_1, then the Rabi frequency is given by:

$$\omega_R = \frac{H_{nm} B_1}{\hbar}. \tag{29.9}$$

Figure 29.2 shows how at resonance the ground- and excited-state populations oscillate at the Rabi frequency in the presence of the perturbing field. Since the difference $\Delta\rho = \rho_{22} - \rho_{11}$ oscillates as $\exp(i\omega_R t)$ the terminology for the π and $\pi/2$ pulses should be clear from Fig. 29.2. The area A_R of a pulse that is resonant with a two-level system is simply the integral of $\int \omega_R dt$, where we now assume that the magnetic field B_1 varies with time during the pulse.

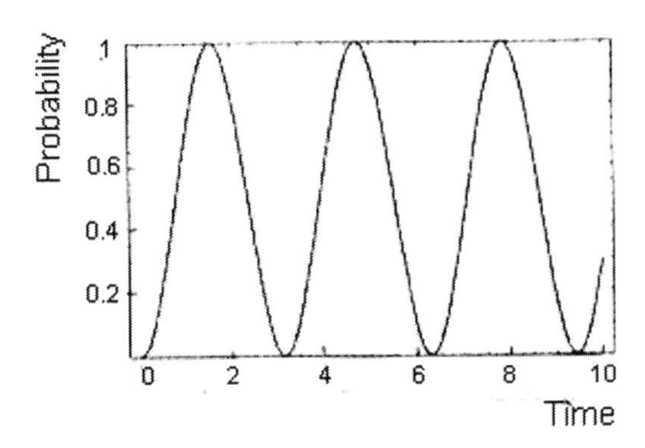

Fig. 29.2. Inversion of the ground and excited states by resonant excitation at the Rabi frequency.

Until now we have viewed this two-level system as being completely coherent: In the absence of external perturbations a complex state with density matrix ρ_{nm} remains entangled or coherent forever, the phase of the state cannot be lost. Likewise, assume that if we put the system into the place where $\rho_{22} = 1$, it will stay there forever in the absence of external perturbations as a totally inverted population. However, in the real world, there are "soft" interactions that, although they do not remove energy from the two states, scramble the phase relationship between u_1 and u_2. Also there are "hard" interactions that remove energy from the system and relax it the ground state so that the population inversion decays. The loss of phase coherence is called transverse relaxation in NMR terminology, and the loss of population inversion is called longitudinal relaxation. The time it takes a system to dephase (but not lose energy!) is called T_2, and the time it takes the system to lose energy is called T_1. The equation governing the density matrix evolution versus time now becomes:

$$\dot{\rho}_{21} = -\left(i\omega_{21}\rho_{21} + \frac{1}{T_2}\right)\rho_{21} + \frac{i}{\hbar}H_{21}(\rho_{22} - \rho_{11}) \tag{29.10}$$

$$\dot{\rho}_{12} = \left(i\omega_{21}\rho_{21} - \frac{1}{T_2}\right)\rho_{21} + \frac{i}{\hbar}H_{21}(\rho_{22} - \rho_{11}) \tag{29.11}$$

$$\dot{\rho}_{22} = \frac{-\rho_{22}}{T_1} - \frac{i}{\hbar}(H_{21}\rho_{12} - \rho_{21}H_{12}) \tag{29.12}$$

$$\dot{\rho}_{11} = \frac{\rho_{22}}{T_1} + \frac{i}{\hbar}(H_{21}\rho_{12} - \rho_{21}H_{12}). \tag{29.13}$$

These are complex equations; if they are written in terms of magnetization rather than the more generalized density matrices, they are called the Bloch equations, but the important thing to remember is that T_1, the longitudinal relaxation time, connects ρ_{22} and ρ_{11} through energy-losing collisions, while T_2 is the dephasing time due to soft elastic collisions and is energy-conserving. An important note: In an NMR experiment, the line width $\delta\nu$ of the individual nuclear resonances is given by:

$$\delta\nu = \left[\frac{1}{T_1} + \frac{1}{2T_2}\right] \tag{29.14}$$

although in practice $T_1 >> T_2$. It is surprising that although T_2 does **not** represent loss of energy but rather loss of quantum mechanical phase coherence, it still greatly influences the line width.

29.3 Fourier Transform Free Induction Decay (FT-FID) and Spin Echoes

The full power of the technology we have developed here, namely, the application of $\pi/2$ or π pulses to a two-level system, comes from analysis of the system response to a pulse of radiation of known area. An excellent introduction to this technique in biological physics can be found in Wüthrich's book [5]. The most basic experiment is to apply an RF pulse of area $\pi/2$ pulse with magnetic field B_1 along an axis (typically the **x**-axis) that is perpendicular to the static field axis (typically the **z**-axis). The result of this $\pi/2$ pulse, as we show in Fig. 29.3 is to rotate the magnetic polarization of the sample from the initial **z**-axis to the **y** as if viewing a frame of reference that is rotating at the Larmor precessional frequency. This rotated spin magnetization then relaxes back at rate $1/T_1$ to the initial position.

The rotated magnetic polarization is detected in a coil oriented at right angles to both the static field and the RF field, and it detects the rotated magnetization as an RF signal oscillating at the Larmor frequency. If the RF pulse is not a pure sine wave of frequency ω_o but rather is a broad band of frequencies sufficient in bandwidth to span the range of chemical shifts δ the nuclei experience in the sample, then the reradiated RF energy caused by the Larmor precession of the spins contains a coherent sum of all the radiating dipoles. This coherent radiation can be detected, and by Fourier analysis discussed in previous chapters, the individual emission lines can be determined in frequency space.

Note here that we have assumed that T_2, the spin dephasing time, is infinite, and that the magnetic field is perfectly homogenous! In fact this is never true, if it were true NMR would not be very interesting. There exist exceedingly powerful techniques using multiple pulses of differing areas to refocus the spin polarization due to dephasing processes. The basic technique is called spin echo spectroscopy. We hasten to point out here that this is a very complex subject and a major industry in NMR technology; our aim here is only to

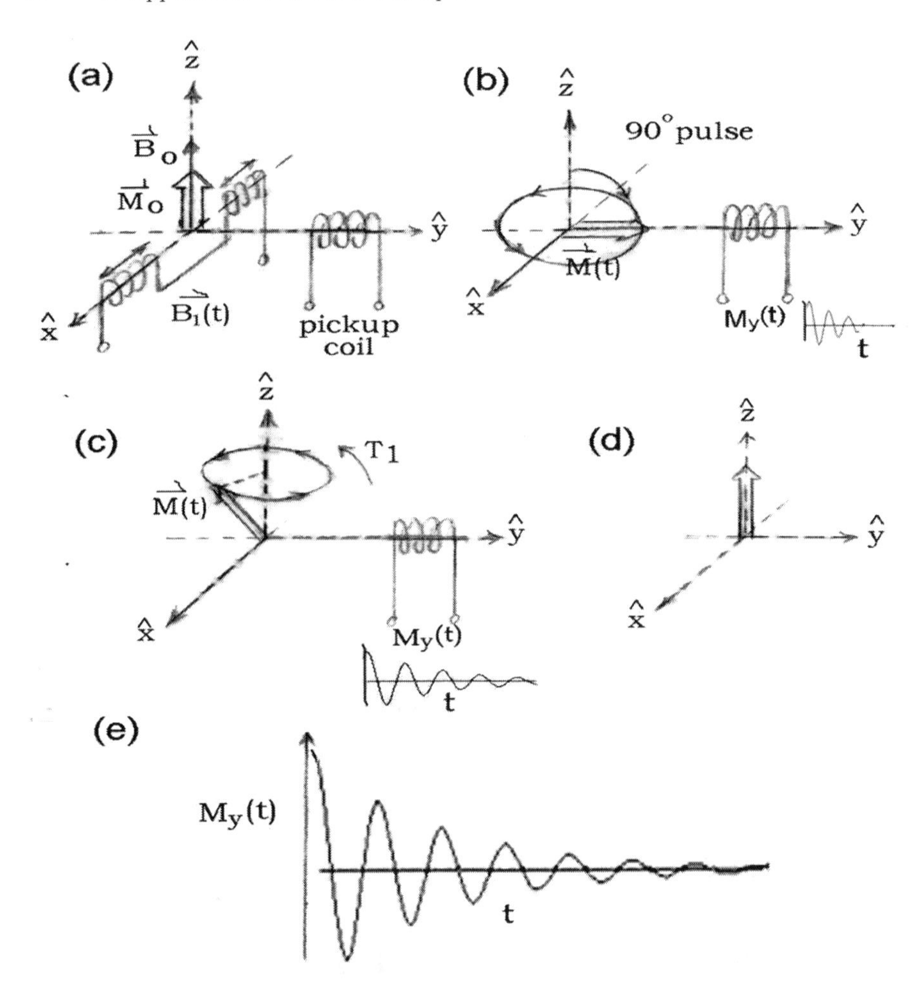

Fig. 29.3. Origins of the free induction decay: (a) The static field $\mathbf{B}_o$ creates a static polarization $\mathbf{M}_o$. A coil along the **x**-axis creates an oscillating field $\mathbf{B}_1$ at the Larmor frequency ω_p, while a pickup coil along the **y**-axis detects any time-dependent magnetization along the **y**-axis. (b) The $\mathbf{B}_1$ pulse first rotates the magnetization $\mathbf{M}_o$ by $\pi/2$ radians. The $\mathbf{M}_y$ polarization rotates at the Larmor frequency ω_p in the $\mathbf{x} - \mathbf{y}$ plane. The radiated Larmor signal by the precessing spins is detected by the pick-up coil. (c) Due to energy removing effects, the magnetization **M** begins to relax back to its equilibrium position along the **z**-axis. The signal in the pickup coil decays. (d) The magnetization has relaxed back to its equilibrium position $\mathbf{M}_o$ along the **z**-axis. (e) The free induction decay envelope at a $\pi/2$ pulse.

briefly sketch the basic idea. Once you understand that the effect of T_2 is to dephase the spins of a system without changing the occupation of the ground and excited states, and that π pulses invert the polarization of a system the

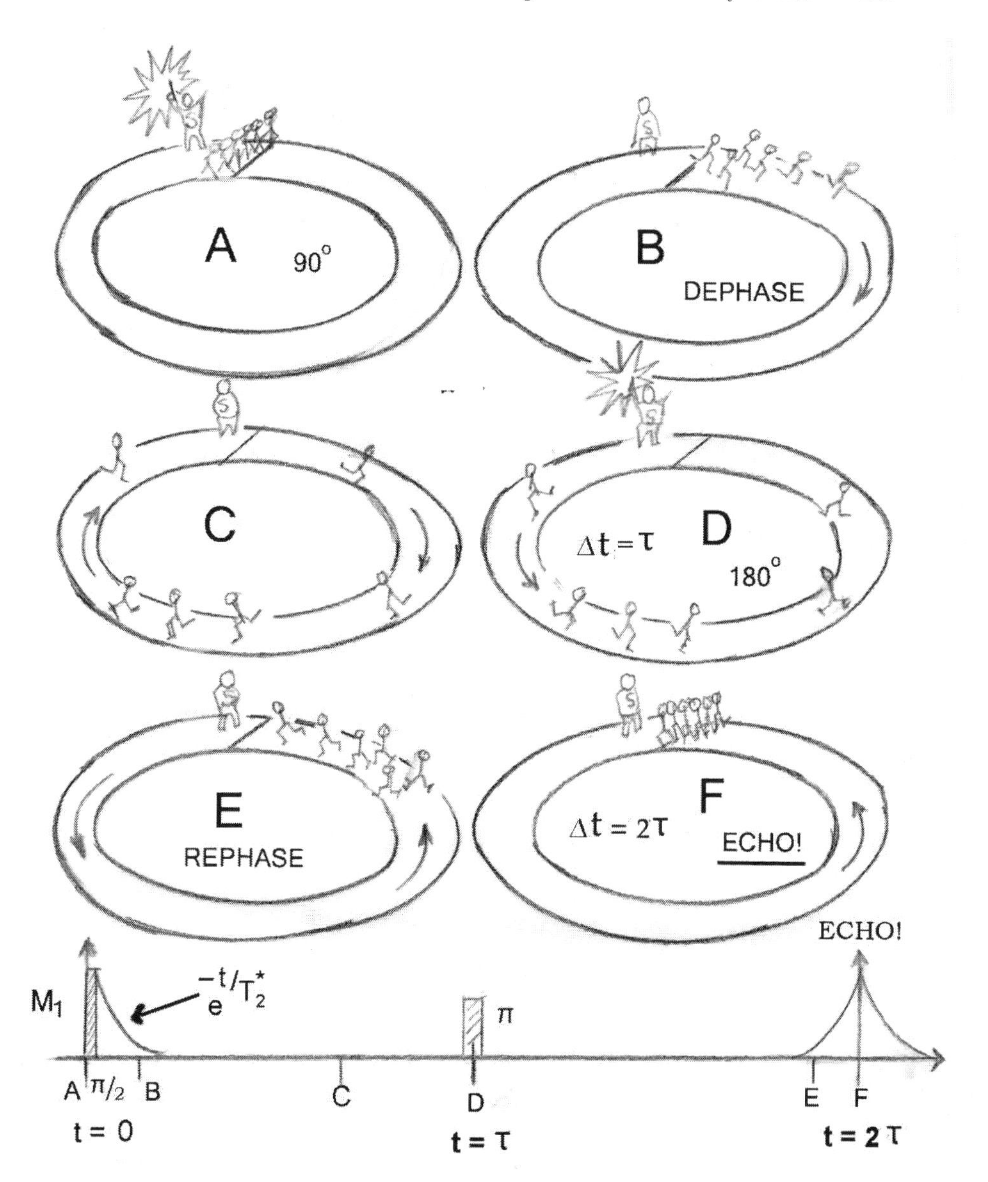

Fig. 29.4. This cover illustration of the November 1953 *Physics Today* shows spin echos in terms of runners. The schematic at the bottom shows the magnetic polarization versus time for a $\pi/2$ pulse followed by a π pulse.

spin echo concept becomes easy to grasp. There was a famous illustration of the spin echo concept by its inventor, Erwin Hahn [6], in *Physics Today* [7] that wonderfully illustrates how spin echos can occur. Figure 29.4 reproduces that illustration.

In the spin echo, two pulses of resonant frequency excite the sample with a delay τ between them. The first pulse has an area $A_R = \pi/2$ and can be viewed as taking the population ρ_{11} and rotating the polarization so that now the polarization is perpendicular to the applied magnetic field direction $\mathbf{B_o}$ as in FID. As before, there is an FID signal. However, due to T_2 the spins dephase and the FID signal disappears as coherence is lost. However, if the dephasing is not due to incoherent processes, but rather is due to differences in the Larmor frequencies of the participating spins, then the dephasing is not irreversible but can instead be reversed by effectively running time backward. Although Superman was able to reverse time by reversing the spin of the earth, an easier way to do this in the lab is by irradiating the system with a π pulse. The reflection of the spins in the y–z plane due to the π pulse effectively reverses the evolution of the dephasing bundle of spins, and at a time τ later then the π pulse, and 2τ from the initial $\pi/2$ pulse the spins refocus and a spin echo appears. The amplitude of this echo is dependent on how much true irreversible spin dephasing actually occurred over the time interval 2τ; this dephasing is a function of the spin neighbors of the different sites.

We now see that depending on the pulse sequence you can probe the coherent dephasing times (T_2) or the longitudinal relaxation times (T_1). Dephasing times can be measured by a $\pi/2$ pulse followed by a π pulse, and measuring the spin echo that is a coherent response of the spin ensemble. In the parlance of NMR spectroscopy, the $\pi/2$ pulse prepares the system by rotating the equilibrium magnetization into the $x - y$ plane, while the π pulse mixes the system. The time τ_1 between the preparation pulse and the mixing pulse is called the evolution period, and the time τ_2 after the mixing pulse is called the detection period. The spin echo is a transverse, coherent process and a function of both T_1 and T_2. Note that in Fig. 29.4 we used the terminology T_2^* instead of T_2 for the transverse relaxation time. T_2 is the transverse magnetization dephasing time which WOULD be measured if the applied magnetic fields were completely homogenous. T_2^* is the MEASURED apparent dephasing time due to applied magnetic field inhomogeneities. $T_2^* < T_2$ always. T_2^* is not a true dephasing relaxation in that the dephasing due to magnetic field inhomogeneties can be reversed with a π pulse as long as there must insignificant diffusion during the waiting time τ, while T_2 is a true irreversible loss of coherence and cannot be recovered by a π pulse.

Longitudinal relaxation processes are **not** coherent processes but instead represent the relaxation of the spin ensemble back to thermal equilibrium due to dissipative, irreversible processes that take energy out of the system. They can be measured by a π pulse that flips the magnetization to be opposite to the steady-state thermal magnetization along the z-axis, followed by a $\pi/2$ pulse which puts the magnetization in the $x - y$ plane. The amplitude of the free induction decay observed after the $\pi/2$ pulse is measured as a function of τ_1, the evolution time between the preparation pulse and the mixing pulse.

29.4 Spin-Spin Interactions and Protein Structure Determination

We briefly discuss how spin-spin interactions can be used to determine the 3-D structure of complex biomolecules such as proteins in solution, without the need to crystallize molecules. We have described the coherent rephasing of the ensemble of spins to form the spin echo, which is a cute quantum mechanical trick but would seem to be of little use. However, in 1951 Hahn and Maxwell [8] discovered something very strange in a spin echo experiment carried out on hydrogen atoms (protons) in dichloroacetaldehyde, $CHCl_2CHO$. Clearly, the two protons in dichloroacetaldehyde see slightly different chemical environments, which lead to chemical shifts $\delta\omega_p$ in their Larmor precessional frequencies ω_p on the order of about 1.5 $\times 10^{-6} \times \omega_p$. This means that in the spin echo the slightly different ω_p's of the two inequivalent hydrogens will combine coherently and "beat" against each other at a beat frequency $2\delta\omega_L$, which of course is proportional to the static magnetic field. No mystery there. However, when the experiment was actually performed, as shown in Fig. 29.5, something remarkable was observed: In addition to the relatively rapid beat frequency $2\delta\omega_p$ due to the differing chemical shifts, a much slower (0.7 Hz) and external magnetic field–**independent** beat frequency ω_J occurred!

It was quickly realized that this external magnetic field B_o independent term must be due to spin-spin interactions, and that the frequency ω_J was connected to how strongly the spins interacted with each other. Note that this was a coherent effect since it was seen through a beating signal. Of course it had been expected that spins would interact with each other through some sort of a dipole-dipole coupling H_{JJ}, but it had been thought this term would average to zero in solution. The surprise that $\omega_J \neq 0$ is related to the surprise in X-ray diffraction that thermal motions did **not** broaden the sharpness of the X-ray interference terms (which would have made X-ray diffraction useless for large-scale protein structures), but rather simply decreased the intensity of the spot brightness through the Debye-Waller factor $W(\theta)$, as we discussed in Chapter 11. In the case of NMR of molecules, a similar situation occurs: The spins are correlated though the bonds hold a molecule together and they do not locally motionally average to zero.

Spin-spin interactions can be coherent or incoherent processes; in Fig. 29.5 we see a coherent beating of the spins. If the spin-spin interaction preserves the coherence of the preparation pulse RF the spin polarization dynamics are described by the off-diagonal density matrix elements ρ_{mn} discussed earlier. This kind of coherent interaction is called a scalar interaction in the NMR literature, and the Hamiltonian that describes it can be written in the form:

$$H_{ij} = \mathbf{S}_i \mathbf{J}_{ij} \mathbf{S}_j. \tag{29.15}$$

The coupling tensor $\mathbf{J}_{ij}$ is independent of B_o, as seen in the original discovery by Maxwell and Hahn, and is due to coupling of the spin polarization through

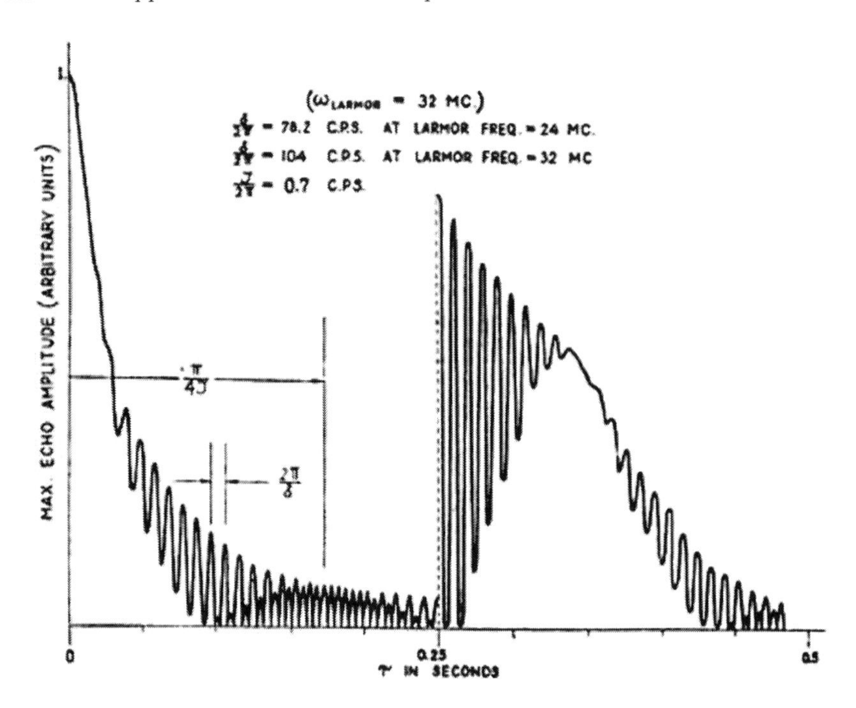

Fig. 29.5. Echo envelope plot for protons in dichloroacetaldehyde. The break in the plot at $T = 0.25$ sec indicates continuation of the plot in the region of small echo amplitude, but multiplied by a factor in order to make the plot readable. MC = megacycles, called MHz today. Taken from [8].

the electronic orbitals. It is a function of the number of bonds, the nature of the chemical bonds, and through the Karplus relation [9], which gives the correlation between J-coupling constants and dihedral torsion angles ϕ:

$$J \sim A\cos^2\phi + B\cos\phi + C. \tag{29.16}$$

Coherent correlation spectroscopy can be done using a variety of pulse sequences; in the case of Fig. 29.5 the sequence was a $\pi/2$ followed by a π pulse. The evolution times t_1 and the observing times t_2 can be converted into frequencies ω_1 and ω_2; the dimensionality of the technique used is given by the number of pulses and hence the number of different frequencies that characterize the times between the pulses. Two-dimentional COSY imaging means that coherent correlations have been made using a two pulse sequence.

The coherent, field-independent scalar interactions give information about the topology of the neighboring amino acids in a protein but not the distances between them. The nuclear overhauser enhancement effect is an incoherent process, but it gives distance information between spins fairly directly [10]. The NOE effect is due to the nuclear magnetic dipole moment μ_i of an atom interacting through space with another magnetic dipole moment μ_j. The field of the ith dipole adds to the local field that the jth nucleus experiences and changes (enhances) the longitudinal relaxation rate, or at least that is a grossly simplified way to view what happens. Since the magnetic field of a dipole falls off as $1/r^3$, the NOE coupling strength D_{ij} through space falls off as $1/r^6$ and thus provides critical distance information.

We mentioned in the discussion about free induction decay that if the $\pi/2$ pulse was sufficiently broad in bandwidth, all the nuclei with various chemical shifts could be excited. In NOE, a different tack is used. An initial pulse of *narrow bandwidth* is used, so that only a narrow subset of the chemically shifted nuclei are excited by the pulse. If, for example, the net area of this narrow-frequency-band pulse is a $\pi/2$ pulse, the local magnetization of that subset if spins is rotated out of the $\mathbf{B_o}$ direction, an evolution period is then used for the spins to interact with each other changing the chemical shift $\Delta\delta$ that those spins exert on their neighboring spins that are not in resonance. This prepulse (or magnetization conditioning pulse) is then followed, in one of many possible scenarios of pulse sequences, by a broad-band $\pi/2$ pulse, and the shift in the chemical shifts for all the resonances is determined. This gives rise to a row of shifted frequencies that are off-diagonal to the diagonal group that is specifically irradiated. While one could do a quasi-2D spectroscopy by moving the narrow selective $\pi/2$ pulse through the 1-D spectrum of the protein, a more powerful way to do this is to make all the pulses nonselective (broad band) and use Fourier transform techniques to create a true 2-D map of off-axis magnetization interactions. This 2-D spectroscopy is called NOESY (pronounced nosey), and stands for nuclear overhauser enhancement and exchange spectroscopy.

Figure 29.6, taken from Hahn's excellent Nobel Prize lecture in 1992, shows how COSY and NOSEY techniques complement each other in determining structures of molecules in solution using coherent and incoherent methods, while Fig. 29.7, also taken from Hahn's 1992 Nobel Prize lecture, shows a structure for the molecule antamanide determined by COSY and NOESY.

We are just describing the tip of the tip of the iceberg here, this is a complex, but important field. Because of the deep complexity, NMR experts are to be avoided at parties lest they tell you of their favorite pulse sequence, although avoidance of NMR people at least at the graduate student level is easy because they cluster in the kitchen to be near food and avoid social contact. Two-dimensional NMR techniques are extremely advanced, and there are many different variants on the theme of pulse sequences. In practice, in order to construct a map of all the connectivities between all the amino acids even in a small protein many different pulse sequences must be used. Even high-order cross-coupling effects and multiple quantum transitions can be used to attack the problem of the structures of large biomolecules in solution [11]. Very recently, this technique has been used to find the structure of a protein inside a living cell, a truly astonishing achievement [12].

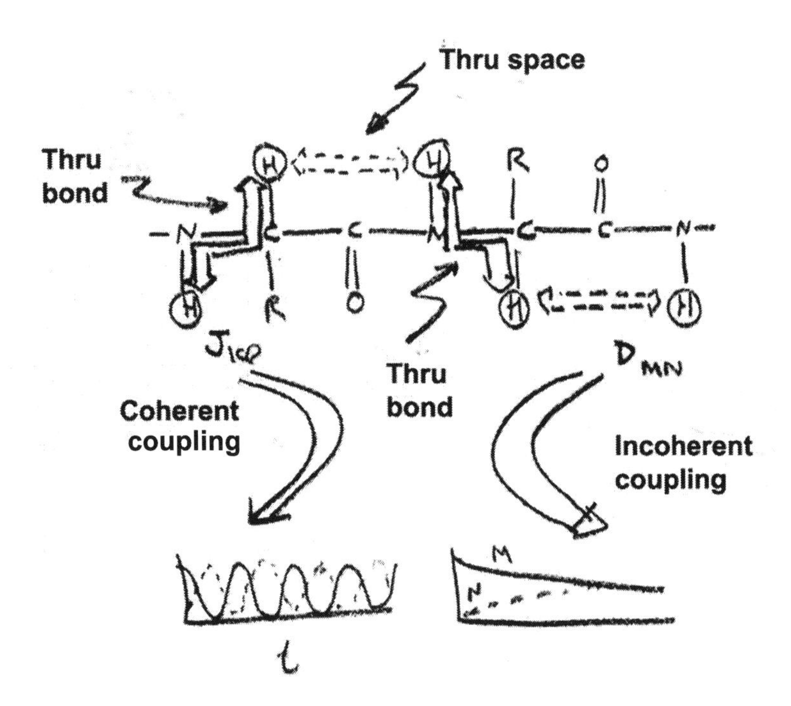

Fig. 29.6. The two-spin pair interactions relevant to structural determination in NMR. The through-bond scalar interaction J_{kl} (dashed lines) gives rise to coherent transfer of spin population (COSY), while the through-space dipole-dipole interaction D_{mn} (dotted lines) causes cross-relaxation between spins S_m and S_n (NOSEY).

29.5 Probing the Protein Conformational Landscape with NMR

Of course, this book is mostly about the dynamics of proteins, in particular conformational dynamics, not just the time-averaged structure of proteins. We have consistently discussed the interplay between structure, conformation dynamics, and function. What does NMR tell us about the conformational structural dynamics of proteins in solution and the shape of the energy landscape? A lot, it turns out, although at present the field is still very much in its infancy because of the difficulties involved in interpreting the complex coherent and incoherent signals obtained using N-dimensional NMR techniques. Two recent reviews [13, 14] offer a glimpse of what is happening.

The physics behind the conformational dynamics sensitivity of NMR using the N-dimensional coherent and incoherent techniques we have discussed should be pretty clear at this point. The time scales of conformation accessible to NMR are enormous, from seconds (simply monitoring changes in cross correlations after a perturbation) to picoseconds (T_1 relaxation time dynamics). However, if all one wants to do is measure the "dynamics of proteins"

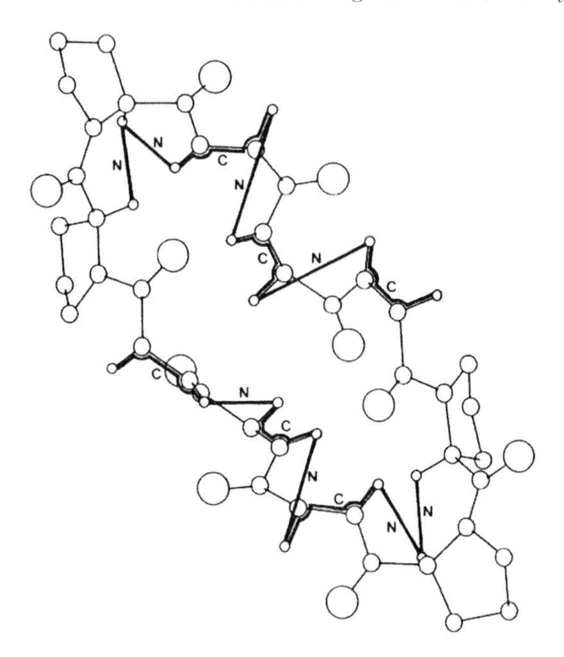

Fig. 29.7. Assignment of backbone protons in antamanide by the combination of COSY (C) and NOESY (N) cross peaks.

then the quest is open-ended and not that interesting in terms of biology: The question is, can NMR access the free-energy landscape we have talked about, and can it connect between the free-energy landscape and protein function in a systematic way with crisp questions and answers?

The key concept to a "free-energy landscape" is the existence of a distribution of conformations of the protein that are quite distinct from one another, speciated if you will, as spin glass physics predicts as we discussed in Chapter 3. These conformations, if they are biologically relevant, must be thermally accessible as was discussed in Chapter 11. Since they are thermally accessible, the lines of the NMR resonances will be affected by the changes in the chemical sifts of the resonances in the various conformational states. Note that in X-ray crystallography the thermal motions are incoherent and average to zero, decreasing the intensity of the diffraction spot (the Debye-Waller factor) but not broadening it. Conformational substates, which give rise to different positions of atoms in the lattice will broaden the diffraction peaks, but the very act of crystallization selects out certain conformations and the full conformational distribution is missing. This is one of the weaknesses of X-ray crystallography.

But in NMR, the protein is in solution and the full range of conformational substates should be accessible and observable as a function of temperature. Perhaps the recent work that most cleanly grasps this fundamental fact is the work of Dorthee Kern [15], which has directly addressed the free-energy landscape issue. The NMR spectrum of a molecule that has multiple accessible and discrete conformational states that give rise to discrete chemical shifts is a function of the rate at which these conformational changes occur. Figure 29.8, taken from [16], shows how even another conformation that has a small Boltzmann probability of being occupied can have a significant effect on the line shape of a transition depending of the ratio of the conformation exchange rate k_{ex} and the chemical splitting of the two conformational states A and B, $\omega_A - \omega_B$. Clearly, this technique is sensitive to conformation exchange rates in the hundreds-of-microseconds to 10-ms time range, since this is the approximate range of the chemical splittings $\Delta\omega$ when expressed in period units.

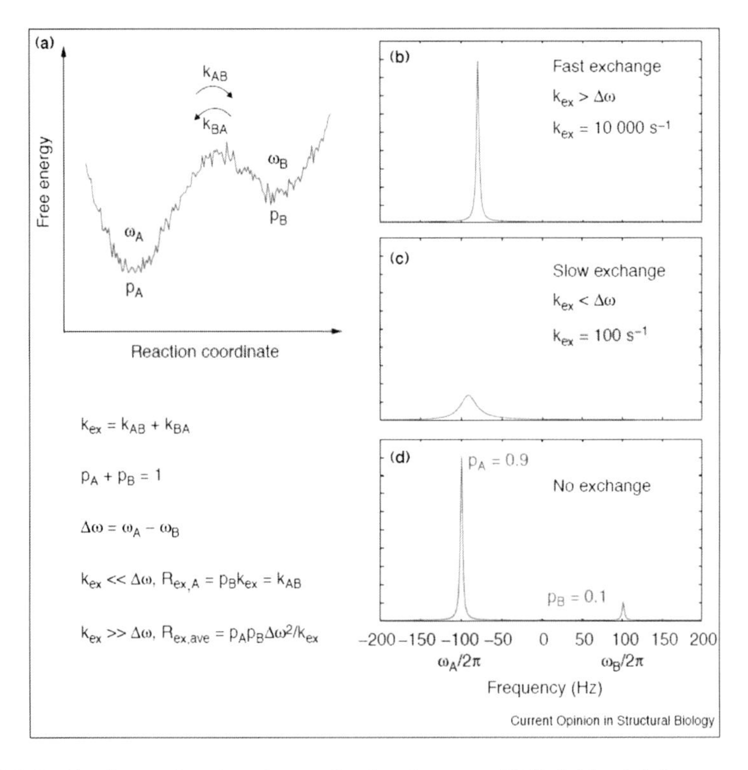

Fig. 29.8. Conformational exchange in the absence of RF fields: (a) Representation of an exchange process between two states with relative populations p_A and p_B, and resonance frequencies ω_A and ω_B. Depending on the relative magnitudes of the exchange rate (k_{ex}) and the chemical shift difference ($\Delta\omega$), the rate is denoted fast ($k_{ex} > \Delta\omega$), intermediate ($k_{ex} \sim \Delta\omega$), or slow ($k_{ex} < \Delta\omega$). The exchange contributes to the resonance line widths, which are proportional to the transverse relaxation rates. Taken from [16].

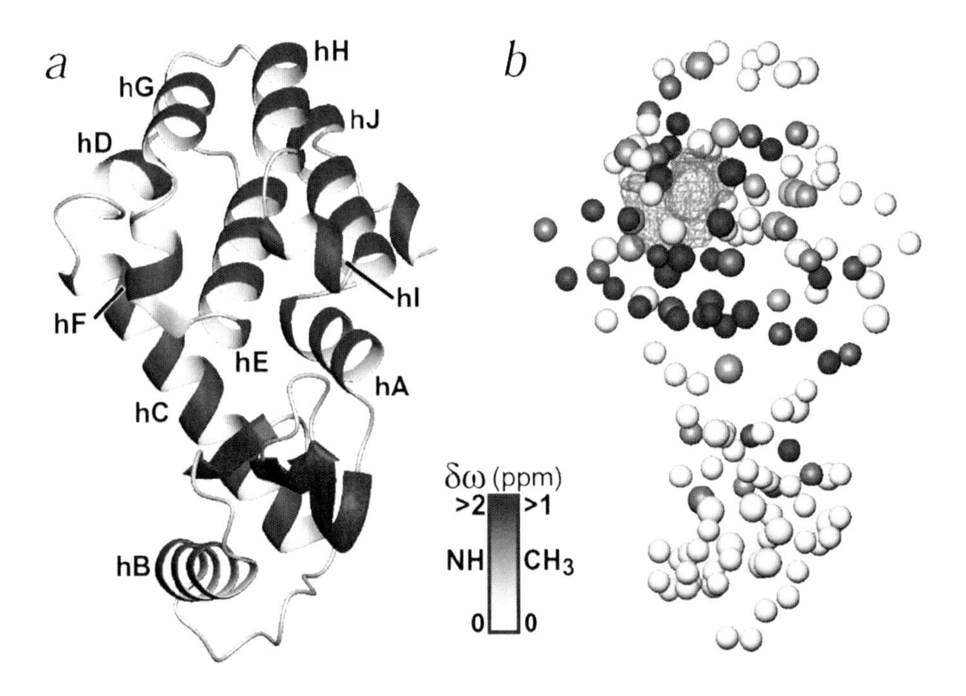

Fig. 29.9. (a) Structure of the T4 lysozyme L99A variant. (b) Identical view, indicating the positions of methyl groups (larger spheres) and backbone amides (smaller spheres) included in the 25°C analysis. The small spheres are color-coded according to values of $\Delta\omega$, with white corresponding to no change in chemical shift and dark blue corresponding to larger chemical shift differences between ground and excited states (>2 ppm for ^{15}N and > 1 ppm for ^{13}C). Large values of $\Delta\omega$ highlight significant structural changes between the G and E states. Taken from [17].

We won't go into the details, which are quite technical, of how it is possible to use longitudinal relaxation rate dispersion to measure exchange broadening and hence determine conformational exchange rates. We will instead close this chapter with results obtained for the enzyme T4 lysozme [17]. Analysis of the longitudinal relaxation rate dispersion at an atomic level enabled Mulder et al. to determine the chemical splitting between two different conformations of the enzyme, at the millisecond time scale. Figure 29.9 shows the result of the conformational changes in T4 lysozyme: changes in the chemical shifts $\Delta\omega$. It shows where they are, but it doesn't show what the actual physical changes are in the position of the atoms, nor how this changes the chemical reaction rate. We have a long way to go.

References

1. C. P. Slichter, editor. *Principles of Magnetic Resonance*, 3rd edition. Springer Series in Solid-State Sciences. Springer Verlag, New York, 1992.
2. K. V. R. Chary and G. Govil, editors. *NMR in Biological Systems*. Springer Verlag, New York, 2008.
3. P. W. Milonni and J. H. Eberly, editors. *Lasers*. John Wiley & Sons, New York, 1988.
4. I. I. Rabi. Space quantization in a gyrating magnetic field. *Phys. Rev.*, 51:652–4, 1937.
5. K. Wüthrich. *NMR of Proteins and Nucleic Acids*. John Wiley & Sons, New York, 1986.
6. E. L. Hahn. Spin echos. *Phys. Rev.*, 80:580–94, 1950.
7. E. L. Hahn. Free nuclear induction. *Phys. Today*, 6(11):4–9, 1953.
8. E. L. Hahn and D. E. Maxwell. Chemical shift and field independent frequency modulation of the spin echo envelope. *Phys. Rev.*, 84(6):1246–7, 1951.
9. M. Karplus. Vicinal proton coupling in nuclear magnetic resonance. *J. Amer. Chem. Soc*, 85(18):2870, 1963.
10. A. W. Overhauser. Polarization of nuclei in metals. *Phys. Rev.*, 92(2):411–5, 1953.
11. G. M. Clore and A. M. Gronenborn. Two–, three–, and four–dimensional NMR methods for obtaining larger and more precise three–dimensional structures of proteins in solution. *Ann. Rev. Biophys. Biophy. Chem.*, 20:29–63, 1991.
12. D. Sakakibara, A. Sasaki, T. Ikeya, J. Hamatsu, T. Hanashima, M. Mishima, M. Yoshimasu, N. Hayashi, T. Mikawa, M. Walchli, B. O. Smith, M. Shirakawa, P. Guntert, and Y. Ito. Protein structure determination in living cells by in-cell NMR spectroscopy. *Nature*, 458(7234):102–10, 2009.
13. R. Ishima and D. A. Torchia. Protein dynamics from nmr. *Nat. Struct. Bio.*, 7(9):740–3, 2000.
14. A. Mittermaier and L. E. Kay. Review–new tools provide new insights in NMR studies of protein dynamics. *Sci.*, 312(5771):224–8, 2006.
15. E. Z. Eisenmesser, O. Millet, W. Labeikovsky, D. M. Korzhnev, M. Wolf-Watz, D. A. Bosco, J. J. Skalicky, L. E. Kay, and D. Kern. Intrinsic dynamics of an enzyme underlies catalysis. *Nature*, 438(7064):117–21, 2005.
16. M. Akke. NMR methods for characterizing microsecond to millisecond dynamics in recognition and catalysis. *Current Opinion in Structural Biology*, 12(5):642–7, 2002.

17. F. A. A. Mulder, A. Mittermaier, B. Hon, F. W. Dahlquist, and L. E. Kay. Studying excited states of proteins by NMR spectroscopy. *Nat. Struct. Bio.*, 8(11):932–5, 2001.

30

Neutron Diffraction

X-rays are scattered by atomic electrons, neutrons by atomic nuclei. The two techniques are not in competition; they are complementary. Because neutron sources are many orders of magnitude weaker than X-ray sources, it is likely that neutron scattering will not be used for routine structure determination but will be important for the determination of the position of the crucial hydrogen atoms. The following discussion is brief; details can be found in a number of books, reviews, and papers [1]–[5].

30.1 Neutron Scattering

The geometrical aspects of X-ray and neutron diffraction are the same, but the interactions between the projectile and the scatterer are different. Photons interact predominantly with the atomic electrons, and nuclear scattering can be neglected in diffraction work. Neutrons interact with the nucleus via the nuclear force and with the magnetic moment of unpaired electrons via the electromagnetic one. The magnetic interaction permits investigation of magnetic materials; we will not consider it here and restrict the treatment to the nuclear force case. A complete theory, starting from the properties of the nucleon-nucleon force, is too complicated. Following Fermi, a *pseudo-potential*, $V(\mathbf{r})$, is introduced,

$$V(\mathbf{r}) = \frac{2\pi\hbar^2}{m}\, b\, \delta(\mathbf{r}) \,. \tag{30.1}$$

Here, $\delta(\mathbf{r})$ is the Dirac delta function and b the *scattering length*, which has the dimension of length. If a potential is attractive enough to produce a bound state, b is positive. If the potential is only weakly attractive, b is negative. If the potential is repulsive, b is again positive.

Equation (30.1) applies to the scattering of slow neutrons from a single nucleus. The cross section for the scattering can be obtained by inserting $V(\mathbf{r})$ into the Born approximation with the result

H. Frauenfelder, *The Physics of Proteins*, Biological and Medical Physics, Biomedical Engineering, DOI 10.1007/978-1-4419-1044-8_30,

$$f = -b$$

and

$$\frac{d\sigma}{d\Omega} = |b|^2. \tag{30.2}$$

The scattering from a crystal with one atom per unit cell is described by a slight generalization of Eq. (30.1). The pseudo-potential becomes the sum of the individual contributions from the nuclei at the positions $\mathbf{r_i}$. The computation gives the scattering amplitude from one nucleus as (cf. Eq. (25.12))

$$b\, e^{i\mathbf{q}\cdot\mathbf{r}/\hbar}.$$

Comparison with Chapter 25 shows that neutron and X-ray diffraction are similar, the atomic form factor f_i is replaced by the scattering length b for neutrons:

$$f(\mathrm{X-rays}) \leftrightarrow b(\mathrm{neutrons}). \tag{30.3}$$

If there are a number of nuclei in each unit cell, the scattering amplitude becomes

$$F(\mathbf{q}) = \sum_s b_s\, e^{i\mathbf{q}\cdot\mathbf{r_s}/\hbar}. \tag{30.4}$$

Following the arguments given in Chapter 25 yields the final result

$$\frac{d\sigma}{d\Omega} = \left| \sum_r e^{i\mathbf{q}\cdot\mathbf{r}/\hbar} \right|^2 |F(\mathbf{q})|^2 + \left(\frac{d\sigma}{dr}\right)^{incoh}. \tag{30.5}$$

The first term in Eq. (30.5) describes coherent scattering, in which the contributions from the various nuclei interfere. This term yields the important information. The second term is isotropic and describes incoherent scattering; it produces the undesirable background.

30.2 Comparison: Neutrons *versus* X-Rays

The strengths and weaknesses of X-ray and neutron diffraction become obvious when neutron and X-ray scattering amplitudes are compared. Table 30.1 shows that the variation of the coherent cross section or the scattering length for neutrons with Z is not monotonic and is smaller than the corresponding variation for f. It is therefore possible to see hydrogen or deuterium with neutrons. Hydrogen has the advantage that its scattering length is negative; it gives more contrast to other atoms. It has the disadvantage that the incoherent cross section, which gives rise to an undesirable background, is very large. Depending on the problem, it is therefore often convenient to deuterate a crystal.

Table 30.1. Scattering amplitudes.

	X-Rays	Neutrons		
	$F(\mathbf{q}=0) \equiv Z$	b (fm)	σ^{coh} (barns)	σ^{tot} (barns)
Hydrogen	1	-3.8	1.8	82
Deuterium	1	6.5	5.4	7.6
Carbon	6	6.6	5.5	5.5
Nitrogen	7	9.4	11	11.4
Oxygen	8	5.8	4.2	4.24
Sulphur	16	3.1	1.2	1.2
Iron	26	9.6	11.4	11.8

30.3 Experimental

Two types of neutron sources are used for neutron diffraction studies, reactors and particle accelerators. Reactors produce fission neutrons, which are moderated ("cooled down") to be useful for the scattering studies. At spallation sources, neutrons are produced by bombarding a metal target with pulses of high-energy proteins [6]. Because neutron sources are much weaker than X-ray sources, the Laue technique is often used. Moreover, since the phase problem is the same for neutrons and X-rays, the initial phases for the refinement are taken from X-ray diffraction. Despite the "weakness" of the neutron sources, neutron diffraction has led to impressive results, as is evident from selected papers [7]–[9].

References

1. G. E. Bacon. *Neutron Diffraction*, 3rd edition. Clarendon, Oxford, 1975.
2. A. A. Kossiakoff. Neutron protein crystallography. *Ann. Rev. Biophys. Bioeng.*, 12:159–82, 1983.
3. B. P. Schoenborn and R. B. Knott, editors. *Neutrons in Biology.* Plenum Press, New York, 1996.
4. H. B. Stuhrmann. Unique aspects of neutron scattering for the study of biological systems. *Rep. Prog. Phys.*, 67:1073–1115, 2004.
5. B. P. Schoenborn and R. Knott. Neutron sources. In M. G. Rossmann and E. Arnold, editors, *International Tables for Crystallography, Vol. F, Chapter 6.2.* Kluwer Acad. Publishers, Dordrecht, 2001, pp. 133–142.
6. P. Langan, G. Greene, and B. P. Schoenborn. Protein crystallography with spallation neutrons: The user facility at Los Alamos Neutron Science Center. *J. Appl. Crystallography*, 37:24–31, 2004.
7. F. Shu, V. Ramakrishnan, and B. P. Schoenborn. Enhanced visibility of hydrogen atoms by neutron crystallography on fully deuterated myoglobin. *Proc. Natl. Acad. Sci. USA*, 97:3827–77, 2000.
8. T. Chatake, A. Ostermann, K. Kurihara, F. G. Parak, and N. Niimura. Hydration in proteins observed by high-resolution neutron crystallography. *Proteins: Structure, Function, and Bioinformatics*, 50:516–23, 2003.
9. M. P. Blakeley, A. J. Kalb (Gilboa), J. R. Helliwell, and D. A. A. Myles. The 15–K neutron structure of saccaride-free concavalin *a*. *Proc. Natl. Acad. Sci. USA*, 101:16405–10, 2004.

Index

H. Frauenfelder, *The Physics of Proteins*, Biological and Medical Physics, Biomedical Engineering, DOI 10.1007/978-1-4419-1044-8,